SOUTHGATE VETERANS
MEMORIAL LIBRARY
14680 DIX-TOLEDO ROAD
SOUTHGATE, MI 48195

Complete Shortwave Listener's Handbook

DATE DUE		
APR 07 1998	APR 22 2015	
NOV 27 1999		
JAN 13 2000		
APR 30 2002		
MAR 13 2008		
DEC 16 2008		
DEC 26 2012		
ul 8/5/13		

SOUTHGATE VETERANS
MEMORIAL LIBRARY
14680 DIX-TOLEDO ROAD
SOUTHGATE, MI 48195

A50527665

14

19.95

Other books by Andrew Yoder

Pirate Radio Stations: Tuning in to Underground Broadcasts
Build Your Own Shortwave Antennas—2nd edition
Auto Audio: Choosing, Installing, Maintaining, and Repairing Car Stereo Systems
Shortwave Listening on the Road: The World Traveler's Guide
Pirate Radio: The Incredible Saga of America's Underground, Illegal Broadcasters
Pirate Radio Operations
The TAB Electronics Yellow Pages
Home Audio: Choosing, Maintaining, and Repairing Your Audio System
Home Video: Choosing, Maintaining, and Repairing Your Video System

Complete Shortwave Listener's Handbook
5th Edition

Andrew Yoder

McGraw-Hill
New York San Francisco Washington, D.C. Auckland Bogotá
Caracas Lisbon London Madrid Mexico City Milan
Montreal New Delhi San Juan Singapore
Sydney Tokyo Toronto

Library of Congress Cataloging-in-Publication Data

Yoder, Andrew R.
    The complete shortwave listener's handbook / Andrew Yoder. — 5th ed.
      p. cm.
    Rev. ed. of: The complete shortwave listener's handbook / Hank Bennett, David T. Hardy, Andrew Yoder.
    Includes index.
    ISBN 0-07-913010-0
    1. Shortwave radio—Amateurs' manuals. I. Bennett, Hank. Complete shortwave listener's handbook. II. Title.
TK9956.Y6415   1997
621.384'151—dc21   97-3765
                                                                                           CIP

McGraw-Hill

A Division of The McGraw·Hill Companies

Copyright © 1997 by The McGraw-Hill Companies, Inc. All rights reserved. Printed in the United States of America. Except as permitted under the United States Copyright Act of 1976, no part of this publication may be reproduced or distributed in any form or by any means, or stored in a data base or retrieval system, without the prior written permission of the publisher.

1 2 3 4 5 6 7 8 9 0   DOC/DOC   9 0 2 1 0 9 8 7

ISBN 0-07-913010-0 (PBK)

The sponsoring editor for this book was Scott Grillo, the editing supervisor was John Baker, and the production supervisor was Suzanne Rapcavage. It was set in ITC Century Light by Jana Fisher through the services of Barry E. Brown (Broker—Editing, Design and Production).

Printed and bound by R. R. Donnelley & Sons Company.

 This book was printed on recycled, acid-free paper containing a minimum of 50% recycled, de-inked fiber.

McGraw-Hill books are available at special quantity discounts to use as premiums and sales promotions, or for use in corporate training programs. For more information, please write to the Director of Special Sales, McGraw-Hill, 11 West 19th Street, New York, NY 10011. Or contact your local bookstore.

Information contained in this work has been obtained by The McGraw-Hill Companies, Inc. ("McGraw-Hill") from sources believed to be reliable. However, neither McGraw-Hill nor its authors guarantees the accuracy or completeness of any information published herein and neither McGraw-Hill nor its authors shall be responsible for any errors, omissions, or damages arising out of use of this information. This work is published with the understanding that McGraw-Hill and its authors are supplying information but are not attempting to render engineering or other professional services. If such services are required, the assistance of an appropriate professional should be sought.

Dedication

To Corbin and Bryn with all my love;
You might not become DXers,
but you'll have a hard time escaping it!

Contents

Acknowledgments *xv*

Introduction *xvii*

1 What is shortwave radio? 1
Shortwave history *1*
Why listen to shortwave? *8*
Fun stuff *9*
The future of shortwave radio *12*

2 Choosing a shortwave radio that suits your needs 15
Buying a new or recent used receiver *15*
Features *18*
Receiver Checklist 18
Receiver Feature: Drake SW8 21
Receiver Review: Grundig YB-400 25
Can I Have a Second Opinion? 26
What are all of those knobs for? *27*
Where to buy a receiver *29*
What are the Leading Names in Shortwave Receivers? 30
Major Shortwave Mail-Order Companies 32
Accessories *34*
The Whys and Whos of DSP 35
Conclusion *36*

3 Antennas and more 37
Antennas for portable receivers *37*
Popular Fallacies About Shortwave Antennas 39

Antenna Stringing Tips 40
The self-sticking screen antenna 46
Antennas for tabletop receivers 47
Conclusion 53

4 Longwave radio 55

Propagation 55
Broadcasting 56
Longwave receivers 56
Hobbyist beacons 57
Amateur radio 59
Government activity 59
Natural radio 60
Tuning in 62

5 Mediumwave radio 63

Propagation 63
Mediumwave in North America 64
Mediumwave broadcasting around the world 68
Mediumwave DXing 69
We Love Those Callsigns (WLTC?) 71
Pick a radio band, any radio band 73
Mediumwave DXing necessities 73
What is a DX Test? 80

6 Shortwave listening 83

Where to listen 83
Using that radio 86
Looking Back at 20 Years of Radio 88
DX friends 96
Finding and eliminating radio interference 104
DXpeditions 105
South Pacific Union of DXers Inc. 106
Locations 106
Antennas 107
Supplies 109
Excursion results 110
Australian DXpedition Results 110

7 International shortwave broadcasts 115

How to use this chapter 115
Albania 116
Algeria 116
Angola 116
Antigua 117

Argentina *117*
Argentine Antarctica *117*
Armenia *118*
Ascension Island *118*
Australia *118*
Australia (Radio Australia) *119*
Austria *120*
Azerbaijan *120*
Bahrain *121*
Bangladesh *121*
Belarus *121*
Belgium *121*
Benin *122*
Bhutan *122*
Bolivia *122*
Bosnia-Hercegovina *122*
Botswana *122*
Brazil *123*
Bulgaria *123*
Burkina Faso *123*
Cambodia *123*
Cameroon *124*
Canada *124*
Canada (Radio Canada International) *125*
Central African Republic *127*
Chad *127*
Chile *127*
China *127*
Colombia *128*
Congo *128*
Cook Islands *128*
Costa Rica *128*
Costa Rica *129*
Cote d'Ivoire *130*
Croatia *130*
Cuba *130*
Cyprus *131*
Czech Republic *131*
Denmark *132*
Dominican Republic *132*
Ecuador *132*
Egypt *133*
Equatorial Guinea *133*
Eritrea *134*
Estonia *134*
Ethiopia *134*

Finland *134*
France *135*
French Guiana *136*
Gabon *136*
Georgia *136*
Germany *136*
Ghana *137*
Greece *137*
Guam *138*
Guatemala *139*
Guinea *139*
Guyana *139*
Honduras *139*
Hungary *140*
Iceland *140*
India *140*
Indonesia *142*
Iran *142*
Iraq *143*
Ireland *143*
Israel *143*
Italy *144*
Japan *145*
Jordan *146*
Kazakhstan *146*
Kenya *146*
Kirbati *147*
Kuwait *147*
Laos *147*
Latvia *147*
Lebanon *147*
Lesotho *148*
Liberia *148*
Libya *148*
Lithuania *149*
Madagascar *149*
Malawi *149*
Malaysia *149*
Mali *149*
Malta *150*
Mauritania *150*
Mexico *150*
Moldova *150*
Monaco *151*
Mongolia *151*
Morocco *151*

Mozambique *151*
Myanmar *152*
Namibia *152*
Nepal *152*
Netherlands *152*
Netherlands 153
New Zealand *153*
Nicaragua *154*
Nigeria *155*
North Korea *155*
Northern Mariana Islands *156*
Norway *156*
Oman *157*
Pakistan *157*
Palau *158*
Papua New Guinea *158*
Paraguay *159*
Peru *159*
Peru: Musical Transmitters 159
Philippines *160*
Poland *160*
Portugal *161*
Qatar *161*
Romania *161*
Russia *162*
Rwanda *162*
Sao Tomé *163*
Saudi Arabia *163*
Serbia *163*
Seychelles *163*
Sierra Leone *163*
Singapore *164*
Slovakia *164*
Soloman Islands *165*
Somalia *165*
South Africa *165*
South Korea *166*
Spain *167*
Sri Lanka *167*
St. Helena *167*
St. Helena (Radio St. Helena) 168
Sudan *169*
Suriname *169*
Swaziland *169*
Sweden *170*
Sweden 170

Switzerland *172*
Syria *173*
Tahiti *173*
Taiwan *173*
Tajikistan *173*
Tanzania *174*
Thailand *174*
Togo *175*
Tunisia *175*
Turkey *176*
Turkey (Voice of Turkey) *176*
Uganda *177*
Ukraine *177*
United Arab Emirates *177*
United Kingdom *178*
Uruguay *179*
USA *180*
United States (Voice of America) *185*
Uzbekistan *186*
Vanuatu *186*
Vatican *187*
Venezuela *187*
Vietnam *187*
Yemen *187*
Zaire *188*
Zambia *188*
Zimbabwe *188*

8 Weird and oddball radio listening *189*

Pirate radio *190*
Clandestine radio *200*
The Clandestine Granddaddy of Central America *206*
More information *215*
Numbers stations *216*

9 Utilities monitoring *219*

What can you hear on the utilities? *220*
The legality of listening to utilities *225*
Equipment *225*
Receiving utility QSLs *231*
For more information *234*

10 CB radio *235*

What is CB radio? *235*
Roots *236*

CB radio today *237*
Community Radio Network (CRN) *240*
Outbanders *241*

11 Amateur radio **245**

Functions of amateur radio *246*
Beginning of Amateur Radio *246*
Getting into ham radio *248*
Ham radio and the SWL *249*
Callsigns *254*
Codes *255*
Other interesting aspects of amateur radio *256*
Sending QSL reports *261*
Amateur radio online *263*

12 FM and TV DXing **265**

Propagation *265*
Equipment *269*
Antennas *270*
DXing *272*
QSLs *275*
FM and TV resources *275*

13 Radio-related collections **279**

Value *279*
QSLs *279*
Pennants *281*
Tapes *282*
Stickers *284*
T-shirts *284*
Photos *284*
Radios *285*
Visiting Radio's Past *289*
Radio magazines and bulletins *294*

14 Computers and radio **299**

Receiver-control programs *301*
Computerized receivers *304*
Shortwave receivers of the future? *306*

15 Reporting and verifications **307**

Report preparation *307*
Report data *311*
Codes *314*
Postage *316*

xiv Contents

    Time *317*
    Tape-recorded reports *318*
    Foreign-language reports *319*
    Enclosures *326*

A Useful shortwave tables 327

    Worldwide time chart *327*
    Commonly used SWL abbreviations *329*
    Language abbreviations *330*
    RST method of signal reports *331*
    Q signals *332*
    The NASWA radio country list *334*
    NASWA station-counting rules *335*
    Practical tips for using the NASWA Radio Country List *336*
        Africa *337*
        Antarctica *341*
        Asia *341*
        Europe *344*
        North America *346*
        Oceania *347*
        South America *348*
        Totals *349*

B SWECHO FidoNet BBS list 357

    Using BBSs *358*

C Utility frequencies 371

Index *395*

Acknowledgments

This book could not have been written without the help of: Roland Phelps, Stacey Spurlock, John Baker, Lori Flaherty, Melanie Holscher, Aaron Bittner, David McCandless, and Jen Priest of TAB/McGraw-Hill; Scott Grillo and Steve Chapman of McGraw-Hill; Richard I. and Judy Yoder; Charles and Diane Haffling; John Stephens and Lynn Hollerman of the International Radio Club of America; Georgia Morgan of Drake; Fred Osterman of Universal Radio; Kim Andrew Elliot of the Voice of America; Mike Witkowski; Harry Helms and Jeanette Barr of HighText Publications; Joe Carr; Jonathan Marks from Radio Netherlands' Media Network; Marius Rensen; Dave Onley of the South Pacific Union of DXers (SPUD); Tom Sundstrum of TRS Consultants; Bob Grove of *Monitoring Times* and Grove Enterprises; Bill O. Rights of Radio Free Speech; Tony Leo of Radio St. Helena; Fred Kohlbrenner; Rich D'Angelo and Harold Cones of NASWA; Don Moore; Ulis Fleming of *Cumbre DX*; Gary Matthews; Viamao DX Clube; G4VZO; Lance Johnson, K1MET; Walt Williamson, W3FGQ; Henrik Klemetz; Milan Hudecek of Rosetta Labs; Sander Schimmelpenninck; Maty Weinberg of *QST*; Radio Animal of WKND; Levi Iversen; Gabriel Ivan Barrera; Jan Fisher; Toya Warner; Barry Brown; Brandon Artman; Joseph Benjamin; Stephen McGreevy; Charles P. Hobbs; Richard E. Hankison; Harald Kuhl; Nick Grace C.; John Cruzan of the Free Radio Network; and Rob Bellville, N1NTE.

Levon Ananikan of the Voice of Armenia; Ignas Yanam of Radio Eastern Highlands; Al Mayers of WJR; Tiny Hammers of Monitor Radio; Jeff White of WRMI; Joe Bernard of RFPI; Bill Early and Karen Zeck of Trans World Radio; Frederica Dochinoiu of Radio Romania International; Ursula Fleck-Jerwin and Sarah Hartley Edwards of Deutsche Welle; Simon Spanswick of the BBC; Aida Hamza of UAE Radio from Abu Dhabi; Ken McHarg of HCJB; Pilar Salvador of Radio Exterior de Espana; Ronald Grunig of Swiss Radio International; Wafa Ghawi of Radio Damascus; Richard W. Jones and Richard Dentici of WEWN; Alden Forrester and Adrian Peterson of Adventist World Radio; Ashraful Alam of Radio Bangladesh; Michael Reuter of KJES; Arie Schellaars of Radio Australia; Mike Osborne of KNLS; Charlotte Addler of Radio Sweden; Tina Hammers of WSHB; Helga Dingova of Radio Slovakia International; Noeleen Vorster of Channel Africa; Alfredo E. Cotroneo of Nexus IBA; Wolf Harranth of Radio Austria; Woinshet Woldeyes of Radio Ethiopia; Juan Mort of XERMX; Fre Tesfamichael of Voice of the Revolution of Tigray; Lourdes López of Radio Havanna Cuba; Gary Hull of The Voice of Hope; Chris Fleck of HRPC; Jose Castañeda of Radio Maya; William Rodriguez Barreiro of Radio Bahái; Yolanda Marcó E. of Radio Santa Cruz; Masmoudi Mahmoud of Radio SFAX, Charles Poltz of WLIS; Sabino Llamo Chavez of Radio Satelite; and Paul Art of Voice of the Rock.

Introduction

You know the routine: Get up, eat breakfast, and read the newspaper. Drive to work, and listen to the AM or FM radio morning DJs. Catch the drivetime program on the way home from work. Eat supper, and relax in front of the tube. It's a way of life for millions of people.

However, millions of others catch alternative media outlets. The present attention-grabber is the Internet. Everyone is talking about the Internet, and the possibilities are truly exciting on the so-called "information highway." With a $2000 computer and $20 per month, you can chat with people around the world, download files, buy things, etc.

You might not often hear about it, but one of the most popular "alternative" media outlets in North America is the shortwave radio. Unlike the Internet, shortwave radio is free: no tolls, no tariffs, and no surcharges. Just like AM and FM radio and TV broadcasting, shortwave radio is free; just buy a radio and catch the action!

With shortwave radio, you can hear broadcasts from around the globe: news from Germany, folk music from Turkey, soccer from England, and politics from Korea. You can hear real-life pirates, spies, and drug smugglers. You can eavesdrop on the people who make news: Air Force 1 and 2, the military, airlines, ships and fisherman, etc. If you feel left out from all of the excitement, you can participate . . . on either the Citizens Band or with an amateur radio license.

For most people, the expected drawbacks to shortwave are: the price of a receiver, the complexity of shortwave equipment, and the feeling that you must be a slave to the radio. Not so on all counts. Shortwave radios are relatively inexpensive and easy to obtain. Radio Shack sells portable models, and this book lists the addresses of many reliable mail-order dealers. It is easier to use most shortwave radios than to program a VCR or to use a home theater stereo amplifier/receiver. Last, with this book and an excellent magazine, you can pop on the shortwave and catch a program. It's almost like using the *TV Guide* to watch TV shows.

If you have been into shortwave listening for years and you are just picking up this book to update your information (like picking up the most recent *TV Guide*), I hope that you will enjoy this 5th edition of *The Complete Shortwave Listener's*

Handbook. Unlike the previous four editions, this one is a complete overhaul; we live in an all-new world from 1973 (when the first edition was written), and this is an all-new book.

If you have any feedback concerning this book, either positive or negative, please feel free to drop me a line (but please don't ask for my recommendations on different receivers!). Because of the large influx of mail, I can't promise a response for all letters, but I do read all correspondence and I will try to respond.

Andrew Yoder
P.O. Box 642
Mont Alto, PA 17237
Internet e-mail: ayoder@cvn.net

1
CHAPTER

What is shortwave radio?

If you bought this book or picked up a copy from the local library, you obviously have an interest in shortwave radio. Maybe you have heard a station in the past, or maybe you've only seen it as a backdrop in a movie, such as *The Killing Fields*, where news from the BBC played a key role at several points and where the Voice of America provided mood music. Unless you are a regular listener, your most important questions are probably "What is shortwave radio?" and "Why should I listen to it?" If you listen to shortwave and know where to tune, you can hear almost anything.

Shortwave history

The original purpose of radio was to eliminate the telegraph wires. The telegraph consisted of electrical impulses sent through long sets of wires by means of a telegraph key. It was used to send Morse code from one telegraph center to the next. There, a relay would change the electrical impulses into clicks (Morse code consists of long clicks—dashes—and short clicks—dots). The telegraph operator at the receiving center would translate the code on paper, then pass on the news to the desired recipient of the message.

Telegraph systems were both expensive and difficult to maintain. The poles had to be made from trees cut from the forest, moved, and erected. Then, miles of heavy-gauge wire (to withstand wind and ice) had to be strung between them. Severe weather, fires, and frontier warfare caused lapses in communications that lasted for days. The initial construction, maintenance, and loss-of-service costs were massive. Nonetheless, the telegraph systems were used throughout North America in the middle of the 19th century.

The creative minds of that time were not satisfied with the performance of the telegraph system. Ideas started brewing. In the 1860s, James Clark Maxwell predicted that waves could be propagated in free space. From that point on, scientists rushed to conduct the first successful wireless test. In the 1880s, Alexander Graham Bell successfully patented the telephone. In 1888, Heinrich Hertz published proof of such transmissions, which covered a grand total of 25 feet. By the end of the century, the telephone began to replace the telegraph. However, although the telephone represented a great leap in technology, it, like the telegraph, required plenty of wires and poles. The bottom line was high cost and no portability. Even though the results of the wireless experimentation in the 19th century were limited, the pioneers of the technology were driven by the lucrative potential.

The early experimenters grappled with discovering how to generate the frequency, how to receive a strong signal, and how to transmit with a large amount of power. These were the days before integrated circuits and even transistors. Imagine the value of the *audion*, invented by Lee DeForest, which could detect and generate radio signals. DeForest built several fortunes from this invention. Edwin Armstrong, perhaps the greatest inventor in radio history, invented the regenerative receiver, the vacuum tube oscillator, the superheterodyne receiver, FM (frequency modulation) broadcasting, and multiplex FM transmissions. Armstrong, DeForest, and several others were involved in lawsuits that stretched throughout most of the first half of the 20th century. The strain of the legal battle was so great that it killed Armstrong; he committed suicide on January 31, 1954. These stories, and many others from the early days of broadcasting, are portrayed in much greater detail in *Empire of the Air* by Tom Lewis.

Very early in the 20th century, large businesses and government agencies began experimenting with the new wireless technology. At that point, its viability was still relatively unproven. This all changed on April 15, 1912 when the *Titanic* sank in the icy North Atlantic. The sinking was first reported via radio. Rescue vessels rushed to the aid of the perishing. Soon after the *Titanic* story reached the front pages of newspapers around the world, everyone was clamoring for the *life-saving* technology that was now a necessity.

The radio signals from the *Titanic* were copied by a number of different commercial radio operators, including David Sarnoff, a Russian immigrant and young radio hobbyist from New York City. Like many of the radio operators of the day, he learned Morse code from a commercial telegraphy company. Naturally, much of the telegraphy lingo and styles for sending code transferred. The operators developed special terminology and abbreviations, such as the *Q codes*, to save time when sending messages. The Q codes consisted of three letter blocks, starting with the letter *Q*, that represented a common term in radio operations. For example, *QSO* means a two-way radio conversation, *QTH* means location, *QRM* means man-made radio interference, and *QRN* means natural radio interference. The Q codes are used as either terms or as complete questions. For example, you can ask "What is your QTH?" or simply "QTH?" A few other interesting developments were the use of sending "VVV" as a test letter sequence, calling "CQ" whenever you want to talk with someone else, and sending "hi," which represents laughter.

In 1906, Reginald Fessenden made the first-known broadcast of voice and music: his program of prerecorded classical and Christmas music, announcements about the experiment, and Christmas wishes dumbfounded radio operators across the Northeast. At that time, no one had ever heard voice or music via radio. In 1915, the now-immortal *QST* magazine began publishing a newsletter for amateur radio operators. For the next five years, amateurs began experimenting with sending voice communications and even entertaining local listeners with music and community news. KDKA from Pittsburgh, Pennsylvania began regular official broadcasting operations on October 27, 1920, when it covered the Presidential election between Warren Harding and James Cox. KDKA was the first station licensed by a government for broadcasting, although a number of other stations (mostly amateur radio operators) had broadcasted before that time.

At this time, hundreds of bright kids were spending their allowances on building crystal radio receiver sets and spark-gap transmitters (crude transmitters that shot a spark across a gap, much like a spark plug) that basically just created blasts of radio interference! In the teens and 1920s, the radio receivers were of poor quality, and the transmitters splattered all over the bands. This, and the many different services vying for the best frequencies on the radio spectrum, made the band a real mess.

During the early years of radio, no one really understood the way the signals were propagated. They assumed that the only signals emitted were from the ground wave. As a result, the authorities believed that the radio spectrum was useless below about 2000 kHz (Fig. 1-1). So, the U.S. government gobbled up the mediumwave and longwave bands, and graciously allowed amateur operators to use the shortwave bands for transmitting. Little did they know that shortwave frequencies are much

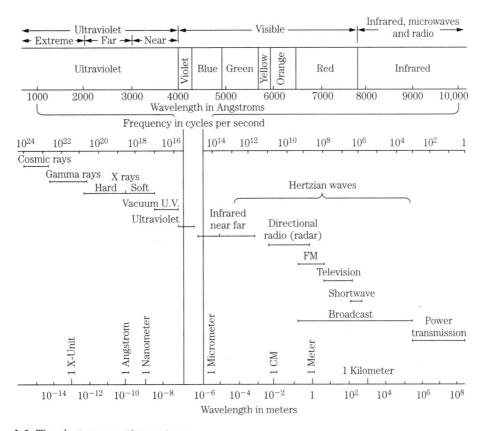

1-1 The electromagnetic spectrum.

more valuable for long-distance communications! If it wasn't for this misunderstanding, the typical broadcast band that you listen to now might have been on shortwave.

Joe Carr has an excellent story in his book *The Secrets of RF Circuit Design—2nd Ed.* He wrote about a young amateur radio operator from Ohio who was off at college. He regularly used the mediumwave band for two-way communications (this was before the days when the MW band was licensed only for broadcasting). When he arrived home from break, he was horrified to discover that his brother had torn down his old antennas, in favor of the smaller, "worthless" shortwave antennas. The older brother grudgingly tried the system on 40 meters; his surprise was quite great when he called CQ and the first response was from France!

Throughout the 1920s, military, broadcasting, and amateur radio stations were all using different frequencies within the mediumwave band. Some frequencies were allocated to different services, and some were used by several services simultaneously. Needless to say, the overlapping of various services caused much resentment. The U.S. government created the Federal Radio Commission in the 1920s to control the rapid growth and conflict within the industry. (In most other countries, government bodies were also created to control radio, but these services were under much greater control by their respective governments. In fact, private ownership of radio stations was not allowed in many countries and even amateur radio was banned in some countries.)

Because of the great number of U.S. hobbyists and commercial radio operations, the airwaves were also plagued with a virtual "airwave anarchy" for a time. Stations kept increasing power to wipe competitors off the air, stations fought for advertisers, and some operators jammed others. The bills that instituted the Federal Radio Commission didn't have enough teeth, and the Federal Communications Commission was created in 1934 to straighten out the radio spectrum. The new FCC closed a number of unlicensed broadcast stations and worked to limit interference on the bands—although many critics claim that the FCC was primarily used to empower the big-money media conglomerates.

Of course, the public went wild over radio in the 1920s and 1930s. For a while, it seemed like every hotel and hardware store operated a radio broadcast station. At this time, radio networks were just forming. Most stations had to produce their own programming. The broadcasts were live, the equipment was unreliable, and the announcers were totally inexperienced. Regardless, the general public loved radio, and most of the people in the United States and Canada spent a month's (or more) worth of wages for a receiver. Most of the receivers from the late 1920s and 1930s were contained in beautifully crafted black walnut veneer or stylish art deco cabinets. Families gathered to listen to programs in the evening (Fig. 1-2).

Before long, radio stations began producing dramatic presentations and live sports coverage. Always out to discover new talent for programs, the stations would find vaudeville performers and hire them for radio productions. The stations with the best ratings and programs were often approached by nearby stations to trade or purchase these shows. As a result of the program exchanges and purchases, many radio networks were created. Before long, some programs reached national audiences, and the radio actors and actresses attained celebrity status. One of the early radio celebrities, Vaughn de Leath, was given a celebrity radio birthday party with

1-2 Crowding around the 1936 model, deluxe receiver from Midwest radio.

plenty of singing and salutations. The next day, she received two truckloads of gifts, including five angora cats from her adoring listeners!

While most of radio technology was still young, stations were naturally concerned with how well they were being received by listeners, especially because many station engineers were amateur radio operators. To discover how well they were being heard, by whom, and where, stations began offering verification cards (known as *QSL cards* by the Q-code lingo) and letters to listeners who sent in reports of how well the station was heard (called *reception reports*). These reports included such information as the time, date, frequency, and signal strength and information about the broadcast. Many of the cards were simply printed in mass quantities, but many stations also offered stamps. If the listener sent an accurate reception report, he or she would receive a special stamp, in return. The listeners usually placed these stamps in a special radio stamp collector's book.

With the world enchanted by the single technology of radio, it's little wonder that some people were caught up with the medium, not the message. In those days, fads ruled. Many listeners were more interested in their collections of QSLs and radio stamps than they were with the station programming. It wasn't just a few people that got into long-distance radio listening (known as *DXing* from the Morse code lingo). Radio clubs sprang up in small towns across the country. Newsletters were sent to and from virtually every corner. Even newspapers and radio stations helped sponsor radio clubs and listening contests. One such club, the *Newark News Radio Club* survived into the 1970s.

In this era of mediumwave listening, distance was equal to difficult-to-receive stations. If you lived in Walla Walla, Washington, you would be very excited if you heard stations from Halifax, Canada or Honolulu, Hawaii; they were considered great "DX." So, imagine what happened when shortwave was "discovered" by the broadcasters and the listeners! Before long, the shortwave bands were included on most of the standard consumer radios of the day. Many receivers even featured the names of the stations right on the dials: Rome, Lisbon, London, Berlin, and so on.

Suddenly, the listeners in North America could tune in stations from Germany, England, Japan, Sweden, etc. with ease. So, by the 1930s, shortwave was all the rage for DXers and QSL collectors. It was great for program listeners, too. The USSR (Fig. 1-3), Germany, and Italy began preparing their propaganda machines—the likes of which the world had yet to see—for world domination. With the strong signals that could be heard from stations around the globe, the whole world could sit back and listen to war preparations, including Hitler's speeches, on a regular basis. The 1930s' European unrest, as evidenced by the radio propaganda, was also featured in numerous magazine and newspaper articles of the day.

At the beginning of World War II, listeners were glued to their radios. A popular U.S. historical scenario features families with shocked expressions sitting at the edge of their seats on December 7, 1941, the day that Pearl Harbor was attacked.

Instead of increasing the popularity of shortwave radio, World War II buried it. The Nazis, who controlled nearly every point in the European continent, banned listening to any broadcasts—except those authorized by the Nazi party. Even, the United States worried that the advanced propaganda of the European nations would be effective (Fig. 1-4). Although radios with shortwave bands weren't confiscated,

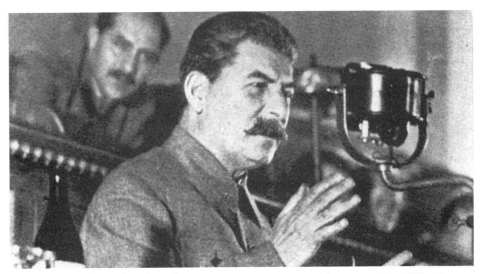

1-3 Stalin made regular international broadcasts before, during, and after World War II.

1-4 Propaganda announcers were prepared to broadcast during any emergency. Imagine what this guy must have sounded like on the air!

the bands were pulled from standard consumer radios. Radio repair shops were instructed to dismantle the shortwave bands from radios that came in for repair. I have one typical broadcast radio (which has a sticker from a local radio shop, dated 1940) that has wires snipped, that rendered the shortwave bands useless. Shortwave was essentially forgotten. After the war, the bands were dropped by the broadcast radio manufacturers altogether.

In Europe, most radio listening was banned, but during the war, the die-hard enthusiasts still listened to and passed on information from the BBC (British Broadcasting Corporation). It was relied upon for accurate news of the war—for many, it was almost like a source of life. So, after the war, shortwave continued as a viable feature of European radio, and the bands were retained on many receivers. Although shortwave was not as popular as mediumwave and, later, FM, it still attracted a large European audience.

The countries that were not particularly involved in World War II—such as Central America, South America, Central and Southern Africa, and internal Asia—were also more remote geographically. The wide coverage area provided by shortwave radio made it a necessity (as opposed to mediumwave and, later, FM broadcasting, which had a much more limited "footprint"). Shortwave broadcasting has been very popular from its inception in these countries and has only recently been losing ground to broadcasting in other radio bands. Most of these geographical regions use the lowest shortwave broadcast bands (120, 90, and 60 meters) as secondary mediumwave bands. To guarantee that these countries would have clear frequencies and not be pushed out by the Cold War propaganda megapowers, an international agreement was signed to prevent stations in North America and Europe from transmitting in those bands. Because many of the stations in the 120-, 90-, and 60-meter bands are located near the Equator, they are often referred to as the *topical bands*.

For most of the post-World War II period, shortwave broadcasting and listening audiences have been steady, but very slowly declining. The most damaging blow in Europe to shortwave was probably caused by FM broadcasting, which, because of its much higher fidelity, also cut into mediumwave listening audiences. In the early days, governments controlled the radio in many nations of the world. In those days, if they wanted to listen to anything different, people in these nations turned to shortwave to hear broadcasts from countries with lighter restrictions. Later on, most countries allowed independent broadcasting, mostly on mediumwave and FM, causing the shortwave audiences to diminish.

For the most part, except for the program content, very little has changed in shortwave broadcasting and listening over the years.

Why listen to shortwave?

One catchy tagline that I read on a few computer bulletin board messages is: "Shortwave is the original information superhighway." Correct! Shortwave radio has been the best source of news and information for more than half a century. How do I know this, and how do I know that it hasn't yet been overshadowed by the Internet? Simple. Just read the international news stories in a newspaper or watch the network news.

When a passenger jet crashed in Colombia in 1996, the U.S. news media cited "Colombian radio" (probably one of the Caracol network stations) as a source. Throughout the civil war in Yugoslavia, the media regularly mentioned "Croatian radio" (Hrvatski Radio) and "Serbian radio" (Radio Yugoslavia) as sources. During the civil war in El Salvador in the 1980s, the newspapers were quoting "rebel radio" (Radio Farabundo Martí) almost daily.

These preceding examples are just minor samples of how shortwave radio is used as a source for the news media. As much as I enjoy the Internet and as close as we are to the 21st century, the newspapers and TV news media isn't yet saying "According to sources on the Internet . . . ." It's still "According to national radio . . . ." This information is passed on because it is considered so helpful; many news organizations employ a staff of journalists to monitor shortwave stations around the world for late-breaking information.

In addition to the excellent news on shortwave, it is also one of the best sources of cultural and geographical information. Do you know where Senegal is; the difference between communism, taoism, and fascism; or how Andean folk, African hi-life, and reggae music are all different? Shortwave radio can enlighten listeners on all of these topics.

In an era when black and African studies are common, it amazes me that these classes apparently are never integrated with shortwave radio. At least a dozen stations from different African countries can be heard nightly in North America. These stations play plenty of contemporary and folk African music and air lots of news and editorials, from many different cultural and political stances.

For those interested in international politics, political science, and current events, the shortwave radio, once again, is a hard-to-beat source of information. Although you could read articles in *Time* or *Newsweek*, what better way is there to learn the different subtleties in political events than to listen to the same newscasts from different government broadcasters. As of this writing, you can get an interesting twist on the conflict between North and South Korea by listening to the news from those two countries, along with the news from the Voice of America and China Radio International. Speaking of North Korea, Radio Pyongyang is one of the last outlets of old-fashioned communist rhetoric still left on the airwaves. There's no better way to learn about Marxist communism than to listen to a few droning editorials about the next five-year plan or about the "glorious comrades of the revolution."

With a bit of creativity, learning from shortwave doesn't have to stop there. For example, you could get interesting cross-cultural recipes from the Radio Slovakia International cooking program, easily learn different styles of music from around the world, or brush up on the foreign language that you learned in high school. The possibilities are virtually limitless.

Fun stuff

Much of what's on shortwave is educational, but it's also fun. For example, although listening to the hardline propaganda on Radio Pyongyang is a great way to learn about communism for a political science class, it's also a blast! As with any hardline propaganda from a totalitarian nation, the announcers often go to absurd lengths to prove how much better they are than their enemies or competitors.

The music on shortwave is really interesting. Local folk music is very popular and you can hear it from almost every nation in the world. The popular music from the different countries is really interesting and often appetizing, by Western standards. That's especially important in the United States and Canada, where music from the other continents rarely receives radio airplay. Another fun aspect of music on shortwave is local songs on distant stations. For example, I've heard Cindi Lauper, Olivia Newton-John, and Dr. Hook on the Soloman Islands Broadcasting Corporation, Linda Ronstadt on Radio South West Africa, "Happy Birthday" (in English) on Radio Satélite and Radio Ancash (both of Peru), Billy Ocean on Radio Noumea (New Caledonia) (Fig. 1-5), Metallica on Radio Ukraine, Mr. Mister on Radio Capital (Venezuela), and Lionel Ritchie on Radio Nacional (Venezuela).

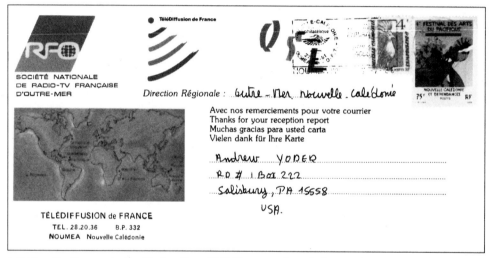

1-5 Despite its remote Pacific location, RFO Noumea played plenty of music from France and other Western nations.

"Weird stuff" has seen the biggest growth in shortwave listening in North America since the late 1980s. Shortwave is one of the only places that you can hear pirate (unlicensed) broadcasters, spy radio stations, clandestines (stations that try to overthrow a government), etc. These days, it seems that most of the new serious shortwave listeners are most interested in catching unusual broadcasts—especially from the pirates. I have to admit that I became involved in shortwave listening this way. My goal was to hear some pirates; it was only when I wasn't listening for pirates that I listened to a number of licensed stations. For more information on these unusual broadcasts, see Chapter 8, "Weird and oddball radio listening."

Much of the "strange" radio programming is actually licensed. With the fall of communism in the countries that formerly comprised the USSR, suddenly dozens of high-powered former Radio Moscow shortwave transmitters are available for hire at a low, low price. These transmitters were leased out to the Radio Aum Shinrinkyo organization, which was later accused of the nerve gas attacks in the Tokyo Subways.

At the same time, the FCC is licensing a number of stations that are playing the "We'll sell chunks of airtime to just about anybody" game. So far, we've collectively been subjected to hearing convicted cult terrorists, populists, socialists, fascists, patriots, nationalists, and a lot more (Fig. 1-6). For more information, see the next section and Chapter 7, "International shortwave broadcasts."

ALTERNATIVE SHORTWAVE RADIO BROADCASTS
Traditionalist/Constitutionalist/Nationalist/Populist/Sovereign/Patriot

PROGRAMMING FOR MARCH 1996

PROGRAM	HOST	MHZ	EST TIME	DAY
Hour Of Courage	Ron Wilson	15.685	7:00 AM	M - F
Hour Of Courage	Ron Wilson	6.040	7:30	
Freedom's Call	Bo Gritz	12.160	11:00	
Hour Of Courage	Ron Wilson	12.160	Noon	
Hour Of Courage	Ron Wilson	9.495	1:00 PM	
For The People	Chuck Harder	9.495	2:00	
Newswatch Magazine	David Smith	9.495	5:00	
The Norm Resnick Show	Norm Resnick	9.475	5:00	
Point Of View	Marlin Maddoux	9.475	6:00	
The Norm Resnick Show	Norm Resnick	5.065	6:00	
Voice Of Liberty	Rick Tyler	5.065	7:00	
Blueprint For Survival	Steve Quayle	7.435	7:00	
Viking International	Mark Koernke	9.955	7:00	
Newswatch Magazine	David Smith	3.315	8:00	
Protecting Your Wealth	Mike Callahan	5.065	8:00	
The Law Loft	Ludlow/Harris	7.435	8:00	
Protecting Your Wealth	Mike Callahan	3.315	9:00	
The Prophecy Club	Stan Johnson	5.745	9:00	
Hour Of Courage	Ron Wilson	5.745	9:30	
The John Bryant Show	John Bryant	2.390	10:00	
Radio Free America	Tom Valentine	5.065	10:00	
Radio Bible Hour	Harold Smith	3.315	11:00	
Hour Of Courage	Ron Wilson	17.510	11:00	
The Prophecy Club	Stan Johnson	3.315	11:30	
Voice Of Liberty	Rick Tyler	3.315	Midnight	
Hour Of The Time	William Cooper	5.065	Midnight	
Point Of View	Marlin Maddoux	2.390	Midnight	
Hour Of Courage	Ron Wilson	5.065	1:30 AM	
Freedom's Call	Bo Gritz	5.065	2:00	
Focus On The Family	James Dobson	5.065	3:00	
The Prophecy Club	Stan Johnson	5.065	3:30	
World Of Prophecy	Texe Marrs	9.475	Noon	Saturday
The Fisherman's Net	Richard Eutsler	12.160	5:00 PM	
Taking America Back	Richard Eutsler	12.160	6:00	
Hour Of Courage	Ron Wilson	5.065	7:00	
World Of Prophecy	Texe Marrs	5.065	8:00	
The American Way	Andrew Gause	5.065	9:00	
Hour Of Courage	Ron Wilson	5.745	9:00	
Newswatch Magazine	David Smith	3.315	9:00	
Open Bible Dialogue	Joseph Chambers	3.315	10:00	
World Of Prophecy	Texe Marrs	9.475	11:00	
Scriptures For America	Pete Peters	3.315	Midnight	
World Of Prophecy	Texe Marrs	9.475	Noon	Sunday
World Of Prophecy	Texe Marrs	9.475	5:00 PM	
Full Disclosure Live	Glen Roberts	5.065	8:00	
World Of Prophecy	Texe Marrs	7.355	9:00	
World Of Prophecy	Texe Marrs	3.315	10:00	
Extraordinary Science	J.W. McGinnis	5.065	11:00	
World Of Prophecy	Texe Marrs	9.475	11:00	
Scriptures For America	Pete Peters	3.315	Midnight	
We Hold These Truths	Paul Parsons	5.065	Midnight	

24 hour satellite: American Freedom Network; Galaxay 6, Transponder 14, Audio 5.8, Vertical Polarity Wideband

1-6 A list of "traditionalist/constitutionalist/nationalist/populist/sovereign/patriot" radio programs aired on shortwave, published by a patriot organization.

The future of shortwave radio

For a decade or so, I've heard doomsayers proclaiming the imminent death of shortwave. Some of these prophets were more caught up in the technological "keeping up with the Joneses" than pragmatism, but some were on the mark with the potential for a gloomy end to shortwave radio.

By looking at the numbers, experts have some reason to believe that shortwave operations will soon dry up and blow away, like a tumbleweed in New Mexico. The three big competitors to shortwave broadcasting and transmitting are satellites, cable TV/radio, and the Internet.

Satellite broadcasting is expected, by some, to be the eventual replacement for international shortwave broadcasting. However, because of the cost of manufacturing and sending the repeater into orbit, it is very expensive. Also, except for DBS satellite TV reception, satellite broadcasting has not become popular enough to outweigh the costs. Satellite radio broadcasting is becoming more popular in Europe, but it hasn't yet reached far enough beyond the hobbyist ranks. Still, many international shortwave broadcasters are also airing their programs via satellite to "keep their bases covered." Perhaps it is naive to believe that shortwave will outlast satellite radio; however, in North America, satellite radio has totally flopped, but shortwave listening is growing steadily.

The shortwave radio vs. satellite radio battle rages stronger with utility transmissions. As mentioned elsewhere in the book, utility transmissions include navigational beacons, ship-to-shore traffic, press service transmissions, military communications, etc. One of the hardest-hit of these branches on shortwave is the press agencies transmissions. For decades, the latest news stories were transmitted by UPI, AP, TASS, Reuters, etc. via shortwave radioteletype. The press services used shortwave because it covered the world and it was an easy way for their radio and newspaper affiliates to receive the information. However, it's much faster and more reliable to transmit the data via satellite, where there are no static crashes to disrupt communications. Today, the big-name press services are gone from shortwave, but some from smaller countries are left. Although some of the most popular utility services (for shortwave listeners to tune in) have departed, most of the "utes" (utility stations) will probably remain because of the lower cost and ease of transmitting mobilely (no satellites to target and transmit to/receive from).

Cable TV/radio is a subset or application of satellite broadcasting. The programs are aired via satellite. A cable TV company receives and retransmits them—either as audio on a TV channel or on an FM radio frequency. The listener subscribes to the cable TV service and either tunes in the proper channel on the TV or connects the cable to an FM receiver and tunes to the frequency where the satellite program is being retransmitted. It's a good scheme, but few cable companies air satellite radio programming, fewer people know that it exists, and fewer still would be motivated or technically competent enough to try hooking up the cable to an FM receiver. Another drawback of this system is that you can only listen to the stations while you are in your own house—not in your car, at work, at someone else's house, outside, etc. The bottom line is that, although the service can be somewhat effective for a few people, cable TV/radio will never be strong competition for shortwave radio.

The Internet has the hype and the riches of companies like Microsoft behind it. It is the future of communications, but will it spell the end of shortwave radio? Probably not; at least not in the near future. Currently, less than 10% of the people in the United States, one of the richest and best educated (per capita) countries in the world, are on the Internet. With such a small presence in the world and such a high cost and heavy learning curve for new users, the Internet will not be very competitive with shortwave broadcasting worldwide . . . not in the next few decades, possibly not ever. Right now, the Internet is most competitive with the mail and telephone service (e-mail and Internet phone vs. mail, telephone, and fax), print media (World Wide Web vs. magazines and newspapers), and advertising (World Wide Web ads and commercial e-mails vs. print media ads, TV ads, billboard ads, and commercial direct mail).

The previously mentioned mediums all have their own specific strengths and limitations. Shortwave radio is no different. Although its services probably will not be eliminated entirely anytime soon, that doesn't mean that it is the best medium for all applications. Still, for simple and inexpensive worldwide broadcasting and two-way or point-to-point communications, shortwave is unbeatable. It's a cost-effective means to reach thousands of listeners as well as to communicate to land from a ship.

In addition to normal broadcasts and communications, shortwave is perfect for unlicensed (pirate), clandestine, and spy broadcasts because of its low cost and long range. So, even if shortwave isn't the most-listened-to medium in the world, it *is* the most exciting. Shortwave radio might have lost the glitter that its early inventors so fervently sought after, but it's still as mysterious as it was during Reginald Fessenden's first broadcast.

If you think that shortwave might interest you, but you are worried that there'll be nothing to listen to in a few years; don't worry, just get a radio. If the idea of kicking back and listening to news, sports, editorials, politics, and music from different cultures entices you, you'll find plenty of satisfaction. If you are intrigued by sitting by a radio with headphones on, scanning for sometimes mysterious signals from around the world, this is the hobby for you.

Now, you only need a radio!

2
CHAPTER

Choosing a shortwave radio that suits your needs

I have been asked by numerous people over the years whether they can hear shortwave radio stations on their "regular radios." Initially, it seems absurd. "Of course not!" However . . . many people don't realize that the radio bands cover "from dc to light" and the frequencies that we commonly use (the so-called AM, FM, CB, and TV bands) are just tiny slices out of the entire radio pie . . . err . . . spectrum.

The shortwave band is a big slice out of the radio spectrum. It can, in turn, be divided into smaller bands. For example, the CB band is part of the shortwave spectrum, as are a number of the amateur radio bands, international broadcasting bands, "fixed" bands, maritime bands, etc. Unlike a standard "AM" or "FM" broadcast radio, a shortwave radio will cover a number of radio bands and a vast frequency range.

With this in mind, yes, you do need a special shortwave ("communications") receiver.

Buying a new or recent used receiver

Portable or table-top?
For decades, if you wanted a shortwave radio, you went to an amateur radio dealer and purchased a hulking mass of steel with more knobs than "Saturday Night Live" has unpopular cast members. Chances are that the radio weighed between 30 and 80 pounds and could emit enough light from its tubes to read and write by.

Over the years, portable shortwave radios have been developed, but most of the pre-1980s models were *luggables*—somewhat heavy radios that you could grudgingly drag off and listen to outside your home. The faux leather-encased Zenith Transoceanic models are excellent examples of luggables. Most of the Transoceanics weigh in at more than 30 lbs. They are finely crafted and were considered quite portable in their prime, between the 1930s and the 1950s. The Transoceanics are well-known for their high-quality, solid build, and status appeal (they were used by everyone from royalty to movie stars). Even so, except for the nostalgia, you wouldn't want to take a Transoceanic on a trip.

Throughout the 1950s and 1960s, only a few good-quality portable shortwave radios were built. Amateur radio was becoming mobile, but these radios were combination transmitter/receivers—heavy (by today's standards), expensive, and only covering the amateur bands. Most of the shortwave portables from this time period hit the international market in the late 1960s and early 1970s, when several Japanese companies introduced stand-up transistorized radios. Unfortunately, most of these radios were just AM/FM models, with shortwave apparently added as an afterthought. The pioneers of the truly portable shortwave receiver were Sony, Grundig, and Panasonic, all of which introduced dedicated (those that focus on shortwave reception) portable shortwave receivers in the 1960s and 1970s. This development created an avalanche of portables that started in the 1980s and has continued through the 1990s.

In essence, if you want to pick up a radio that travels well, you are limited to purchasing a radio that has been manufactured within the past 15 years. Since that time, a number of small portable radios have been designed. Because of the technological advances over the years, increasingly better radios have been placed in smaller boxes. Even some of the table-top radios can be operated from a 12-volt dc power supply, suggesting that the manufacturer expects you to take them along to remote areas. In fact, if the design trends continue, it appears that, some day, all shortwave radios will be at least somewhat portable.

Your needs

One of the best ways to choose between a portable and a tabletop receiver is to evaluate your budget, your needs, and your traveling habits (Fig. 2-1). If you want a new receiver and are on a tight budget, you *will* get a portable. Even the least-expensive tabletop receiver currently on the market costs approximately $600 (U.S.). Unless you have a professional interest in shortwave or have a great job, that's a lot of money. The prices rise steeply from there. Some shortwave tabletop receivers cost over $4000 (U.S.)! Another possibility is buying a used or classic tabletop receiver; more information on these topics is included later in Chapter 13 (Fig. 2-2).

If you are still undecided about which type of receiver to purchase, think about your needs and how often you travel. If you rarely travel, you might as well buy a tabletop receiver. Most of the recent tabletop models can be hauled around anyway. However, if you often travel on business and you need to pack your radio away in a briefcase or a carry-on bag, go with a small portable. That way you won't miss any listening opportunities.

My first receiver was a used Yaesu FRG-7 tabletop/luggable. This receiver was about as big as a minitower computer and weighed about 25 pounds, *but* if you stuck in 6 or 8 D-cell batteries, you could drag the thing around outside or wherever and listen to shortwave for an hour or two (until the batteries died). At the time, I bought the radio because it was cheap, of decent quality, and readily available. I was in high school and working as a dishwasher, so I couldn't afford a better receiver. Also,

Choosing a shortwave radio that suits your needs 17

2-1 The American-made Drake R-8 is a top-notch shortwave receiver that looks more like an expensive stereo component.

2-2 August Stellwag's excellent arrangement uses many modern and 1950s-era shortwave receivers.

portable receivers were somewhat new then, and you paid a higher price for that portability. So, I stayed with the FRG-7 for a number of years.

A few years after college, I noticed that the FRG-7's technology was showing its age. Although its sensitivity wasn't bad, the selectivity (more on these terms later) was terrible, and it had no digital readout. I was also traveling more to visit parents, in-laws, and friends. Time for a new receiver. I bought a portable. Although I didn't really have much more spending money than I did during my dishwashing days, the prices of good portables were dropping. I picked up a used Sony ICF-2010 for a great price. The 2010 was, and still is, one of the best portables on the market. Its small size allowed me to take it anywhere.

A few years later, I also obtained a number of huge military and communications radios from the 1950s and 1960s. I used the large radios at home, and the 2010 became the traveling rig.

Two years ago, I moved up to my first tabletop radio from the digital era, a used Kenwood R-5000. Although I still travel frequently, most of my trips are brief visits to family and friends' homes, so I have enough room to pack the R-5000. It's much larger in size and weight than a handheld portable, but I would rather lug around a larger (but still very portable) radio than sacrifice the much better performance.

I would not, however, be likely to include it with my camping gear. Even lower performance ratings can't hurt a tiny portable's value when the trunk is full or the backpack is heavy.

Features

The cost of a portable receiver ranges from about $40 to almost $600 (U.S.). Fortunately, the shortwave radio market is rather small and only a few dozen different models are available. However, the sizes, styles, quality, and features obviously vary considerably. A good receiver isn't very important if you only want to catch the news on the Voice of America. If that's the case, you can probably get by with a very inexpensive receiver. Even if you just want the cheapest receiver to perform the easiest tasks, you should still check out the different features so that you can find a receiver that will best suit your needs. The following paragraphs cover some of the qualities and features of shortwave radios that you should examine before choosing a receiver.

Receiver Checklist

- Is the receiver within your price range?
- Will it perform up to your requirements?
- Does it cover enough frequencies (preferably 500 to 30,000 kHz)?
- Does it have SSB reception or a BFO?
- What is the smallest frequency step that the receiver can display? 1 Hz? 10 Hz? 100 Hz? 1 kHz? 5 kHz?
- Is it small enough to take everywhere?
- Can it operate from dc power?
- Is it built well enough to handle some physical abuse?
- Does it have wide and narrow filters?
- Does it cover the FM or air bands?
- Does the radio's audio sound good?
- Does it have a digital frequency readout?

- How many filters does it have?
- Can it receive shortwave in the FM mode?
- Is the tuning rate adequate?

Sensitivity

Sensitivity is the capability of a receiver to pull in weak signals. If the receiver has poor sensitivity, weak stations will not be audible on this receiver (strong signals will still be audible on a set with poor sensitivity). If sensitivity is a real need for you, look at the receiver specifications, which are often printed in the catalogs of the major shortwave mail-order companies. Sensitivity is measured in microvolts, which might seem like a complicated measurement; however, there's an easy way to check the sensitivity: the lower the number, the better. For portable receivers, you can't get much better than the Drake SW8 and the Lowe HF-150, which both have an SSB sensitivity of 0.5 microvolts.

Selectivity and filters

Selectivity is the capability of a receiver to choose between different radio signals. For example, a very strong signal might be on 6010 kHz and another strong signal might be on 6020 kHz. On a receiver with poor selectivity, both stations would interfere with each other. A receiver with excellent selectivity will have no problem separating signals that are just a few kilohertz apart.

The selectivity of a receiver is based on the filters that it contains. The filters allow a small segment of the radio spectrum to pass through and be heard from your speaker. Most people without technical backgrounds just assume that when you tune a radio to a frequency, the radio will just "play" whatever is on that frequency. However, radio signals are very wide, often occupying as much as 12 kHz for amplitude-modulated (AM) signals, which are typically used for shortwave and mediumwave broadcasting.

Amplitude-modulated signals consist of a carrier and two audio sidebands. The carrier is just a "blank" signal. You can hear the carrier whenever there is no audio on the station. A good time to hear a carrier is when a station is signing on or off, between commercials, or during long pauses in a talk show. The sideband signals (upper and lower sideband) are on either side of the carrier; they contain the audio. Typically, the total signal power is 50% carrier, 25% upper sideband, and 25% lower sideband. As a result, a strong radio station often sounds best when you tune just above or below the center (strongest) part of the signal. At the center, you will receive more of the carrier than the audio from either of the sidebands. However, if you tune to either side, you will receive lots of signal from one of the sidebands.

The width of the signal and the audio filters have a lot to do with how the radio station will sound in your speaker or headphones. So, how do the filters apply to real life?

If a really strong station is operating with a 10-kHz wide signal and your receiver has a 9-kHz filter, the audio will probably sound really good. However, if you are listening to a weaker signal that is only 5 kHz wide with the same receiver on the 9-kHz filter, the results will be less spectacular. The audio will be crisp and clear, but an extra 2 kHz of noise or splatter from other stations will be audible on either side of the signal, making it very difficult to copy even though the sound is not muffled at all. For best results on receiving this signal, a 3- or 4-kHz filter would be used.

On the other hand, if a 3- or 4-kHz filter was used on the first, 10-kHz-wide signal, the station's audio would be much more muffled and unpleasant than if the 9-kHz filter was used.

So, what to do? If the filter is very wide, very strong signals will sound great, but weak signals and signals with adjacent-channel interference will be basically unlistenable. On the other hand, if the filters are too narrow, weak signals will be received well (but not with good audio), but strong signals will also have muddy audio.

Most inexpensive radios—especially portables—have one intermediate width (usually about 4 or 5 kHz wide). The best portable receivers, such as the Grundig YB-400 and the Sony ICF-2010, have a "wide" and a "narrow" filter position so that both strong and weak signals can be received. The tabletop receivers have as many as five different filter positions for the most flexible receiving conditions.

Many portable receivers have only one or two different filters because of their high cost. New filters each typically cost anywhere from $75 to $200 (U.S.). A handful of good filters alone could cost several times the cost of a new portable receiver. Considering that filters are the "windows to the radio spectrum" and that their cost is prohibitive, the filter is one of the most important components that influences your satisfaction with a radio and its compatibility with your budget. When choosing a radio, be sure to check out what types of filters are included. Some tabletop radios companies—such as the Japan Radio Company (JRC), Kenwood, and Icom—sell their receivers like cars: basic models will contain fewer or lesser-grade filters; better or more filters are available as options.

The "wave of the future" for radio filter technology is digital signal processors (DSPs). The Watkins-Johnson HF-1000 has an entirely digital front end; instead of traditional crystal or mechanical filters, the filtering is all processed digitally through integrated circuits. The HF-1000 can be set to more than 55 different filter widths! Using traditional filters, they alone would probably cost over $10,000 (U.S.). Instead, as of this writing, the HF-1000 priced in at "only" about $3700 (U.S.). Although these prices are tremendous, expect that the technology will become less expensive in the next few years. About five years ago, I heard shortwave experts talking at a convention about how some of the major radio manufacturers would soon be out with digitally filtered portable radios for less than $500. It hasn't happened yet, but such a radio would all but render the competition obsolete.

Image rejection

Image rejection allows the receiver to be exposed to very powerful signals without "overloading." An extremely potent signal, such as that from a local AM broadcast station, might swamp the front end of the receiver. As a result, you might hear "images" on frequencies other than the one that the station broadcasts on. Because of the small size and emphasis on low cost, portable receivers traditionally have been notorious for poor image rejection. I have had some receivers that, when connected to a longwire antenna, swamped on a local AM broadcast signal and I could hear that signal in the background wherever I tuned. Very annoying. However, the image-rejection problems have been corrected (or at least improved) in many of the newer portables. Image rejection in tabletop receivers typically ranges from very good to excellent.

SSB/BFO

The *single-sideband (SSB) mode* is a special type of transmission that is much more efficient than traditional AM broadcasting. As mentioned in the section about selectivity and filters, each sideband only consists of about 25% of the total signal that is transmitted from an amplitude-modulated radio station. With SSB transmission, the theory is that, if the carrier and the one sideband are unused, all of the audio signal can be concentrated on that one sideband.

It works tremendously well. However, the fidelity is drastically diminished. Because of the reduced fidelity, only a few shortwave broadcast stations operate in SSB, but nearly every amateur operator uses this mode. The other problem with SSB transmitting is that, when you listen to an SSB station in the AM mode, it will sound like some cartoon ducks—except less intelligible. To hear the SSB stations, you need a receiver that either has the SSB mode or has a BFO (beat-frequency oscillator). Receivers with the SSB mode are often easier to tune than those with just a BFO, so this should be a consideration.

Receiver Feature: Drake SW8

Unquestionably, one of the best portable receivers on the market is the Drake SW8. Here are some of its features:
- Covers 0.5–30 MHz, 87–108 MHz (FM band), and 118–137 MHz (airband)
- Receives AM/USB/LSB on 0.5–30 MHz and FM on 87–108 MHz
- Three built-in bandwidths (6.0, 4.0, and 2.3 kHz)
- AM synchronous detector for improved quality of AM signals
- Built-in whip antenna
- Requires either D cells or 110-Vac power
- High dynamic range
- FM stereo for headphone operation
- Operating parameters displayed on the front timer
- Backlighting can be shut off to conserve battery life

The SW8 is excellent for all types of shortwave reception: the audio quality is superb, the tuning is smooth and is displayed in 100-Hz steps, and images from strong signals are rejected—even if large antennas are connected. Not only is the SW8 very pleasant to listen to, but the "hard-to-catch" signal reception is so good that I was the first person in North America to report one station (among other good catches), thanks to the SW8.

The downsides of the high-quality, well-built Drake SW8 are size and cost. The SW8 is approximately the same size as a typical modern tabletop shortwave receiver, and the weight is about 10 pounds (lighter than most tabletop models, but much heavier than most portables). The price is about $600 (U.S.), which is about three times the price of a typical portable but is still anywhere from $350 to $600 cheaper than comparable tabletop receivers. However, the cost is not prohibitive for the serious listener—especially if you need to pick up and go from time to time.

Digital or analog readout

Most adventurous people like to explore by going somewhere new, but not by feeling lost wherever they are. For this reason, digital readout (which prints out the numbers on an LCD) is much more popular than analog tuning (which moves a pointer in front of a few rows of frequencies). Because of the small size of most analog portables, few numbers can be written in the dial space, and you wind up guessing at your frequency (±10 or 15 kHz). Fortunately, with the low cost of components for digital tuning, very few analog shortwave receivers of any type are made any more.

One step beyond digital tuning is *Station Name Tuning*, which is available on several Sony portable receivers. With this system, you can tune to some of the popular frequencies around the shortwave bands and the name of the station will also automatically appear on the LCD readout. In the Sony ICF-SW77, 94 of these memories are preset at the factory, and you can program up to 162 different stations into the memories. The station name tuning is an unnecessary extra for me because I always listen to and tune the radio with a few shortwave newsletters, magazines, or books handy.

Tuning rate

One of the problems with low-cost digital receivers is that most of them don't tune continuously across the bands. Some of the small, portable, digital receivers only tune in 1- and/or 5-kHz steps. Neither of these types of receivers are useful for hearing amateur radio stations, and you might have some problems with hearing some of the broadcasters, such as the Voice of Greece on 7448 kHz or Radio Copan International on 15674.6 kHz. These stations could fall between the tuning steps of the receiver and be difficult or impossible to receive.

For the SSB stations, you might even have some problems with the digital receivers that tune in 100-Hz blocks. If the amateur is transmitting between the 100-Hz steps of your receivers, such as on 7240.15 kHz, the signal will sound "ducky," no matter where you tune. This can be a real problem when listening to some of the pirate broadcasters that transmit music in SSB.

Some of the better receivers tune finer than you can see. For example, 100-Hz readout is common on the better portables, but some of these tune in increments down to 10 Hz or so. As a result, you can tune in SSB stations much better on these receivers (such as the Drake SW8).

Unfortunately, few of the shortwave mail-order companies list the tuning rate in their catalogs. If you plan to spend a little more on your portable and want to have an easier time tuning in the SSB stations, be sure to call the information line and ask a customer service representative.

Coverage

Most shortwave receivers cover the entire shortwave band (1800 to 30,000 kHz), plus the mediumwave (known as the *AM band* in most of North America) band (500 to 1800 kHz), and some of the longwave band (usually anywhere from 10 kHz, 150 kHz, or 200 kHz is covered up to 500 kHz). Some receivers also cover the FM broadcast band (known as the *VHF broadcast band* in many parts of the world), and a few cover the aero band (108 to 136 MHz), and some cover the aero and public service bands (108 to 172 MHz) with an optional converter.

Most of the tabletop receivers cover the entire shortwave band, plus a few (sometimes mediumwave, FM, aircraft, or some of the VHF utility bands are included). However, many of the portable receivers have an incomplete shortwave range. Many only tune down to about 3500 kHz and up to about 21 MHz. Others only tune the shortwave broadcast bands. Many old receivers covered the amateur radio bands exclusively; these are only useful to amateur radio operators or radio nostalgists. Beware of these "incomplete" radios!

By contrast, a few very old receivers covered very large portions of the bands continuously. Several 1940s- through 1960s-era Hallicrafters radios could receive from 500 kHz to 108 MHz in AM, FM, and SSB (BFO) modes! The famous military

1950s-era Hammarlund SP-600 covered from 540 kHz to 54 MHz, although that receiver only had AM and SSB (BFO) modes, so the 30- to 54-MHz areas (which primarily are now occupied by stations using the FM mode) are basically useless.

The concept of communications receivers covering into the VHF or UHF bands had been dead for decades. However, the idea returned to life in the early 1990s when some scanner manufacturers added shortwave bands to their scanners. In the mid-1990s, AOR and Bearcat revolutionized the radio-monitoring scene by manufacturing shortwave/scanners that could cover from 500 kHz to as high as 2000 MHz. Future shortwave receivers will probably also cover a good chunk of the "scanner" bands.

Size and weight

Even though some of my old 60+-pound radios are excellent, they just don't cut it for traveling; they're a bear just to clean around, let alone carry. All of the "portable" shortwave receivers are at least relatively portable, but otherwise the size varies greatly. The largest of the portables, the Drake SW8, is more like hauling a VCR with a handle than a portable radio. The SW8 is basically a tabletop radio that has been built tough for traveling. I wouldn't use it for business trips, but it's great for taking to a remote site for serious listening.

The funny thing is that although the SW8 is considered a portable, the tabletop Kenwood R-5000, Icom R-71A, and Yaesu FRG-100 are all smaller! When you purchase a radio and size is a consideration, check out its physical dimensions so that you know just how small it is! Of the current crop of shortwave receivers, only those in the professional or semiprofessional bracket are truly nonportable. Most have a handle on one side and little plastic feet on the other. The Drake R8A is probably the least portable of the common hobby-level tabletops. It can operate from 12-volts dc, but it's more like dragging along a piece of stereo equipment.

Most of the other high-grade portable receivers are close to the size of a large book. The Grundig Satellit 700, for example, is 12.25" × 7.25" × 3". The Satellit 700 is an excellent receiver. The only smaller (barely) comparable receiver is the Sony ICF-2010. The next general size bracket for portable radios is similar to that of a small, paperback novel. A few in this bracket are the Sangean ATS-606, Grundig YB-400, Panasonic RFB-45, and Sony ICF-SW7600G. These radios are all near 7" × 4" × 2". One of the smallest portables around is the shirt-pocket Philips AE-3905, which is only 2.75" × 3.75" × 1.25".

Some people like to quote the old "size isn't everything" adage. Well, in regard to shortwave portables, size *is* everything. The smaller, the better. However, because of the high cost of technology and the cutesy novelty of small electronics, you have to pay for the tiny size with both higher prices and lost performance. To put the quality of one of the large portables into a novel-sized package, the price will increase. Also, in spite of the companies' efforts, the performance won't be as good, either. To get what you want, you have to make sometimes painful tradeoffs: "This size is too big, but I don't have the money for this feature, so I'll settle for this . . . ."

Receiver Review: Grundig YB-400

The YB-400 is one of the highest-rated small portables because of the high performance that the user gets from such a small, solid, inexpensive box. Here are some of its features:
- 0.54–30 MHz AM and shortwave, and 87-108 MHz FM (stereo on headphones)
- 40 memory presets
- Direct-entry keypad
- Multifunction LCD readout
- Dual clock/timer and alarm modes
- Selectable tuning steps (1- and 5-kHz steps)
- Telescopic antenna and external antenna jack
- Adjustable BFO for SSB reception
- Only 1.51 pounds
- Low price

The YB-400 is very sensitive for such a small receiver. In fact, I was able to hear several low-power SSB pirate stations on the YB-400 with just the whip antenna. The stations were drifty, yet it was easy to "track" the drift with the BFO control. Furthermore, when I used the YB-400 with a longwire antenna, it was not susceptible to overloading and being splattered with images from other strong stations. This radio is very small, sturdy, and perfect for mobile listening.

Of course, although the YB-400 is good for general listening, it isn't a "DX machine" like the Drake SW8. Because of the small package, only a tiny speaker with a small compartment is provided—the sound from the speaker is tinny. However, the cost is only ⅛ of the SW8, and it can travel to many places that the SW8 can never go. Overall, for the size and the price, the Grundig YB-400 is tough to beat.

Batteries

The number of batteries that a receiver requires is closely related to its size and weight. You can assume that a large receiver will require more batteries than a tiny one. At the top end of the spectrum, the Drake SW8 requires six D cells. On the other hand, the tiny Philips AE-3905 requires only two AAA batteries for operation. Most portables require a few AA or C batteries (or a combination of the two). Of course, the batteries add quite a bit to the weight. Some light receivers, such as the Sangean ATS-803A, require a stack of batteries that weigh more than the receiver.

Battery life is also a key consideration with a portable receiver. If your batteries croaked during a safari or a mountain hiking trip, the radio would be useless until you could get back to civilization. Like the tuning rate, battery life is one of those specifications that's tough to find information for. Part of the reason for this lack of information is that the life depends on the type of batteries used and how loud the volume is. Still, it's worth calling to some of the shortwave mail-order distributors and checking to see if they can give you some helpful hints.

Can I Have a Second Opinion?

If you are about to shell out the big bucks on a new receiver, you sure don't want to be stuck with a lemon. What if the ad lied and those knobs do nothing more than spin? What if the $200 portable is just a $20 radio with a salvaged surplus calculator keypad? What if you can pick up more stations on the fillings in your teeth than on the radio you're about to purchase? Peruse the equipment reviews from the following:

- IBS—IBS produces the annual *Passport to World Band Radio* and the in-depth *Radio Database International* white pages technical radio reviews. These reviews are some of the best you can find. They include technical specifications, plenty of hands-on radio articles, a star rating scale, and a price-versus-performance symbol.
- WRTH—For decades, the *World Radio TV Handbook* has been *the* annual source of information on radio and TV broadcasting in the world. The *WRTH* receiver reviews are less "flashy" and witty than their counterparts at *Passport*, but they are also excellent and very thorough.
- Magazines—*Popular Communications* and *Monitoring Times* both feature regular receiver reviews. The amateur radio magazines *QST*, *73 Amateur Radio*, and *CQ* also feature radio reviews. These reviews are typically written by different people, so you might not get a real sense of consistency. However, *Monitoring Times* does feature columnist Larry Magne, who reviews receivers for *Passport* and *RDI*. Nonetheless, you will learn a lot by reading from these sources, and you won't have to wait until the next radio annual is available.

What are all of those knobs for?

Anyone accustomed to simple broadcast band receivers might find the front of a shortwave set intimidating. As Fig. 2-3 indicates, modern shortwave receivers have no shortage of controls!

2-3 A superb receiver, with memories, dial and keyboard tuning, and remote-control capabilities.

Because controls and options vary, the following is a survey of the more popular features, but it is not an exhaustive list.

Display Many modern shortwave receivers use LCD displays, although some gas-plasma displays are used in Icom and Kenwood sets, and analog (dial) readout is used in a number of portable receivers. The LCD and gas-plasma digital panels simply state the frequency tuned, usually in megahertz. Some sets have a dimmer control to dim the light during the daytime and to save battery power. These panels also often contain the clock, type of mode being received, the memory number, the filter position in use, etc.

Tuning knob Just a few years ago, it would have been ridiculous to consider a receiver without a tuning knob. However, with the advances in technology, some receivers only use keypad entry. With keypad-only entry, you type in the frequency that you want to receive. If the station is slightly off frequency, you must either type in a new frequency or tune a little thumb-dial for fine frequency tuning. Personally, I am very uncomfortable using receivers that have keypad-only entry.

Many portable and most tabletop digital receivers use a combination of keypad entry and knob tuning. This is really handy because you can key in a frequency, then manually tune it in better or scan across the band. With keypad entry/knob tuning, you have the best combination of convenience and having a good "feel" for the radio.

Tuning speed Very small shifts in frequency require fine tuning. On the other hand, going from 12000 to 12985 kHz quickly requires very coarse jumps. Many sets have buttons that control how rapidly the display changes when the tuning knob is rotated. A fast setting of 5 kHz jumps also makes scanning much quicker because broadcast stations are generally spaced 5 kHz apart. Tuning speed variations are very helpful when you are seriously exploring or DXing the bands.

Mode For shortwave broadcast reception, the only necessary mode is AM. However, the USB, LSB, and CW modes are all much more common on the two-way (amateur, military, marine, etc.) bands than the AM mode. These modes are all necessary for standard shortwave listening. The more exotic receiver modes are FM (sometimes used for broadcast feeds on shortwave) and FSK or RTTY (for receiving radioteletype).

Filter settings For most portable receivers, there is only one nonselectable filter position or there is a switch so that you can change between the "wide" or "narrow" positions. Tabletop receivers have a switch so that you can flick between the different positions.

S meter The S meter shows the strength of the signals being received at the radio in decibels. Although S meters are helpful, meters are not calibrated similarly on different radios. S7 on one receiver might not equal S7 on another. Some receivers, such as the Sony ICF-2010, have a row of LEDs that run from 1 to 10 and scale the signal strength linearly.

AF gain The audio-frequency gain control is basically just a volume control. This potentiometer controls how much the AF circuits amplify the signal.

RF gain The radio-frequency gain control is much like the AF gain. However, instead of amplifying the AF signal, this potentiometer controls the amplification from the RF circuits. Both the AF and RF gains are used in conjunction to control the volume.

Tone As with ordinary mediumwave and FM receivers, this control alters the tone of the audio from more bassy to more treble-ish.

Automatic gain control (AGC) The strength of shortwave signals often surges up and down; radio signals traveling thousands of miles and ricocheting off the earth, sea, and ionosphere rarely stay at exactly the same strength from minute to minute. An AGC enables the radio to compensate, reducing amplification when the signal comes in strongly and increasing it when the signal fades. Fast, slow, and medium settings govern how often the set samples the incoming signal and how quickly it responds. The fast setting is best with AM, the slow is best with SSB and CW, and the medium is best when you are scanning many different signals and modulation modes.

Noise blanker (NB) The noise blanker will eliminate some types of impulse noise. Sometimes useful in vehicular-mounted radios, most noise blankers do little against the bulk of shortwave interference.

Passband tuning and RF notch filters Used only in the finest sets, passband tuning allows you to narrow the band of radio signals passed into the set; it thus functions as an adjustable filter. An RF notch filter lets you chop out a narrow range of frequencies within those passed on, so as to cut out whines, hums, etc. This can also be accomplished with audio filters, but the technique is more efficient before the signal is amplified.

Receive incremental tuning (RIT) A form of extremely fine tuning, RIT can be very useful for RTTY and SSB tuning, where accuracy is very important.

Squelch The squelch can be set to turn the audio off until a signal of a certain power is received. Thus, it silences the annoying background noise of an "empty" frequency. It is useful for monitoring a single frequency. The threshold turn-on level is adjustable, so everything from very weak to very strong signals can turn on the receiver's audio.

Output Most receivers have a built-in speaker of marginal quality, a socket for headphones, another for a recorder (which is attenuated and does not cut out the receiver's loudspeaker), and a socket for auxiliary speakers (which does cut out the inboard speaker). The last, when patched to the input or tape connections of a stereo system, can produce impressive audio!

Antenna connections Most sets incorporate at least two antenna connections—a PL-259 socket for coaxial cable and push terminals for wire connections. The former is used for low-impedance antennas, such as for dipoles or quarter-wave verticals, the latter for high-impedance antennas, such as longwires.

Memories and scanning Many sets now have the ability to memorize a list of favorite frequencies and pull them up at the push of a button. Most of these receivers have one or more search modes, which can scan all or some of these memories or scan a preset range of frequencies. These can be useful for casual listening (such as having the strongest frequencies for the BBC, Voice of America, Radio Netherlands, etc. in the presets) or specialized listening (such as having the spy or Strategic Air Command frequencies in the presets).

Most memories also control the mode, as well as the frequency, so you can go from 6955 kHz USB to 7355 kHz AM without having to separately switch from the USB to the AM mode. Some receivers allow their memories to be changed by a keyboard, and others require the user to retune the set and punch a Memory In or Enter button for each memory. The Kenwood R-5000 has a really convenient memory access; just put the radio in memory mode and you can change between the different memories by turning the tuning knob.

Other features and controls Many of the other shortwave radio features are related to such things as clocks and timers. Many portable radios have clocks with Time Set, Sleep, and Alarm Set buttons. Some of the portables also feature FM, FM Mono, and FM Stereo switches or buttons.

Where to buy a receiver

If you have looked over some of the shortwave books and magazines, you might already know what receiver you want to buy. Now what? Chances are that the only local store that carries any type of shortwave radios is Radio Shack. Radio Shack carries a full line of portable receivers, most of which are made by Sangean. Some other primarily nonshortwave companies, such as Sony, are getting some of their receivers into such chain stores as Sears. So, with some searching around town (provided that you at least live in a small city) at the various electronics stores and stereo shops, you might locate about half of the available portable receivers.

The problem is that you won't see many of the good receivers. Radio Shack receivers *are* available at every one of their branch stores (Fig. 2-4). A few Sony, Panasonic, and Magnavox receivers are somewhat common because these companies all make consumer video and audio equipment. One of the specialty dealers might decide to pick up a shortwave radio or two, just to see if they'll sell.

2-4 Like many portables, the Realistic DX-390 contains scanning capabilities and a number of presettable memories.

However, some of the equipment manufacturers specialize in shortwave equipment. You won't find their receivers everywhere. Grundig is one of the largest manufacturers of shortwave radios in the world; although they do specialize, some of their radios are available in retail outlets. Sangean radios are most commonly seen under the Radio Shack name. Drake is legendary for innovative, high-quality U.S.-made shortwave and satellite receivers, but their equipment is rarely available outside of specialty shortwave shops and mail-order distributors. Lowe's description is much the same as Drake, except that the company is based in England.

What are the Leading Names in Shortwave Receivers?

You might be flipping through a general-interest magazine and see an ad for a Platypusonic ZR-2000 receiver. It can do everything! Receive all of the shortwave stations in the world without an antenna. Make shortwave sound better than FM. Why, you can hear stations all of the way from the mysterious land of Russia! And because it's made in China, they can offer it to you for only $79.95! Sound too good too be true? It probably is. There aren't many manufacturers of shortwave radios, so tracking the good companies is easy. Although the following companies might have made a few "dogs" in their time, the companies themselves are all dependable and have been building shortwave radios for years:

- AOR
- Drake
- Grundig
- Icom
- JRC
- Kenwood
- Lowe
- MFJ
- Panasonic
- Philips/Magnavox
- Radio Shack
- Sangean
- Sony
- Yaesu

Another drawback of picking up a receiver from a local store is that it will probably be close to the list price. The major shortwave mail-order companies sell so many radios that they can offer portables for anywhere from $10 to as much as $200 off the list price. The only exception to this rule is Radio Shack, which generally offers their radios at good prices and sometimes has excellent sales. Of course, because Radio Shack only sell their own receivers, you have to be yearning for a Radio Shack receiver. If you are, be sure to check all of the local stores for special sales and especially for the closeouts. During the closeout sales, receivers are often sold for 50% or less off the original price!

If you're as cheap as I am, you might be on the lookout for a used receiver. However, the prospects of finding one are slim. Used shortwave radios are almost never advertised in the classified sections of newspapers or "shopper-type" papers. I go to plenty of hamfests/computerfests, where all sorts of amateur radio equipment, computers, consumer electronics, and parts are sold. Not many portable receivers even turn up here; however, if you have a hamfest/computerfest coming to your hometown, it's worth a look.

There's a slim chance that you might find the portable receiver that you're looking for by looking through the classified ads in the back of shortwave magazines and newsletters or through the shortwave/amateur sections of computer systems (such as HAM-FORSALE on the Fidonet or rec.radio.shortwave on the Internet). As with anything, there's a chance that you could get ripped off by ordering a radio through the mail from an unknown individual. However, the active shortwave and amateur radio communities are generally small, honest, and tight-knit.

In one case, I saw where someone had ordered a radio from another hobbyist over one of the computer networks. The man who ordered the radio hadn't received the radio after a year, and he started complaining. He couldn't sue for his money back because the two lived in different countries. However, an amateur operator in the same city as the "seller" who had never met the victim, tracked the "seller" by using information from the network. The petty thief had run off with the money. The operator somehow convinced him to return it. I have no idea what he said that made such an impression. This kind of helpfulness and accountability is very rare in most other hobbies.

Other than these rather infrequently used offerings, your only hope for a used portable is to get a trade-in radio from one of the specialty shortwave mail-order companies. Most of these companies service their equipment and offer brief "return-for-any-reason" periods and 30-day parts-and-labor warranties. Because of their reputation and because most of these companies have been in business for decades, you can purchase a used receiver from them, without fear.

Major Shortwave Mail-Order Companies

Looking for a particular shortwave radio that you can't find anywhere else? Want the best price for a receiver? Want a radio that's backed with a warranty? Or do you just want a good price on a used receiver? Try one of the shortwave mail-order stores. Although a handful (or two) exist, these are the largest:

ACE Communications
10707 E. 106th St.
Fishers, IN 46038
800 445 7717 (tel)
800 448 1084 (fax)

Atlantic Ham Radio
368 Wilson Ave.
Downsview, ON M3H 1S9 Canada
416 222 2506 (tel)
416 631 0747 (fax)

Barry Electronics
540 Broadway
New York, NY 10012
800 990 2929 (tel)

Communications Electronics
POB 1045
Ann Arbor, MI 48106
313 996 8888 (tel)
313 663 8888 (fax)

Com-West
48 East 69th Ave.
Vancouver, BC V5X 4K6 Canada
604 321 3200 (tel)
604 321 6560 (fax)

C. Crane Co.
558 10th St.
Fortuna, CA 95540
707 725 9000 (tel)
707 725 9060 (fax)

Dubberley's on Davie
920 Davie St.
Vancouver, BC V6Z 1B8
Canada 604 684 5981 (tel)

Electronic Equipment Bank
323 Mill St. NE
Vienna, VA 22180
800 368 3270 (tel)
703 938 6911 (fax)

Gilfer Shortwave
52 Park Ave.
Park Ridge, NJ 07656
201 391 7887 (tel)
201 391 7433 (fax)

Grove Enterprises
POB 98
Brasstown, NC 28902
800 438 8155 (tel)

Lentini Communications
21 Garfield St.
Newington, CT 06111
800 666 0908 (tel)

Norham Radio Inc.
4373 Steeles Ave. W.
N. York, ON M3N 1V7 Canada
416 667 1000 (tel)
416 667 9995 (fax)

Tucker Electronics & Computers
1717 Reserve St.
Garland, TX 75042
800 527 4642 (tel)
214 348 0367 (fax)

Universal Radio
6830 Americana Pkwy.
Reynoldsburg, OH 43068
800 431 3939 (tel)
614 866-2329 (fax)

Accessories

An amateur radio operator will often have a basement full of radios and a pile of little boxes that adjust levels, match levels, couple different boxes together, switch between antennas, etc. The way these guys accessorize, you'd think that even Mary Kay could make a killing off of them. Fortunately, portable shortwave radios are so simple that you don't really need to accessorize.

Antenna tuners Antenna tuners aren't particularly useful for shortwave-listening applications; they are most useful when transmitting, to "match" the antenna to the transceiver. For more information on amateur radio equipment and operations, see Chapter 11, "Amateur radio."

Audio filters These plug into the auxiliary speaker connection of the set and work on the audio input. Most incorporate an audio bandpass filter. Speech frequencies are generally about 500 to 2500 Hz. Cutting off frequencies above and below this range can eliminate a lot of whine and hum, while leaving the speech relatively unimpaired. Most audio filters are adjustable to narrow this bandpass, cutting out further noise at some expense of intelligibility. Most also incorporate a notch filter to eliminate more interference. Most of these audio filters are analog; they are now being rendered obsolete by the more flexible and effective DSPs.

Converters Converters attach between the antenna and the receiver. They enable the radio to receive signals from other bands. The most common converters are those for shortwave radios that allow them to receive VHF frequencies (for amateur radio or public service bands) or those for standard mediumwave car radios that allow them to receive shortwave stations. Converters were most popular in the 1950s and 1960s, when radios were large and expensive. Even in the 1990s, however, converters are still somewhat common.

Computer interfacing In the 1980s, shortwave radio manufacturers anticipated the technological innovations of the personal computer revolution. At this time, some began manufacturing optional computer interfaces for their tabletop receivers. With the computer interface, the receiver can be controlled by special software. Most of the programs allow you to set up a programmable database, automatically turn on and tune in different stations, control the receiver remotely via modem, and much more. This topic is covered in more depth in Chapter 14, "Computers and radio."

DSPs *DSP (digital signal processing)* units are a broad category of computer-related electronics that receive an analog signal, change it into computer binary code, process that code with various algorithms, and change it back into an analog signal. DSPs are very useful for eliminating white noise, removing signals, limiting signals, and filtering audio (Fig. 2-5).

DSPs have been used for decades, but they have just recently hit the consumer market because of the high cost of computer processing chips. In just the past few years, the new advancements in miniature computer chips have allowed more research into consumer products and parts that cost less. Now, DSP "black boxes" are becoming common in shortwave listening.

Most motels, apartment buildings, and other public buildings are infested with radio interference from fluorescent lights, neon signs, and computers. Although the interference-reducing circuits in these DSPs can't entirely eliminate this type of in-

2-5
The JPS NTR-1 DSP is excellent for removing heterodynes and wideband noise from shortwave broadcast signals.

terference, they can remove some of it. JPS Communications makes several units that are effective against this type of noise and also against *heterodynes*—high-pitched squeals that are the byproduct of several overlapping radio signals.

 However, beware of most DSPs! The DSPs that are available in the shortwave market are typically from the amateur radio market. These DSPs are intended for better reception of narrow Morse code and SSB signals. As a result, the bandwidth is much too narrow for standard AM audio. In most cases, I found that using a DSP filter made the audio of even a clear AM broadcast sound muffled, muddy, and unpleasant for listening. Presently, only JPS Communications builds some external DSP units specifically for shortwave listeners. If you are uncertain about which DSPs are good for general shortwave listening, ask a dealer to show you which models are made for broadcast listening or ask if the DSP has bandwidths between 4 and 6 kHz wide.

The Whys and Whos of DSP

Why do you want DSP?
- Reduce random noise
- Provide better filters than your receiver has
- Remove heterodynes

Who makes DSP units?
- JPS Communications
- MFJ
- Timewave
- Ramsey
- Radio Shack

Two innovative companies have introduced DSP into their shortwave radios: SGC and Watkins-Johnson. SGC offers several DSP functions that are integrated into an amateur/marine transceiver. Watkins-Johnson built an almost completely digital receiver that features programmable filters and 1-Hz digital readout. Although it seems doubtful that either of these companies will make a portable receiver any time soon, it will be interesting to see how long it will be before digital and DSP-integrated portables are on the market (thus reducing the need for DSP "black boxes").

DSPs aren't for everyone. They generally cost upwards of $150, and they are about the size of a hardcover novel. So, if space and cost are big issues with you, you're better off without one. However, if you're looking for performance, a DSP might perfectly complement a high-grade portable or tabletop receiver.

Preamplifiers and preselectors Most preselectors are actually tuned RF amplifiers. They preamplify a weak radio signal but are tunable so that radio signals on other frequency ranges are not amplified. They can thus improve both sensitivity and selectivity. However, they make tuning much more difficult, and they are basically useless with the recent tabletop radios that already have excellent sensitivity and selectivity. However, preamplifiers and preselectors are very helpful with old, classic tube receivers that are lacking in both sensitivity and selectivity, yet are historical pieces that should have a permanent spot in the shack.

Surge suppressors Up until the dawning of the personal computer era, voltage surge suppressors were uncommon technical pieces. Today, however, you can pick one up at any mall or shopping plaza. Lightning strikes a mile (or closer) away, snapped power lines, and even the startup of appliances with motors can generate voltage spikes or surges within your electrical lines.

The typical North American 115-volt house voltage can surge to several thousand volts for a few thousandths of a second. Far too brief to start fires or otherwise pose a danger, these spikes can nonetheless wreck the latest solid-state electronics. Some surges are powerful enough to leap across the gap in the set's on/off switch and cause damage—even if it is turned off.

Unplugging the set when it is not in use is the best protection, but you can hardly leave it unplugged while you're using it. Surge suppressors, available for as little as $20, plug between the set and the wall outlet. They divert the surges to ground while passing the wall voltage to the radio.

 Beware! Not all power strips are true surge suppressors. If you have an expensive radio that you want to be protected, be sure that the power strip is a true suppressor. Many power strips also are called "surge suppressors," but only buy the ones that will insure your equipment in the case of damage. The others provide no guarantees and really, no protection.

Conclusion

One of the most difficult aspects of shortwave listening is simply choosing what equipment is best for you and then finding it! With just a little bit of research, such as calling for information and looking through catalogs, you can choose a receiver that will perform well and will be easy to use. Now, you need an antenna.

3
CHAPTER

Antennas and more

If you want to hear anything on the airwaves, you've got to provide a way for that radio signal to get into your radio. If you want to fish, you have to have a fishing rod, enough fishing line, a hook, and some bait. If you want to listen to shortwave, you simply need an antenna. Shortwave radio antennas are something like using a fishing rod, but you never have to worry about whether or not the signals are "biting," if you will cast into a knotted mass of reeds, or if you're going to reel in an angry snapping turtle.

One look at an engineering textbook on antenna design and construction, though, and you will find a mass of equations printed on thousands of onion-skin pages. If you wanted be a rocket scientist, you would be at the Goddard Space Center. Shortwave antennas really aren't too difficult to install and use—especially when you're traveling somewhere. In spite of anything you might read in a communications magazine, newsletter, or book, the bottom line is getting a chance to listen to that radio. If you think that you must string up a 32-element log-periodic antenna at 60 feet anytime you want to listen to the shortwave, you'll never hear anything (Fig. 3-1).

The antennas in this chapter are for portable and tabletop radios. If you have a portable receiver, just read the next section; it should cover everything that you need to know for good reception. However, because most portable antennas are simple compensation versions of standard, fixed antennas, you might want to read the entire chapter if you are a new listener and have a tabletop receiver. Portable antennas are acceptable to use with tabletop receivers, but the reception probably won't be quite as strong.

Antennas for portable receivers

Take a gander at a portable receiver, and you'll see that it has a retractable chrome whip antenna. It's got more segments than a millipede, and it's long enough to disembowel a giraffe. You should be able to hear broadcasters from Tibet with this monster, right? Unfortunately, *you* are probably a better antenna than the mighty chrome behemoth. If you want to check it for yourself, turn on the radio, wet your fingers, and touch the antenna. Chances are that the signals will improve. You're a

3-1 Who wouldn't like to spend a few hours receiving with the curtain antenna at the Swiss Radio International relay station in French Guiana?

better antenna than the one on the portable, but don't bother mentioning it on your resume; you're still not a very good antenna, and jobs as antennas don't pay well.

So, if the whip antenna on a portable doesn't do its job, what then? It's easy to string up a quick antenna that will greatly improve reception. The cost is almost nothing, and the time spent on installing these antennas is also just above nil. Even with excellent portable radios, such as the Grundig Yacht Boy 400 and the Sony ICF-2010, using only the whip antenna results in a real loss of performance (about like coupling a $1000 stereo to surplus 89¢ speakers). These two shortwave radios will perform on the same level as many tabletop radios that cost much more. Your radio is only as good as your antenna; even the most basic shortwave program listeners should at least do some limited experimentation with external shortwave antennas.

However, if you forget your antenna or don't have any wire handy (not everyone carries a spool of wire in their purse or suitcase), you can use the whip antenna with limited success. In a worst-case scenario, don't sweat it, worrying about trying to string up another antenna. If you're out in Las Vegas with the bingo club and stringing up a decent antenna will require too much time, then don't do it. The key here is to be able to hear some news or other interesting programming when you can, not to win a Nobel Prize for Shortwave Technology on the Road.

Popular Fallacies About Shortwave Antennas

1. *Indoor antennas are ineffective.* As long as the building is not metal or metal-structured, they should work fine.
2. *Loops in antennas will cause large signal losses.* Loops might cause an antenna to snap, but they won't make a noticeable difference in the signal.
3. *The antenna must be as high as possible.* This is true for transmitting, but not for receiving—especially in the low shortwave frequencies and in the AM broadcast band.
4. *Insulation or tarnish will prevent the radio signal from entering the antenna.* Corrosion might ruin connections, but neither it nor tarnish will ruin an antenna.
5. *Only specific-frequency antennas work well.* There is a case for this, but random-wire and longwire antennas are very effective, nonetheless.
6. *TV antennas are fine for shortwave reception.* They're okay in a pinch, but you'll notice a big difference between a TV antenna and any standard shortwave antenna.

The important thing at this point is to get back to the basics. An antenna is necessary to receive signals. In the case of the portable radios with the built-in whip antennas, an external antenna should improve reception. Just how much the reception should improve depends on what it's worth to you to provide the auxiliary antenna.

For example, when I go to my in-laws' house over the holidays, it's not worth it to me to install an elaborate antenna. I want to do some listening, but I don't want the time that I would spend setting up an antenna to cut into the holiday festivities. So, I settle on a low-class wire antenna that requires about 5 or 10 minutes of installation time. It's not great, but it's much better than just using the whip antenna. At other times, when I plan to do more listening and when I have more time, it's worth it to spend more time and install a better, more complicated antenna.

Likewise, a travelling businessperson would not likely be able to build a proper shortwave antenna and use it on a regular basis. The setup time, space, and weight required would be prohibitive. Besides that, chances are that he or she would be planning only to catch the news or programming from one of the large, easily heard shortwave stations. This being the case, an antenna that could raise the signal level from fair to good is all that would be necessary. That is exactly what this chapter is all about: simple, quick antennas that can fill the minimum requirements for listening.

Connections

It's very simple to attach antennas to the new breed of portable radios. Some portable radios (such as the Grundig Yacht Boy 400) contain built-in jacks for plugging in external antennas for shortwave, some (such as the Drake SW8) contain built-in jacks for plugging in external antennas for the radio bands other than shortwave, but most (such as the Sony IC-SW30) don't contain any jacks for external antennas.

Through the 1950s, 1960s, and 1970s, if you wanted a decent shortwave antenna, you had to build it yourself. The radio manuals didn't mention how to do it, and there were few available publications for the shortwave listener. Then, the approach seemed to be that shortwave was just something to listen to while you prepared to be a radio amateur. In that time, you just read amateur radio books, experimented with building antennas and falling out of trees, and waited until that glorious day when you could graduate to being a ham operator. Amateur radio can be a great hobby for experimenters and people who like to talk. However, shortwave broadcast listening is also highly informative, and it's a lot of fun. By the late 1980s, some receiver companies recognized the real potential of shortwave listening, and they started selling complete shortwave kits: an easy-to-use portable shortwave receiver with a guide that explained what shortwave is, and a quick-install reel antenna.

The reel antenna is one of the handiest shortwave implements in years. Rather than stringing an antenna or suffering with the weak signals from the whip antenna, the reel allows you to almost effortlessly put up a decent little antenna. You can take it down and put it away in less than a minute.

Antenna Stringing Tips

1. *Never string antennas where someone might trip or fall against them.* Avoid this even if you are by yourself in a motel room. You might forget about the antenna and hurt yourself or your radio.
2. *Avoid metal obstructions.* Either get your antenna outside or against the windows of a metal-structured building (such as nearly all motels

and large modern buildings), or the signals won't even reach your antenna.
3. *Ask permission to install large antennas.* Some mom-and-pop motels or inns have windows that open to a courtyard—perfect for a 100' (or longer) antenna. Ask for permission to install any such antenna, or you will be held accountable if it causes any damage.
4. *Use the room furnishings to support your antenna.* When stringing wire around a room, loop it around pictures, over door frames, curtain rods, etc. This way, you won't damage the walls with nails or tape.
5. *Never use rocks to support wire antennas in trees.* Chances are good that you'll knot the rock in the tree and leave it dangling—a real health hazard. A better deadweight is a sock containing a few ounces of sand.
6. *Use a fine gauge of insulated wire for homemade antennas.* Large-gauge wire is heavy and bulky.

If you don't have a reel antenna for traveling, but you want an antenna other than the whip (and you don't want to hire someone to act as your antenna), you can either buy or build an antenna. Reel antennas are inexpensive and are available from shortwave dealers and Radio Shack, so buying tips are unnecessary. However, the following sections cover some simple antennas that you can build. For most of these antennas, you only need wire and a few easy-to-find parts. Soldering is an option in a few cases, but there are enough choices here that you won't even have to solder.

The basic approach for connecting an external antenna to a portable receiver would be to strip about 2" of insulation off the end of a wire (most anything from #8 to #30, although the most popular antenna gauges are from #14 to #22) and to wrap the stripped end around the end of the receiver's whip antenna (Fig. 3-2). If you want to impress your friends, you can use the proper term for this type of connec-

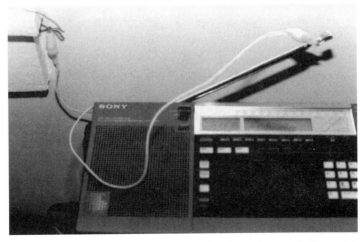

3-2 Clip leads connect the antenna to the whip antenna of a Sony ICF-2010 receiver.

tion and tell them that the antenna is *inductively coupled* to your radio. One problem with this method of antenna coupling is that it's easy to end up with an incomplete connection. It is really annoying to have to rewrap the end of your antenna around the radio's whip antenna every time you tune to another station, just to make sure that the connection is complete.

A solution to the connection problem would be to pick up a pack of *clip leads* (also known as *test leads*) from an electronics store, such as Radio Shack. Each lead is a different color and has a shielded alligator clip on each end. These leads are really handy to have around to make emergency connections, to hook up several receivers to the same antenna, etc. It's a good idea to purchase a pack of these whether or not you want to use this as your main antenna connection arrangement. With the clip leads, you only have to connect one alligator clip to the antenna wire and connect the other clip to the receiver's whip antenna. Easy!

To make the most permanent antenna connection for a portable radio, solder a connector on the end of the antenna. For radios with an external antenna jack, this would probably either be a ¼" RCA plug or a ⅛" plug (like the kind used on headphones for portable stereos). Make sure that you check what type of jack your radio has; otherwise, you might accidentally solder the wrong one on your antenna. For radios without the external shortwave antenna jack, a standard alligator clip is the best connector to solder to the antenna. I would rather use clip leads to connect than plugs because I'm prone to accidentally ripping them off or breaking them. However, some people prefer to solder the plugs.

The wire

The *wire antenna* is the most basic form of antenna. The wire might not do everything well, but it does everything nonetheless. What I am calling the wire antenna is simply just a piece of wire, configured in whatever way. Unlike most standard antennas, the wire has no transmission line (used to carry signals from your antenna to your receiver), such as coaxial cable. This antenna functions as both the antenna and as the transmission line.

Like most forms of horizontal and vertical antennas, the wire can be almost any length. However, a longer, relatively straight version of the wire would generally be considered a *longwire antenna*. When does the wire become a longwire? I would hesitate to call a 5' piece of wire a "longwire" antenna. Some sources say that any antenna that's one wavelength long or longer is a longwire. Under this criteria, my 125' antenna would be considered a longwire above about 7500 kHz, and it would be considered a simple wire below about 7500 kHz. This sort of variable definition seems overly bulky and confusing for standard shortwave-listening purposes. Personally, I would say that anything over about 100' is a longwire antenna.

The receiving characteristics between a 100' longwire and a 1000' longwire are quite different, so if you're writing a reception report to a station or describing your antenna to someone, it's best to state the length of it and the direction that it is pointing. Wire antennas of less than 100' in length are much less directional in the way that they receive, so the direction in which shorter longwires are pointing is not as important for others to know.

As I mentioned, the receiving characteristics of a long longwire antenna are much different than those of a shorter wire antenna. The longer it is, the better it will receive from each end of the antenna. Likewise, the longer it is, the less signal can be picked up from off the sides of the antenna. This phenomenon occurs according to the proportion of antenna length to the wavelength (frequency) of the station that you're listening to. An extremely long longwire antenna might be a most incredible antenna for someone who only listens to a few small segments of the world.

For example, one weekend some friends and I got together to listen for low-power European shortwave pirate stations. We strung a longwire antenna that was somewhere between 1200' and 2000' in length (according to the campground map, the antenna was approximately 1700' long). The antenna was pointed to the northeast, right at Europe. As a result of our efforts, we heard several European pirate broadcasters with low power (only about 100 watts) and many North American pirates, including one low-power AM pirate from New York City. In the case of the New York City AM pirate, it was booming in with excellent fidelity. I decided to try hearing the station with about 30 feet of wire a few hundred feet away. Nothing. The massive longwire made the difference between hearing nothing and hearing an excellent signal. On the other hand, we found that many of the North America stations had better signals with a simple inverted V antenna because it was much less directional. As you can see, each antenna type has its advantages and disadvantages. It's all a matter of choice and taste.

 One of the few drawbacks with today's crop of miniature portable shortwave sets is that the tiny electronics contained within the case are susceptible to electric currents—even signals within the radio. I've heard a handful of stories about one excellent portable. Because of the quality of the radio, many listeners DX with it. As a result, the longwire is a logical choice to connect to this radio for DXing. However, if the signal into the receiver is too strong, it can overload the front end of the receiver and POOF!, a semiconductor or two is rendered useless. You won't even see smoke, hear a blood-curdling explosion, or smell molten plastic. You'll only notice a significant drop in signal strength. Although the radio is repairable, the $50 (or so) bill per pop should be prohibitive. So, the warning is: *The longwire antenna could put your portable radio at risk. Be careful!* Just how long is too long? I haven't gone over 400' long with my portable receivers (except the Drake SW8, which is built to handle stronger signals). Maybe I'm conservative with my antennas, but I like my radios and I like them to be working!

Thus far, the only version of the wire that I've covered in this section has been the outdoor longwire. The wire can also be used indoors, and it can be configured into different shapes. These antennas, which must be easy to install and remove, are the key types for portable shortwave radios on the market today.

As stated earlier in this chapter, you shouldn't worry about little details, such as antenna resonance or impedance, with these indoor makeshift antennas. The big concern is just to get a length of wire up to "grab" as much signal as possible from the air. Crude as that might seem, it sure beats not having an antenna at all!

I have had moderate success with these types of miniature indoor antennas. I am the type of person who enjoys listening to music, news, or sports while working, so they have been particularly useful. For this type of simple program listening, these antennas often made the difference between success or failure at being able to understand the BBC news over the sound of running water at the sink while I washed dishes. As you might expect, these indoor wires are poor DX antennas—I have had a few good catches while using them, but these were rare and more based on extraordinary propagation conditions than on good antenna performance.

The reel antenna

One of the most practical new developments in shortwave-listening antenna technology (or gadgetry) is the *reel antenna*, which was mentioned earlier. The reel antenna is basically a handy version of the indoor wire that is neat and easy to carry. The actual wire antenna is held inside of a reel. The other end of the wire has a small clip (somewhat similar to the type used to clip on chip bags), which you can clip on to the end of a whip antenna.

I received a commercial version of a reel antenna from one of my friends. I love having this antenna because I can wind it in immediately after a brief program-listening session in the kitchen. Because the regular wires were sometimes a pain to install and wrap up, I would normally just leave them strung up. This was always a problem because the wire usually crossed the kitchen cabinets and every time someone would reach in for a plate or a glass, they would be in danger of setting off my booby trap. The reel antenna has ended nearly all of these problems for me.

The reel antenna has one flaw: it's very short and it works for only the most basic listening. I wanted to experiment with constructing longer reel antennas—antennas that could actually be used for light- to heavy-duty shortwave listening.

The first problem was finding an appropriate reel. I didn't really want to modify a commercial reel antenna because it seemed to defeat the ingenuity of the experiment. I thought about using a fishing reel (fishing reel antennas have appeared in the hobby press for several decades), but I felt that it would be too bulky and that it would probably unwind into a mass of knots while I was taking it on a trip. My best guess for an appropriate reel was a standard chalk line, like those used in house construction.

I don't know how many people have old chalk lines sitting in their basements and garages, waiting to be converted into portable antennas. However, I would guess that few people do. A new chalk line isn't terribly expensive. In fact, I picked mine up for about $4—much cheaper than the price of a commercially produced reel antenna.

The huge advantage of using a new chalk line is that it hasn't yet been filled with chalk dust. If you choose a used chalk line to convert into a reel antenna, you must rinse it out several times with water. Otherwise, every time you unwind the antenna, the wire will be covered with chalk dust and bits of dust will spill out inside the house. Using a dusty chalk line for a reel antenna might work well for a vacuum cleaner ad on TV, but the resulting mess sure isn't fun in real life.

To build my reel antenna, I first pulled all of the string out of the chalk line reel until the reel was completely empty. Then, I removed the screws in the chalk line case and pulled it apart. The string was easy to get to on the plastic spool, so I just cut it off with a pair of scissors. Next, I got the wire that I was planning to use, ran it through the spool, and knotted it. I used standard #20 stranded hookup wire with rubber insulation. After winding about 50' of wire onto the reel, I cut off the wire, stripped about an inch of insulation off the end and soldered on an alligator clip. Then, I screwed the case back together (the case can be screwed back together anytime after you knot the wire inside of the spool).

I often string up longwire antennas when I visit my parents over the winter holidays. Invariably, I end up with a massive knot in my wire about 75' from the house, and I spend about an hour attempting to untangle the wire with numb hands in 10° F weather. After building the reel antenna, a lightbulb popped up over my head, and I imagined how convenient it would be to have a longwire reel antenna.

For the longwire reel antenna, I used very light gauge (approximately #26) enamelled wire. I chose this type of wire because it has a very small diameter and quite a bit of it could be held on a chalk line reel (Fig. 3-3). Also, the insulation is very hard and slick and is less apt to jam up the reel.

3-3 A reel antenna being assembled.

I was surprised to find that I could fit well over 300' of wire on the reel without having any winding problems. However, one day when the temperature was about 25°, I was putting up the long antenna. The wire pulled off the spool inside of the chalk line case and locked up. I broke the chalk line while trying to straighten the mess. Enraged, I unceremoniously finished off the rebellious antenna by smashing it with a big rock. On my next attempt at reel antenna construction, I used less wire. It stays within the contours of the spool and works great.

The procedure for building the longwire reel antenna is the same as that for building the standard reel, except that I didn't use an alligator clip on the end. I was afraid that the clip would break off because of the light gauge of the wire, so I tied the end of the wire around the old end of the chalk line (to prevent the antenna from unwinding off the reel inside the case). Then, I moved down the wire about an inch and carefully scraped off the enamel insulation for about an inch. After the antenna was up, I connected an alligator clip lead to the place where the insulation was removed and connected the other end to the whip antenna of the portable shortwave receiver.

The reel antennas perform on par with standard wire and longwire antennas of equal length. The whole benefit of using a reel antenna is merely in the ease of installing and dismantling. In my winter experiences with the second longwire reel, I found that I could wind it up in about 5 to 15 minutes (in the latter case, it ran through trees and into the woods, was covered with snow and ice, and it was several hours after sunset). I was pleased. I find the prospects of spending more time keeping warm while listening and less time freezing in snowstorms rather exciting!

The self-sticking screen antenna

The wire antenna and the reel version of the wire are excellent portable antennas. However, both are totally ineffective inside of a metal-structured building, such as a motel, trailer, or office building. In these cases, it would be best to have an antenna installed outside. After working at my job for years, the office was reorganized and I got a room with a window. One of my first actions was to bring in a portable receiver and run an inconspicuous random wire down the side of the building. I used the #26 wire from the last project and tied it off behind a "No parking" sign so that it would be safe. No one would ever see it . . . or so I thought. At the first ice storm, my thin wire had grown to an inch in diameter with ice! Fortunately, no one told me to remove the antenna. A friend of mine tried the same experiment, but his antenna was quickly torn down and whisked away.

The bottom line is that using an outdoor antenna from a public building could cause you problems with your job or you could wind up with a lawsuit (if, for example, someone trips as a result of an antenna that you installed at a motel).

The only good location for an indoor antenna in a steel-structured building is at a window, where only glass is between the antenna and the great outdoors. A 3' wire antenna is impracticably short at shortwave frequencies, so that rules out the standard wire and vertical antennas. One of the best makeshift antennas for this awkward predicament is a piece of window screen.

Window screen antennas have occasionally appeared in hobby publications over the past decade or two. In these publications, the window screen was normally assumed to be in the window. However, it seems that in most of the cases where this type of antenna would be necessary, there would be no window screen in the window. Also, with the "wonders of modern technology," most of the newer window screens are made out of plastic, not metal. Because plastic does not conduct electricity, plastic window screens cannot be used as antennas.

Essential to the window screen antenna is (obviously) a piece of window screen. In the ratio of size and performance versus convenience, a best bet for the size of this antenna would be between $1' \times 2'$ and $2' \times 3'$.

Try the local hardware store or construction supply center for window screen. Most of these stores sell various grades of window screen on large rolls. Once again, make sure that you use metal screen, not plastic screen. Either cut the piece to the size that you want or have it cut to size right in the store by a clerk.

Afterward, you can place duct tape around all of the sides, except for a ½" gap along one of the sides. The duct tape around the edges will prevent you from cutting yourself or other objects while using or transporting the antenna. The small gap in the duct tape is where an alligator clip lead should be placed to connect the antenna to the whip antenna of the receiver. You might instead choose to seal all of the sides with duct tape (not leaving a gap for an alligator clip) and instead solder a wire directly to the screen. I didn't choose this means of connecting the screen antenna to the receiver because the solder connection would be rather fragile. I can just imagine a solder joint getting crushed and breaking in my luggage.

The next step is to find some small suction cups from a craft store. Lately, there has been a proliferation of small suction cups that are used for such things as hanging sun catchers on house windows and hanging Garfield dolls in car windows. Just fasten one of these suction cups on each corner of the antenna, and it's ready to hang!

When I constructed my version of this antenna, I used an old piece of metal screen that I found behind the wall in our laundry room. The screen was old, dirty, and slightly corroded, so I wire-brushed the area where I was going to connect the alligator clip. Cleaning this area of the screen will ensure a good contact between the clip lead and the screen.

As you might expect, the screen antenna is no great performer, as compared to a standard outdoor antenna. However, it does work to some degree of satisfactory, and it is one of the best makeshift antennas if you're stuck in a metal-structured building.

Antennas for tabletop receivers

Dozens of different, fairly common shortwave antenna types are used for different applications. Some of the antennas are chosen for their specific receiving/transmitting characteristics; others are chosen for their size and ease of installation. This section covers two of the most common shortwave antenna groups: the dipole and the longwire.

Wiring cable connectors

Many shortwave receivers provide a connector on the back for a secure antenna connection to the set. This connector is usually the type that matches a common male connector, known as a *PL-259*. The correct connector wiring can be accomplished in five steps:

1. Using RG-58 coaxial cable, remove the black outside jacket insulation for 3", and be very careful not to cut the braided conductor. Remember to slide the collar of the connector down the coaxial cable. I have soldered connectors onto coaxial cable only to find that I had forgotten to put the collar on the coax first. That left me with no alternative but to unsolder the joints, remove the PL-259, and start over.
2. Unbraid down one side of the braided outer conductor. Pull the braid together into one strand, twist it together, and carefully draw it through the hole in the inner sleeve of the PL-259. Then, solder the braid to the shield. Instead of pulling the braid back, you could also leave it on and cut it at the

same place as the inner dielectric insulation. Then, solder the braid solid. The PL-259 will then slide over the end of the coaxial cable and the outside will make contact with the braided shield.
3. Bare one-half of the center conductor by carefully stripping away the white center dielectric just under the braid. Be careful not to nick the center conductor.
4. Push the plug assembly onto the cable, and make sure that the center conductor slides easily into the hollow tip. If the center conductor is longer than the hollow tip, cut that little bit off. Screw the plug to the cable adapter, and solder the braid through the holes that are located at the half-way point on the plug assembly. Solder the center conductor to the hollow sleeve tip, and be certain not to drop any loose solder on the outside.
5. Screw the collar to the plug assembly. The wiring is complete.

One point of contention is the type of coaxial cable used. Like some antenna builders, I prefer the larger RG-8U coaxial cable for better signal strength over long transmission line runs. If you go with RG-8U cable, you will need to use a PL-259 connector that is made for the wider diameter cable. The wiring process is then followed exactly as for the smaller RG-58 cable.

Some of the older, less-expensive shortwave receivers do not use the screw-in type of connector for the antenna, but instead supply the owner with a terminal board with two screw-in connections called *terminal lugs*. One terminal is marked *ground*, and the other is usually marked *antenna*. Coaxial cable can be used with this type of connector by separating the braid and the outer connector into distinct conductors for connection to the back of the receiver. The following steps will produce the best results for this type of cable preparation:
1. Strip the black plastic insulation from the cable for a length of 2". Do not nick the braid.
2. At a point 1¾" from the end of the stripped cable, separate the braid from the center conductor by unbraiding down one side with a pen or a small nail.
3. Twist the braided conductor into one thick piece of wire.
4. Strip the insulation from the center conductor for ½" only.
5. Connect the inner conductor to the antenna terminal. Connect the twisted braid to the ground terminal. The wiring is complete.

The standard dipole and its variations

The standard dipole (and its many variations) has to be shortwave radio's all-time most-popular antennas (Fig. 3-4). This antenna is very simple, performs very well, and can be configured to save space and/or change the reception/transmission pattern. Although some amateurs and shortwave listeners move onto other antennas, most continue to use a dipole for the lower frequencies (below about 13 MHz). The more complicated antennas are quite handy, but below about 13 MHz, they become unwieldy. In these cases, the dipole is a wonderful alternative.

The dipole is a half-wave wire antenna that is fed in the center with a piece of coaxial cable. Because of the two equal-length conductors on either side of the coax-

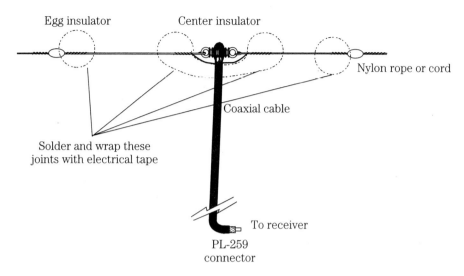

3-4 The dipole antenna.

ial cable, the dipole was known as a *doublet* in much of the older literature. The term *doublet* is still occasionally used—especially by old-timers and those in foreign countries. Because the dipole is fed by 50-Ω or 75-Ω coaxial cable, there are no problems with matching the antenna to your radio. This is a real problem with many other antennas, and the effects are especially dramatic when transmitting. Some, such as the Zepp (so named because they were originally made to dangle out of zeppelins), will allow some of the transmitted signal to get back into the shack and create disturbances (such as cause electrical shocks at the radio, feedback into telephones, etc.).

The materials needed for a dipole are: plenty of coaxial cable, plenty of wire, plenty of small-diameter rope, rosin-core solder, a center insulator, and two egg insulators. When choosing coaxial cable, RG-58 is fine for listening, but use RG-8X for low- to mid-power transmitting and RG-8 for anything over about 750 watts output. The small-diameter rope should be small enough to fit inside the holes of the end insulators, yet strong enough to support the dipole in the air.

Really, any gauge of wire from about #22 up to #6 is fine for dipoles, but the general consensus is that #14 or #16 is the best. Higher-gauge wires are usually used for transmitting. All of the energy from a transmitter is on the very outside of the wire, thus it's best to have as large of diameter as possible. This is known as *skin effect*.

I once heard one story from a broadcast engineer about how he had used more than a kilowatt of power going into a small-gauge dipole. The wire gauge was too small to support the high wattage; it heated up the wire. After a few minutes of transmitting, the antenna wire burned up, and the remains of the dipole fell to the ground!

Measure out and cut the appropriate amount of wire for the band on which you will use. Then, cut this wire into two pieces of equal length. Bare about 3" of one end of each wire. Then, cut an appropriate length of coaxial cable. Attach the PL-259 connector to one end of the coax. Then, remove approximately 3" of the outer sheathing with a knife (be sure not to cut into the wire braid). Then, unbraid down one side of the braided shield. When you are finished, twist the braid together into one thick 2" piece of wire. Next, move down 1" from the end of the coaxial cable and cut off the inner dielectric, leaving 2" of the center conductor exposed.

Tie the bare antenna end of the coax around the middle of the center insulator, and twist some wire around the area where the coax crosses itself. The end is then pulled up so that the center conductor and the braided shield will each reach one loop of the center insulator (Fig. 3-5). Then, pull out each piece of wire, and loop the bared end through the hole in the center insulator. Twist each around itself on the back side of the center insulator. Then, twist the coax braid in with the wire loop on one side of the insulator, and twist the center conductor in with the wire loop on the other side of the insulator. Solder these twisted connections, and make sure that the joints are solid.

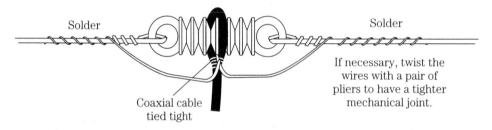

3-5 A close-up of the center connections of the dipole antenna.

Using the same technique that you used for wrapping and soldering the antenna wire around the center insulator, bare the antenna wire at the ends, run each through an egg insulator, twist each upon itself, and solder the connections (Fig. 3-6). Then, tightly wrap electrical tape around all of the exposed joints to prevent them from being weathered and to prevent the solder from joints cracking.

The dipole is then ready to go. The reception pattern of the dipole is essentially a doughnut around the center of the dipole. In order words, it receives and transmits best from angles that are perpendicular to it. The height above the ground makes a rather dramatic improvement in the performance of the dipole, so you should try to get the antenna as high off the ground as possible. Most shortwave listeners have their dipoles fixed at about 20' to 40' off the ground (Fig. 3-7). Radio amateurs often get their dipole up to about 60' off the ground because the height really makes a difference in transmitting effectiveness. When a transmitting antenna is too low to the ground, much of the signal is absorbed by the ground under the antenna. For receiving only, the height difference between 30' and 60' is so inconsequential that the extra height is hardly worth the effort.

Longwire antennas

The classic *longwire antenna* (Fig. 3-8), sometimes called a *wave antenna* or a *Beverage antenna*, is an old-timer that was, in part, developed by H. H. Beverage in

Antennas and more 51

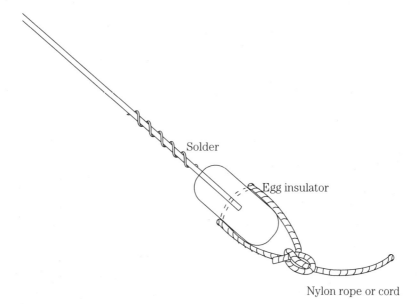

3-6 A close-up of the egg insulator end connections.

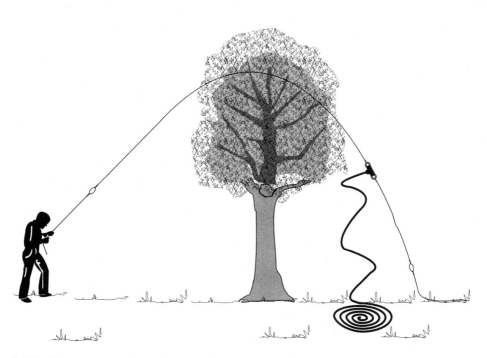

3-7 Pulling an antenna into place.

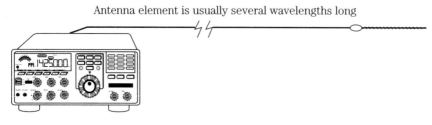

3-8 A simple longwire antenna.

the 1920s. Unlike some of the older antennas, such as the Zepp, the longwire has remained popular throughout the years for a variety of reasons. As stated earlier, the longwire is one of the easiest antennas to build, and it has the effect of being directional. This is very beneficial for the beginner because most other directional antennas are very difficult to build. The longwire is probably the only directional antenna that is easier to build than the dipole.

If electrical noise is a problem in your household, prepare a length of coaxial cable to reach from your receiver to the point at which your antenna will start to head off in a straight direction from your listening post. Twist and solder the center conductor of the coaxial cable to the antenna element. Then, twist and solder the shielded braid to a wire that will connect to a ground stake or a ground system.

It is best to use a thicker gauge of wire for the longwire; otherwise, there will be some losses in signal strength. However, the antenna is so large that lack of signal strength probably will not be a problem, and the cost of several hundred or thousand feet of #14 copper wire might prevent you from ever wanting to build a longwire. For my nonpermanent longwires, I normally just use #26 enamelled wire with success.

Unlike most other antennas, longwire reception is not especially negatively affected by its proximity to the ground. In fact, some of the cold-climate DXers string a few thousand feet of wire across the snow in preparation for a big weekend of DXing. The normal height for a receiving longwire is about 8' off the ground: low enough that it is easy to string, but high enough that no one will be "clotheslined" by the wire. For transmitting/amateur radio work, it's best to keep the antenna as high as possible so that the signal from the antenna won't be mostly absorbed by the surface of the Earth.

An interesting effect of the longwire antenna is that, the longer it is, the more directional it is off the ends of the antenna. If these patterns could be seen from above, they would look like two very large lobes, heading off in opposite directions. The longer the antenna is, the greater the gain and also the smaller the width of the lobes.

Because of the large signal strengths, the longwire is the best antenna to use to hear a tiny signal from a distant country. Because of the very directional properties of this antenna, it is also a favorite for mediumwave DXers, who use the directional properties of this antenna to "null out" strong stations that are on either side of the wire. The more wavelengths long the antenna is, the tighter the lobes become and the more directional the antenna becomes. However, after the antenna is several wavelengths long, some of the smaller lobes that are perpendicular to the two main lobes become stronger. These minor lobes blow the nulling effects of the longwire that are so important at medium wave frequencies. I guess that it's just as well anyway; a four-wavelength long antenna at 540 kHz is 7289' long!

The mediumwave listeners often use a resistor and a ground at the end of the antenna to make the longwire even more directional (Fig. 3-9). By doing this ("terminating" the antenna), the antenna will only receive signals well from the far end—not from both ends of the antenna. To terminate the end of a longwire, solder a 600-Ω resistor at the end of the antenna, then solder enough wire at the end of the resistor to reach the ground. The ground should be either a standard 4' or 6' ground stake or a buried radial system. As stated earlier, you probably won't want to terminate the longwire if you are listening to shortwave because the shortwave bands aren't filled with nearly as much interference as the mediumwave band.

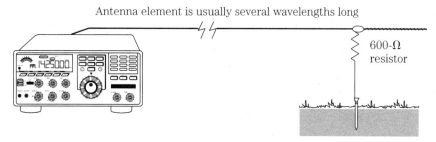

3-9 A terminated longwire antenna.

One advantage of the longwire is that it works well at most any frequency, whether receiving or transmitting. A 308' long half-wavelength longwire that is cut for 1600 kHz will perform well from that frequency and up through the bands. This antenna will also perform quite well even on the FM band and on television frequencies. Longwire antennas are rarely used at these frequencies because, at that many wavelengths, they won't null out stations as well as a yagi or a log periodic. More importantly, you can't hit a button and turn your longwire!

Conclusion

With just a whip antenna, you won't "reel in the big ones." In fact, you'll probably be lucky if you get more than a few nibbles. However, with most any sort of external antenna, your catches will greatly improve.

If these very simple antennas whetted your appetite for experimentation, plenty of other designs are available. For more information on other antennas for home or portable use, see *Build Your Own Shortwave Antennas* by Andrew Yoder. It contains these designs and many others to help you hear as much as possible—and have fun doing it.

4
CHAPTER

Longwave radio

If *shortwave* is a little-known radio band, what about *longwave* radio? Most people in North America have never even heard of longwave radio, and the band is a mystery even to most shortwave listeners and amateur radio operators.

The longwave band, as you might guess, consists of radio waves that are longer than mediumwave and shortwave radio signals. This places the longwave band at lower frequency than either of those bands. Although the longwave band is sometimes broken into several different segments, it is generally considered to run from "0" (dc) to 500 kHz.

Propagation

The longwave band is distinctive in the manner in which the waves propagate. Many people initially become interested in shortwave because the waves "bounce" around the world—almost like magic. Mediumwave and FM are thought to be passé; mediumwave signals seem to be only groundwave (they don't appear to "bounce" anywhere), and FM signals seem to be only line-of-sight. However, if you pay attention to the mediumwave band above 1000 kHz, you will notice some of the same skipping effects that are abundant on shortwave (more on this in the next chapter). However, you might notice that some of the signals above 1000 kHz have traveled further than those at the bottom of the mediumwave band.

With skip distance dropping as the frequency lowers, the bottom line is that there is virtually no skip on the longwave band. At night, the signals behave much like those in the mediumwave band during the daylight hours; during the day, the signals in the longwave band are absorbed by the ionosphere.

Because of the high absorption and lack of skip, radio stations in the longwave band often use brute-force methods to make the signals travel. Although most shortwave broadcast and utility transmitters only output between 1 and 500 kW, longwave stations often use transmitters in excess of 1000 kW (a megawatt)!

You can make a simple deduction here:
1. Longwave stations must be high powered for the signals to reliably travel hundreds of miles.
2. High power costs!
3. For short-range reception, it is much less expensive to use VHF/UHF equipment, where low-power, line-of-sight equipment is inexpensive and whip or directional antennas are small.
4. The result? Few stations operate on longwave.

Broadcasting

In much of the world, there are no broadcasts on the longwave band—with the poor propagation, the huge antennas, and high power required, it's simply not cost-effective. However, in Europe and Asia, it's popular. I haven't talked with any authorities in Europe about this, but I imagine that these frequencies are used because there are so many different countries with different languages in a small space. The mediumwave band is finite. With English, Spanish, Portuguese, French, German, Italian, Dutch, Swedish, Norwegian (and many more!) spoken, there is simply no place for a variety of programming in any one language. Although some of the pressure was relieved after the FM broadcasting bands became widely accepted in the 1970s, longwave is still used by stations that want to cover an international audience, as opposed to merely a local one.

If you live in Europe, you already know about the wonders of the longwave band. It's nearly as common for general listening as the mediumwave band. Unlike in North America, where the longwave band typically only appears on expensive communications receivers, even European car stereos often cover longwave.

The longwave band features a nice selection of countries broadcasting in Europe and the Middle East, but it's not a smorgasbord of radio stations, like the mediumwave dial. Most of the stations on longwave are merely relays of regular broadcasting services that are also aired on the mediumwave and FM dials. For example, the United Kingdom longwave stations on 198 kHz all air the BBC's Radio 4 programming, which is also aired on a number of other frequencies and bands. For more information on longwave broadcasting frequencies, see Table 4-1.

Longwave receivers

For decades, longwave listeners were forced to either modify their own communications receivers, build equipment, or search for military-surplus longwave receivers. This difficulty limited longwave listening in the Western Hemisphere to those who were technically competent and *really* interested in radio. Fortunately, nearly all tabletop communications receivers since the early 1980s have access to all or most of the longwave bands. The typical receiver will cover down to 100 kHz. If you really love longwave DXing and the last 90 or so kilohertz are important, you will probably need to choose one of the three previous options. Military-surplus longwave receivers are still being bought and sold. However, you will probably need to consult

Table 4-1. Select longwave broadcasting stations

Frequency	Station	Country	Power	Service
153 kHz	Radio Algiers	Algeria	1000 kW	
	Radio Romania	Romania	1200 kW	Romania Actualitati
	Radio Rossii	Russia	1200 kW	Radio Rossii
	Deutschland Radio Cologne	Germany	250 kW	Deutschlandfunk
162 kHz	Radio France	France	2000 kW	Radio France
171 kHz	Belaruskaje Radyjo	Belarus	1000 kW	Moscow Ostankino
	Ukrayinsko Radio	Ukraine	1000 kW	Moscow Ostankino
	Moscow Ostankino	Russia	1200 kW	Moscow Ostankino
177 kHz	Deutschland Radio Berlin	Germany	100 kW	DRB
180 kHz	Kazakh Radio	Kazakhstan	250 kW	Kazakh Radio 1
	Radio Rossii	Russia	150 kW	Radio Rossii
189 kHz	Radio Rossii	Russia	1200 kW	Radio Rossii
198 kHz	BBC	UK	500 kW	Radio 4
	Radio Algiers	Algeria	1000 kW	
	Radio Mayak	Russia	250 kW	Radio Mayak
207 kHz	Deutschland Radio Cologne	Germany	250 kW	Deutschlandfunk
	Radiodiffusion-Télévision Marocaine	Morocco	800 kW	Network A
	Ríkisútvarpid	Iceland	100 kW	Channel 1
	Ukrayinsko Radio	Ukraine	500 kW	Ukrayinsko Radio
216 kHz	Norsk Rikskringkasting	Norway	200 kW	Program 1
225 kHz	Polskie Radio	Poland	600 kW	Radio 1
	Radio Rossii	Russia	1000 kW	Radio Rossii
234 kHz	Moscow Ostankino	Russia	2000 kW	Moscow Ostankino
	RTL	Luxembourg	2000 kW	RTL
	Radio Rossii	Russia	1000 kW	Radio Rossii
243 kHz	Kazakh Radio	Kazakhstan	1000 kW	Kazakh Radio 2
	Radio Rossii	Russia	1200 kW	Radio Rossii
252 kHz	Atlantic 252	Ireland	500 kW	Atlantic 252
	Radio Algiers	Algeria	1500 kW	International service
261 kHz	Radio Rossii	Russia	2500 kW	Radio Rossii
270 kHz	Czech Radio	Czech Republic	750 kW	Czech Radio 1
279 kHz	Belaruskaje Radyjo	Belarus	500 kW	Belaruskaje Radyjo
	Radio Rossii	Russia	1000 kW	Radio Rossii

the classifieds in *The Lowdown* or the Internet, or scour hamfests (see Chapter 12), to find the radio of your dreams. They aren't very common.

Hobbyist beacons

The longwave band is such a strange radio "nowhere land" that some radio hobbyists are challenged, not thwarted, by the difficulty of transmitting there. Although the band does not contain an actual amateur radio band in most parts of the world, the

FCC (and communications legislative bodies in other countries) has adopted a section into its rules that is known as *Part 15*.

Part 15 covers the amount of radio signals that can be radiated (either intentionally or unintentionally) by a device in a particular radio band. For example, in the United States, a device can output as much as one watt ERP of signal in the 160- to 190-kHz range. This specification covers everything from intentional transmitters to computers. However, if you are intentionally transmitting, an antenna of no greater than 15 meters (about 50 feet) is allowed in this frequency range.

Many radio hobbyists around the world have built and installed "micropowered" transmitters for this band—these stations are often called *lowfers*. Because 1 watt is a tiny amount of power in this static-filled range and the antenna length can only be a fraction of a wavelength long, the hobbyists must think of any way possible to be heard. Voice transmission is out of the question, unless you want to talk to a friend on your block. Because propagation must be perfect for stations to be heard, most people don't actually make contact via Morse code. Instead, the transmitter sends only a slow, simple self-created callsign over and over. It's not very exciting to me, but it is a neat propagational experiment. It also is very helpful for checking the propagation on the longwave band.

If, per chance, you would want to actively listen to the longwave beacons, you only need a receiver that is capable of covering the 160- to 190-kHz segment and an antenna. A loop antenna works best (see Chapter 5 for more information), but other types of antennas, such as random wires, will perform fairly well for casual listening (they receive too much noise on these frequencies to be a good pick for serious listeners).

Many times, when I tune through this band segment, a number of radio beacons are audible. One reason that I can hear so many beacons at my location is because the population density is relatively high, and a number of amateur radio operators live within 50 miles. At other locations in the world, and even in North America, the 160- to 190-kHz band will be devoid of radio signals. Still, this lack of activity shouldn't cause you to turn off your radio. You're here because you want a challenge, right?

Even though it is handy to know Morse code, it is not necessary for DXing lowfer beacons. The callsigns are repeated so simply and slowly that even people who don't know Morse code can copy the transmissions. The real difficulty is finding their location and address for reception reports. I thought about including a listing of beacon frequencies, callsigns, and addresses, but they change so frequently that the list would be completely out of date by the time that this book was published. Instead, your best bet is to subscribe to a longwave newsletter, such as *The Lowdown* from the Long Wave Club of America (LWCA), which is available from:

LWCA
45 Wildflower Rd.
Levittown, PA 19057

Otherwise, check some of the Internet Web pages for more information.

If you're ready to jump right in and build some Part 15 equipment for the longwave band, a few companies are manufacturing. Try:

Curry Communications
825 North Lima St.
Burbank, CA 91505

The LF Engineering Co.
17 Jeffry Rd.
East Haven, CT 06512

Amateur radio

In addition to the unlicensed Part 15 beacon transmitting, amateur radio organizations around the world are lobbying their respective governments for some spectrum space in the longwave bands.

One of the first countries to pass such legislation was the United Kingdom, which recently allowed amateur radio in the longwave bands—with several important rules. Instead of allocating the 160- to 190-kHz band for amateur radio, the British allocated a "band" that is a whopping 2.8 kHz wide (71.6 kHz to 74.4 kHz)! Although any modulation modes are allowed, many modes, such as AM, would probably produce a signal that is wider than the entire band! The power allowed on this band is a mere 1 watt (ERP).

To make the conditions even more difficult, no mobile or maritime mobile operations are permitted. However, just to prove that the legislators had a sense of humor, they established these operations under a no-interference basis! Considering the tiny amount of power and the limited frequency range, it seems a bit silly to require licensing for these operations. After all, I would imagine that the radio signals are typically only heard a bit further than someone shouting outside. However, if this sounds interesting, check out the *73 kHz* Web page at:

```
http://www.stonix.demon.co.uk/73kHz
```

If other countries begin licensing amateurs to operate in the longwave band, I hope that they increase the power a bit to make the operations somewhat useful. For more information on amateur radio, see Chapter 11.

Government activity

Aside from some LORAN (*LOng-RANge* radio navigation) activity around 100 kHz in the longwave bands, very little government radio communications are audible. LORAN is somewhat like radar; pulses are transmitted and their reflections are timed. It really isn't much fun to listen to, and it's basically impossible to receive a QSL from the transmission.

There is one other type of government radio activity, however, that is common throughout the longwave band. Although it isn't entertaining to listen to, it is fun to QSL. Many Morse code beacon stations operate throughout the longwave frequencies. You might initially think that these are amateurs that have departed from their locations in the 160- to 190-kHz band. Nope, these beacons are operated both privately and by the government; you can hear such signals as AJG on 524 kHz, IIY on 435 kHz, and DMZ on 203 kHz. Before you develop some sort of *X-Files* theory about how aliens are using these beacons for human mind-control experiments, stop. These beacons aren't sinister; they are simply the navigational beacons that airplanes use to find their location, with respect to the airport.

Most of these beacons operate with about 25 watts of power, although sometimes as much as 2000 watts is used. Some of these aviation beacons have been heard over great distances—even more than 1000 miles! Beacon tracker Jill Dybka collects QSLs from these beacons, and she lists many of the addresses on her Web page. Typically, she must prepare a QSL card or letter for herself in advance. Someone from the airport will verify it with a signature or a rubber stamp. So, unlike shortwave and mediumwave broadcast QSLing, you won't receive pretty postcards, pennants, photos, stickers, or magnets. You'll be lucky if you even see some airport's letterhead.

However, if you enjoy radio listening for its own sake and you want to do something that few other people are doing, check it out. Jill's handy Web page is:

http://funnelweb.utcc.utk.edu/jtdybka

One really excellent Web site for aviation beacon DXers is the Navaid page. Among other aviation-related topics, this page has an excellent search/listing page just for these aviation beacons. You can enter a frequency or a callsign, and it will either list all of the stations on the frequency or list the information on that particular beacon (depending on the information that you entered). This site lists tons of information about the beacon in question, such as the power, the location, and even what the Morse code is for the particular letters that it transmits. You can check this one out at:

http://www.cc.gatech.edu/dbl/fly/navaid-info.html

An excellent source for information on longwave beacons is *The Aero Marine Beacon Guide* by Ken Stryker. This book lists more than 7000 U.S. and international beacons by frequency and call. Like some of the other annual frequency listings, this book is copied onto loose sheets so that you can insert it into a binder and update it. For information, write to:

2856-G West Touhy
Chicago, IL 60645

Natural radio

For many years I was under the mistaken assumption that if you wanted to listen to the radio, you always listened to a *man-made* radio signal. Recently, talk of *natural radio* has surfaced within the radio hobby. Natural radio covers any type of radio signals that are radiated by non-man-made objects or materials, such as those caused by the movement of the plates of the earth, planets traveling through space, and propagational effects. Some people even listen to radio "noise." Although "static crashes" from electrical storms can be heard throughout the radio spectrum and some of the planetary radio-astronomy signals are best heard around 19 MHz, most of the natural radio listeners monitor the longwave band.

Obviously, no one listens to this type of radio strictly for the programming or for the QSLs. Most people listen to these sounds for some other application, which usually has very little connection to radio listening. For example, some people have found enough evidence to believe that the Earth's seismic activity can alter radio propagation. They have noticed changes in the levels of noise at certain frequencies in the longwave band (such as 25 kHz and 300 kHz) that correlate with extreme storms and earthquakes. It is hoped that with enough research and experimentation, this radio monitoring will help seismologists and meteorologists accurately predict severe weather and earthquakes.

One book dedicated to earthquake and weather research and experiments is *How to Build Earthquake, Weather, and Solar Flare Monitors* by Gary Guisti. Most of the projects are centered around longwave radio monitoring and construction, although a few monitor such things as vibrations in the Earth and the Sun. This book is very interesting, but it is fairly technical. It is certainly more for a hobbyist than for someone who just wants to be alerted in case an earthquake should strike.

Natural radio also consists of some signals with names that are sillier than the cast of a blue movie. Some of the different signal types that you can hear include: *whistlers*, *spherics*, *tweeks*, *hooks*, *the dawn chorus*, *triggered emissions*, *artificially stimulated emissions*, *hiss*, and *static*. These sounds are all mostly related to a combination of lightning and long-distance propagation. Whistlers, one of the most common of these types of effects, is a descending whistle that lasts for as short as a few milliseconds or as long as several seconds. After decades of mystery, it is now believed that this phenomenon is caused by the radio-frequency signals emitted from lightning strikes. A small projection of this signal shoots thousands of miles into the atmosphere and is refracted to another part of the world, over a signal path that is more than 10,000 miles long! The descending whistle is caused because the higher frequencies of the signal arrive slightly faster than the lower frequencies. Bizarre, huh?

Most whistlers are heard in the lowest portion of the longwave band, down in the 10-kHz region. If you wonder why you can't hear a signal that is obviously within your range of hearing, it's because these transmissions are radio signals, not audio vibrations. So, you can't hear them, just as you cannot hear people talking at the very low frequencies on your LF receiver. However, because noise and "interference" travel so well at the low frequencies, power-line noise is a real problem. Most LF enthusiasts recommend receiving the signals at a location that is at least a half mile away from any power lines. Obviously, this rules out listening at home for more than 95% of the world's population. Most people who are really into natural radio tend to travel out into the middle of nowhere, set up an antenna and a cassette recorder, and let'er rip.

If you want to check out more natural radio on the low frequencies, one good starting point is *The Secrets of RF Circuit Design—2nd Ed.* by Joe Carr. This book covers all sorts of information about receiver, transmitter, and amplifier theory. It also includes design, construction, and repair. Be sure that you get the second edition of this book, which has a chapter with the anomalies on the low frequencies; the first edition doesn't.

Aside from this book, one great source of information on natural radio is the Internet. One of the premier current researchers of the topic is Stephen P. McGreevy, who often travels out to the desert for noise-limited LF DXpeditions. As I am writing this book, McGreevy is in Alaska, performing more natural radio monitoring and research. McGreevy is somewhat of a pop scientist—one of the first radio-scientific types to release an album since Edison. McGreevy's first release, a double CD, is entitled *Electric Enigma: The VLF Recordings of Stephen P. McGreevy*. Although I can't imagine this CD hitting the Billboard hit list, I'd bet that it rates pretty high in the New Age top-100. This CD isn't for everyone, but I've seen reviews from natural radio enthusiasts that describe it as "fantastic." *Electric Enigma* is available for £18.50 (plus postage) from:

Irdial-Discs
P.O. Box 424
London SW3 5DY England

You can check out Stephen McGreevy's Internet Web page at:

`http://www.netcom.com/spmcgrvy`

If you want to hear samples of some of the natural radio sounds (sampled at a lower rate than the audio on the CD), check out his Web page at the University of Iowa:

`http://www.pw.physics.uiowa.edu/mcgreevy`

Tuning in

Listening to longwave radio can be difficult because the frequencies are more affected by natural and most man-made interference. For example, you can watch TV during a thunderstorm (not advisable), and you probably won't see or hear the static crashes on the set. Static crashes will get comparatively louder as a radio is tuned lower in frequency through the shortwave spectrum. Because the thunderstorm noise travels such long distances, the tropical and mediumwave bands are almost useless for heavy-duty DXing during the summer (most strong signals will still be useful in the summertime).

Imagine the summertime noise problems on the longwave band, where it seems that you can't even walk across your lawn without causing radio-frequency interference. In addition to being nearly worthless for DXing and most general listening in the summer, the longwave bands are also destroyed by bad powerline transformers, computer interference, fluorescent lights, touch lamps, neon signs, etc.

To make listening even more difficult, a full-wave antenna for the longwave bands could be a mile long! Obviously, this would rule out heavy-duty longwave listening for most everyone, except a few farmers. Fortunately, loop antennas are excellent performers at the low frequencies, where they pick up less noise than other antennas. They are also very directional. If you want to pick up a longwave loop antenna, try one manufactured by Kiwa, RF Systems, Palomar, or a few others. If you want to experiment with making loop antennas, read *Build Your Own Shortwave Antennas—2nd Ed.* by Andrew Yoder, which features a section on the topic.

If you live in a city or apartment building, you can forget longwave DXing. If you live in Europe, where the longwave broadcasters are nearby, general listening shouldn't be much of a problem, but DXing will still be out of the question if you live in a densely populated area.

For more information on longwave radio on the Internet, see the Longwave Home Page at:

`http://members.aol.com//lwcanews/index.html`

This page features information on most every aspect of longwave listening—from natural radio to amateur beacons.

5
CHAPTER

Mediumwave radio

If you own a radio and live in the United States, there's at least a 90% chance that it covers the AM (530 to 1700 kHz) and FM (88 to 108 MHz) broadcast bands. In many regions of the world, the AM band is known as the *mediumwave band* and FM radio is known as *VHF*. Because of the confusion between the AM band and AM modulation, the "AM broadcast band" is called the *mediumwave band* hereafter throughout this book.

· The mediumwave band is the most popular frequency range in the world, although FM is gaining popularity for its high-quality sound. Mediumwave is so popular because it's the oldest and best-established broadcasting band, and because the signals don't skip over most of the local listeners (like shortwave), yet a relatively small amount of power can cover hundreds of miles (unlike longwave or FM).

I don't need to introduce you to the mediumwave band because you've probably listened to it for years. However, I'll start with some characteristics of the band and mediumwave differences around the world for casual listeners.

Propagation

Long-distance radio reception is possible because of one of the layers of our atmosphere that is known as the ionosphere. The ionosphere is many miles above the earth, where the air is "thin"—containing few molecules. Here, the ionosphere is bombarded by X-rays, ultraviolet rays, and other forms of high-frequency radiation from the sun. This energy ionizes the layer by stripping electrons from the atoms in the ionosphere. The ionized layers of the atmosphere make long-distance radio communications possible.

Scientists have determined that the ionosphere can, in turn, be divided into several layers, but the D, E, and F layers are the ones that affect radio propagation. The D layer is closest to the surface of the Earth. The existence and strength of this layer depends on and is proportional to the sun in the sky. As a result, the D layer gradually grows in strength in the morning, is strongest at mid-day, and it gradually decreases until it disappears by nightfall. Also, the D layer is generally stronger in the summer than in the winter. On the lower frequencies (below about 10 MHz), the D layer will absorb any signals that are transmitted into it. To experience a great example of this effect, tune across the mediumwave band at midday and then do the same at night. At midday, you will hear local stations and maybe a few powerful cross-state stations. However, with nightfall, the band comes alive, and stations from across the country are audible. You have just witnessed the effect of the D layer.

The E layer is much like the D layer, except that it is a bit higher and the layer doesn't absorb all of the radio signals. In fact, instead of just absorbing the signals, the E layer will often bend (refract) the radio waves back toward the earth, thus enabling radio signals to be heard over *much* greater distances than would otherwise be possible. Most of the E-layer skip at mediumwave frequencies results in stations several hundred miles away fading in. For example, at my location in Pennsylvania, a number of broadcasters are audible during the daylight hours from my own state and several others: Maryland, West Virginia, and Virginia. However, near sunset, I can begin to hear stations from New York, Chicago, Boston, etc. This is the E layer in action. However, although the E layer does refract signals back to the Earth, it also absorbs (attenuates) them. This is why the radio signals on the mediumwave band aren't refracted by the ionosphere, bounce off the Earth, are refracted by the ionosphere again, etc. The E layer simply absorbs too much of the signal for it to bounce around the globe.

The F layer makes most shortwave and long mediumwave skip possible. This layer is generally about several hundred miles above the Earth. It remains ionized throughout much of the day and night. However, unlike the other layers (under most circumstances), the F layer will refract signals back to the Earth. Because the F layer is so high above the Earth, signals often skip over very great distances from the F layer—sometimes several thousand miles.

This is a very brief description of mediumwave propagation. For more information on the topic, contact the National Radio Club and Medium Wave Circle (their listings are included later in this chapter).

Mediumwave in North America

In the United States and Canada, literally thousands of commercial stations operate regularly. In this region of the world, commercial radio is king. Very few other types of stations are on the air. The formats on the mediumwave band are somewhat limited; most common are talk radio and oldies, classical, pop, rock, and country music. The mediumwave programming, formats, and target audiences are "safe"—don't expect any revolutionary or "edgy" programming here. The programs are typically intended for older listeners, and you can hear many advertisements for pain relievers, arthritis medications, denture creams, and different types of insurance.

Another characteristic of these stations, in terms of format and type of operation, are distinctive regulations for the amount of power. The frequencies in this broadcast band are assigned three different statuses: clear channel, regional, and local.

Clear-channel stations have very few other stations on the frequency in the continent. They typically cover many states. The typical clear-channel power is the maximum allowed in the United States or Canada: 50 kW. Clear-channel stations have very large audiences and are very well-respected stations, with long histories and good reputations for news reporting. Many are even associated with major TV stations in the large cities. A few good examples of these are: KDKA in Pittsburgh, PA; KYW in Philadelphia, PA; KCBS in San Francisco, CA; WJR in Detroit (Fig. 5-1); and KOA in Denver, CO. Because of international agreements, some clear channels are open for Mexican and Canadian stations; this allows these countries to have less interference locally and allows U.S. listeners to listen to some international programming.

5-1 A QSL from clear-channel mediumwave station WJR in Detroit, Michigan.

Regional-channel stations typically operate with 5 or 10 kW and are expected to cover only a relatively small region—perhaps half of a state. As a result of the smaller range of the regional broadcasters, more stations are allocated to operate on these frequencies. Many regional stations are very popular—even if they aren't major media forces.

Local-channel stations typically operate with anywhere from 250 watts to about 1 kW. Local stations are expected to cover about a county or two. Because of the small range (and smaller audiences), local stations often exemplify bottom-budget radio—broadcasters struggling to survive by airing satellite programs and using announcers right out of college. The local channels are filled with literally hundreds of radio stations.

As mentioned in the previous section, the time of day drastically impacts the propagation of radio signals on the mediumwave band. At night, even a few watts can travel hundreds and even thousands of miles. As a result, the local and regional channels are a mess after dark. I don't think that I've ever even copied more than a few seconds of audio from a local-channel station after dark. The audio sounds like little more than white noise—throw 35 stations into a blender, and an indistinguishable mess oozes out of your speaker. Because these frequencies are relatively useless, the local channels are often called *graveyard channels* by those in the listening hobby or broadcasting industry.

Currently, mediumwave broadcasting in North America is in a state of flux. The FCC has opened the 1600- to 1700-kHz region for broadcasting, known as the *expanded band* (see Table 5-1). As of late 1996, a number of stations were licensed to

Table 5-1. Expanded band allotments

Call	City	State	kHz	kHz
KALT	Atlanta	TX	900	1610
WJRZ	Toms River	NJ	1550	1620
WGYJ	Atmore	AL	1590	1620
WAMJ	South Bend	IN	1580	1620
KRIZ	Renton	WA	1420	1620
WTAW	College Station - Bryan	TX	1150	1620
KQKE	Soledad	CA	700	1620
KHRT	Minot	ND	1320	1620
KJCK	Junction City	KS	1420	1620
WKZQ	Myrtle Beach	SC	1520	1620
KECN	Blackfoot	ID	690	1620
WRRA	Frederiksted	VI	1290	1620
KNBA	Vallejo	CA	1190	1630
WRDW	Augusta	GA	1480	1630
KSVE	El Paso	TX	1150	1630
KCJJ	Iowa City	IA	1560	1630
KHVN	Fort Worth	TX	970	1630
KSHY	Fox Farm	WY	1530	1630
KOQO	Clovis	CA	790	1640
WSYD	Mount Airy	NC	1300	1640
WTRY	Troy	NY	980	1640
WVMI	Biloxi	MS	570	1640
KPHP	Lake Oswego	OR	1290	1640
WKSH	Sussex	WI	1370	1640
KURV	Edinburg	TX	710	1640
KCRC	Enid	OK	1390	1640
WWHL	Cocoa	FL	1350	1640
KLXX	Bismarck/Mandan	ND	1270	1640
KWFM	Tucson	AZ	940	1640
WGOD	St. Thomas	VI	1090	1640
KOJY	Costa Mesa	CA	540	1650
WAOK	Atlanta	GA	1380	1650
WPMH	Portsmouth	VA	1010	1650
KWHN	Fort Smith	AR	1320	1650
KSVP	Artesia	NM	990	1650
KTMT	Phoenix	OR	880	1650
KCFI	Cedar Falls	IA	1250	1650
KTKK	Sandy	UT	630	1650
KKLS	Rapid City	SD	920	1650
KJDM	Elizabeth	NJ	1530	1660
KRCX	Roseville	CA	1110	1660

Call	City	State	kHz	kHz
WGIV	Charlotte	NC	1600	1660
WCHQ	Camuy	PR	1360	1660
WKRG	Mobile	AL	710	1660
WQSN	Kalamazoo	MI	1470	1660
KZOK	Seattle	WA	1590	1660
KRZI	Waco	TX	1580	1660
WMIB	Marco Island	FL	1480	1660
WREN	Topeka	KS	1250	1660
KRKS	Denver	CO	990	1660
KQWB	Fargo	ND	1550	1660
KBLU	Yuma	AZ	560	1660
KHPY	Moreno Valley	CA	1530	1670
WRCC	Warner Robins	GA	1600	1670
WLVW	Salisbury	MD	960	1670
KBTN	Neosho	MO	1420	1670
WTDY	Madison	WI	1480	1670
KHTE	Redding	CA	600	1670
KKEL	Hobbs	NM	1480	1670
KSOS	Brigham City	UT	800	1670
KLOQ	Merced	CA	1580	1680
WHWH	Princeton	NJ	1350	1680
WKTP	Jonesborough	TN	1590	1680
WXTO	Winter Garden	FL	1600	1680
KSLM	Salem	OR	1390	1680
WSFN	Muskegon	MI	1600	1680
KMLB	Monroe	LA	1440	1680
KBNA	El Paso	TX	920	1680
KCOL	Ft. Collins	CO	1410	1680
KBRF	Fergus Falls	MN	1250	1680
WBIT	Adel	GA	1470	1690
WDDD	Johnson City	IL	810	1690
WPTX	Lexington Park	MD	920	1690
KRGI	Grand Island	NE	1430	1690
KFVR	Cresent City	CA	1310	1690
KAPR	Douglas	AZ	930	1690
KSTR	Grand Junction	CO	620	1690
KAHI	Auburn	CA	950	1700
WSVA	Harrisonburg	VA	550	1700
WZNN	Rochester	NH	930	1700
WCMQ	Miami Springs	FL	1210	1700
WEUP	Huntsville	AL	1600	1700
KAST	Astoria	OR	1370	1700
WONX	Evanston	IL	1590	1700
KIDR	Phoenix	AZ	740	1700
KNRB	Fort Worth	TX	1360	1700
KDDR	Oakes	ND	1220	1700

operate in this band, but only two had made their move from their old frequencies within the old mediumwave band. Because no other stations are operating on these frequencies, the stations on these frequencies are a sort of temporary "low-power, clear-channel station." For example, when WJDM took to 1660 kHz with only 1 kW, they were reported from coast-to-coast and around the world. Even though I am only about 200 or 300 miles from the transmitter, other mediumwave stations with that power and the same general locations are difficult to hear at my location because of the tremendous co-channel interference.

Mediumwave broadcasting around the world

Two general scenarios for mediumwave broadcasting exist: commercial radio and government-controlled radio. Commercial radio reigns across nearly every portion of the Americas, plus Japan, Australia, New Zealand, and many of the Pacific islands. Government-controlled radio is the general rule in Africa, Europe, and Asia.

In Africa, Europe, and Asia, radio is, for the most part, controlled by agencies that are organized and funded by the government. Radio is often operated by the government for purposes of totalitarianism and censorship. Since the 1930s, nearly every fascist or communist government has relied on censorship and media control for much of its power. However, such is not always the case. England, for example, has the British Broadcasting Corporation (BBC), which is known for its fair and accurate reporting and for producing some excellent radio and television programs. In this case, some might argue that, if the BBC was disbanded in favor of independent "free" radio broadcasting, the British radio and television quality would be significantly degraded.

Aside from the issue of control, another important difference in mediumwave broadcasting in Europe and Asia (Africa is different in this respect) is that of power. Not "who-controls-what," but rather transmitter output power. Most European and Asian mediumwave stations are international broadcasters. Just compare the maximum power of the U.S. and Canadian mediumwavers (50 kW) to the 2000-kW stations in Jordan, Russia, and Serbia. With that kind of power, it's not a question of whether you can hear the station, but if your popcorn pops in the cupboard and your appliances operate when they aren't plugged in while you're listening.

Because of the relatively small amounts of land that are occupied by most of these countries, the signals from these broadcasters are commonly heard across a dozen or so countries—even into neighboring continents. As a result of this coverage and the many different nationalities and countries that are packed within a small space, many of these mediumwave stations broadcast with international programs in a variety of styles and languages; much different than anything you could receive in the Americas (Fig. 5-2).

Mediumwave broadcasting in Africa varies a bit from that of Europe or North America. The stations are typically relatively low-powered (often less than 25 kW). Only a handful operate from each country, and most are strictly controlled by the ruling government. Thus, very few of the African mediumwave stations are commercially operated, most are the voice of the ruling government, and most broadcast in a variety of languages to satisfy the different nationalities within the country (Fig. 5-3).

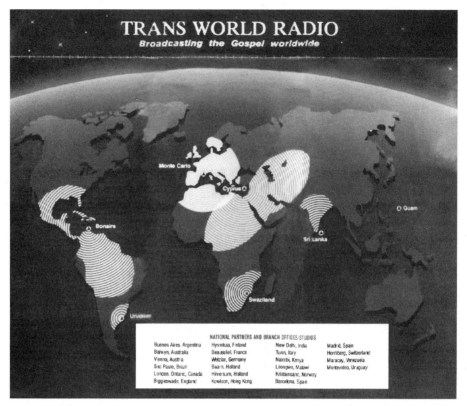

5-2 International broadcasting via mediumwave: the powerhouse transmitters of Trans World Radio.

Mediumwave DXing

On a winter evening many years ago, Larry Godwin of Englewood, Colorado, was methodically checking the mediumwave band for signs of the unusual. As he passed 840 kHz, he noticed a station slightly higher in frequency. After carefully tuning the signal, he noted that it was on 844 kHz and that the programming appeared to be in a Polynesian dialect. Larry carefully adjusted his loop antenna and determined from the loop bearing that the station was, indeed, in the South Pacific.

From a nearby shelf, Larry pulled down his well-worn copy of the *World Radio-TV Handbook*, the most complete listing of broadcast stations in the world. He turned to the Pacific section and started hunting for the information. Soon, he found that the only station on 844 kHz was a 50-watt station on Tarawa, in the Gilbert and Ellice Islands, just south of the Equator and about 6000 miles from Colorado. Everything about the programming, including chimes on the hour, seemed to fit the description in the handbook, except for the listed power. The signal was just too strong for only 50 watts. Reaching for the telephone, Larry called several mediumwave DXers on the West Coast.

5-3 Vatican Radio's 1611-kHz transmitter is one of the better-heard Mediumwave broadcasters.

"I know what I'm hearing, but I can't believe it!" he said. Within the hour, word had spread up and down the Coast. Within a few days, a whole group of DXers logged a new country.

Such events don't happen every night or even every month, but the excitement or challenge of such an event is enough to keep mediumwave DXers tuning the dials, season after season, looking for stations that weren't there yesterday. The stations also often disappear or are covered by other stations as quickly as they originally appeared. For example, you won't find a listing for the Gilbert and Ellice Islands in the most recent edition of the *WRTH*. Another good example of this is the British shipboard pirates that operated from the North Sea. At their peak in the mid-1960s, some of these broadcasters were logged in the United States on occasion. One DXer friend from the National Radio Club told me how he heard the infamous Radio Caroline with very good signals while he lived in Pennsylvania in 1967. I'm sure that he was pretty excited at the time, but now that the days "when pirates ruled the waves" are long past, the event seems to hold magical significance.

This is the main attraction of DXing of any type—the magic that's involved. It's the feeling that results from sitting in bed late at night with only the glow of the tubes (or the glow of a digital display) casting a few shadows around the room. I have talked with many people about radio, and I have been interviewed by dozens of radio stations and newspapers. I can see the difference between those that are just talking about radio to get a story and those who truly love radio. Nearly everyone with a true passion for radio has had that "radio by the bed late at night when I was a kid" experience. Many of the people that I've talked with can't explain the feeling that they had (or still have) when DXing, but many describe such an experience.

At least half of the drive for DXing is this "radio magic." The other portion consists of the challenge of listening for, copying, and receiving QSLs (more on this in Chapter 15) from difficult-to-receive stations from around the world, an interest in geography and world cultures, and a desire to form international friendships or procure friendly acquaintances.

I have talked with some shortwave DXers who said that they would retire from DXing and sell their shortwave radios if they received and QSLed Tristan da Cunha, recognized by many as the most difficult-to-receive shortwave station/shortwave country in the world. I can't imagine snuffing out the magic of the radio just because I had reached some predetermined goal. In these cases, it appears that the passion for radio has been replaced by some sort of cold compulsion. Fortunately, most DXers seem to really love radio for its own merits!

We Love Those Callsigns (WLTC?)*

If you listen to North American radio long enough, you will realize that some of the callsigns actually mean something or say something. A few of the local stations from around where I grew up included WFRB (*Frostburg, Maryland*) and WKGO ("GO-106"). Stations with slogans like this are really common. The "GO-106"-type slogans are a relatively new phenomenon.

In the early days of radio, when the medium was still an experiment, just about any little niche group or business owned a radio station, and many created elaborate slogans to fit their designated callsigns. Instead of hearing "97 Rock," you'd hear a slogan such as KFDC ("*Keeping Ft. Dodge Country*").

Some of the other fun and interesting callsign slogans include: "*Kansas' Finest Hotel*" (KFH), "*Known For Neighborly Folks*" (KFNF), "*Keep Forever Radiating Cheer*" (KFRC), "*Keeping Good Folks Joyful*" (KGFJ), "*Keep Growing Wiser*" (KGW), "*Keeping Listeners Aware in Style*" (KLAS), "*MAy Seed and Nursery Company*" (KMA), "*Music, News, and Sports*" (KMNS), "*We Bring A Program*" (WBAP), "*We Broadcast Better Music*" (WBBM),

WLIS ⇒ WE LOVE INTERVAL SIGNALS

We are pleased to verify your reception of WLIS. Thank you for your report!

DATE:					TIME:	FREQUENCY:
Jan.	1 2 3 ④ 5 6 7				0 0 0 0	0 0 0 0 . 0
Feb.	8 9 10 11 12				① 1 1 1	1 1 1 ① 1 . 1
(Mar.)	13 14 15 16				2 2 ② 2	2 2 2 2 2 . 2
Apr.	17 18 19 20				3 3 3 3	3 3 3 3 . 3
May	21 22 23 24				4 4 4 4	4 ④ 4 ④ . 4
June	25 26 27 28				5 5 5 5	5 5 5 5 . 5
July	29 30 31				6 6 6 6	6 6 6 6 . 6
Aug.					7 7 7 7	⑦ 7 7 7 . 7
Sept.	(1990)	1991			8 8 8 ⑧	8 8 8 8 ⑧
Oct.	1992				9 ⑨ 9 9	9 9 9 9 . 9
Nov.	1993				(UTC)	(kHz)
Dec.	1994				EST	USB LSB

"Wolfe Bank Newspaper and Shoes" (WBNS), "Washburn Crosby Milling COmpany" (WCCO), "Where Coal Meets Iron" (WCMI), "Doctor George Young" (WDGY), "Edison Electric Illuminating" (WEEI), "Eugene V. Debs" (WEVD), "World's Finest Fishing Ground" (WFFG), "World's Greatest Loudspeaker" (WGL), "Where Historic Blackhawk Fought" (WHBF), "Johnstown Automotive Company" (WJAC), "We Love Our Caves" (WLOC), and last, but certainly not least, "World's Happiest Broadcaster" (WHB).

The days of long callsign slogans are nearly, but not quite, gone. Most of them disappeared many decades ago, but the Christian and pirate stations still have a passion for these catch phrases. A few of the current Christian stations include "Heralding Christ Jesus' Blessings" (HCJB), "Your Family Radio" (WYFR), "World Wide Christian Radio" (WWCR), and "World Harvest Radio International" (WHRI).

Some of the pirate callsign slogans include: "We Love Interval Signals" (WLIS), "We Hate Dead Air" (WHDA), "We ate Hot Dogs for Lunch" (WHDL), "World Parody Network" (WPN), "We're KaNine Dog" and "WeeKeND radio" (WKND), "Sea (C) SICk" (CSIC), "KULPsville" (KULP), "Radio eXperiment on 40 Meters" (RX4M), and many, many more. Remember, these callsigns are self-declared, and many of these have been legitimately claimed by commercial broadcasters.

If you have an interest in these strange slogans, read Alice Brannigan's monthly historic broadcasting feature in *Popular Communications*. Better yet, if you are on the Internet, the IRCA Web page features more than 900 of these slogans! Go to:

`http://fly.hiwaay.net/waholler/irca.html`

---

* WLTC is actually on 1370 kHz from Gastonia, North Carolina. To the best of my knowledge, it doesn't represent "We Love Those Callsigns."

Pick a radio band, any radio band

Why mediumwave DXing over shortwave DXing? Most serious DXers have dabbled in both mediumwave and shortwave DXing, but choose one or the other. They do so because the radio spectrum is so large—you can easily spend years tracking down tough catches and getting used to either one of the bands.

Some of the attractions of mediumwave DXing include the large number of possible stations to hear from—a veritable endless sea of broadcasters. A DXer in the United States could have literally thousands of different stations to hear in his or her country alone! Toss in the regular callsign changes and new stations, and the United States' mediumwave scene alone is an ocean of radio stations (Fig. 5-4), changing with every tide. Not only is the local DX bountiful, but the stations are all a cheap fax or a first-class stamp away. Shortwave QSLs, on the other hand, often require an air mail stamp, a report written in the station's broadcasting language (and your attempts at Portuguese, for example, might not be understood by the station), and return postage. Besides, if foreign DX is desired anyway, it can be had on the mediumwave bands. It might be difficult, but it is very possible (Fig. 5-5).

5-4 A great catch for most listeners: KHLO, Hilo, Hawaii.

A few other reasons for DXing the mediumwave band include the cost; with just a very inexpensive radio, a listener can receive many different stations on the mediumwave bands. Although shortwave receivers are becoming less expensive, they are still much less common and much more expensive than a mediumwave receiver of the same quality.

One key to interest in mediumwave DXing seems to be an involvement in domestic broadcasting. Many of the U.S. mediumwave DXers that I know are also professional broadcasters—especially engineers. There's just something about keeping tabs on your industry....

Mediumwave DXing necessities

If you want to get into mediumwave radio DXing, you need a few different resources. Refer back to the beginning of this section with Larry Godwin hearing the station from the Gilbert and Ellice Islands. Larry:

- Was listening to the radio.
- Had a good radio that was capable of accurately tuning stations in that were several kilohertz away from the standard North American channels.
- Had a loop antenna.
- Had good sources of recent information.

5-5
The Australian Broadcasting Corporation's 3LO.

You

This is one of those categories that almost seems unnecessary. Of course, you need to be there in order to listen to the radio. However, it's surprising how many people don't follow this advice. If you want to hear stations, you must listen to the radio.

Too many listeners turn on their radio every now and then and expect to have some great catches every time that they have one of these sessions. First, you must know the band and the stations that are out there before you can regularly make the great catches. It's like spending "quality time" with your family; before you can have the "quality time," you must spend plenty of hours getting to know them and doing things that aren't always a load of fun.

Of course, you shouldn't be a slave to your radio, but if you have DXing in mind, it's best to listen frequently while doing other things. You don't even have to DX, just leave it on in the background and notice when the band changes or when different stations fade in.

The bottom line is that you can have the best radio, antenna, and information in the world, but if you don't get to know the radio bands or spend any time listening, you won't hear much. Chances are that an active DXer with a lesser radio, antenna, and information will do much better.

Equipment

The necessary equipment for mediumwave DXing can be quite simple, although you can bet that you will have better results if you have a good receiver. Most "boom boxes" contain pitiful receivers. Even some of the tabletop and component stereo receivers are poor. Typically, the sensitivity is awful and the selectivity is almost as bad. I can receive only a few stations during the day and about twice as many at night on my portable stereo. The only way that I could expect to hear a distant radio station on that radio would be to travel there and tune it in! In addition, I can hear images across the band, and the audio is worse than what I can get from the 1954 tabletop model that I keep on my dresser!

Portable receivers

One of the best bets for mediumwave DXing is the GE Superadio, available in several different models (such as the Superadio III). This radio is known as the best consumer radio for DXing. Although it looks somewhat like the mono portable cassette players that were popular in the 1980s, it is very sensitive and selective, and its loopstick antenna does a nice job of nulling out stations that cause co-channel interference. Best of all, the Superadios sell for about $60.

In general, communications "shortwave" receivers are better at receiving the mediumwave band than typical mediumwave-only receivers. Many of the older shortwave portables were inexpensive "all-band" radios that you could receive everything with—but nothing well. With the emergence of microelectronics, an excellent receiver can be placed in a tiny package. The lines are blurred now between portable and tabletop shortwave receivers, but in general, most of the portables still don't perform very well on the mediumwave bands.

Mediumwave listening is demanding; the receivers must be incredibly sensitive (enough that you can hear faint stations), yet be resistant to overloading (so that you don't receive images all over the band), and still be selective enough to separate the split-channel stations (more on this later) from the domestic operations. Most portables are so small that they are susceptible to overloading and causing images. This problem is "solved" in other models by the designers, who merely desensitize the radio to prevent it from overloading.

Car stereos

Some listeners like to use car stereos for DXing. This style of listening is a bit peculiar, but many of these radios possess a nice balance of sensitivity and selectivity. One problem with using car head units for DXing (Fig. 5-6) is that many are programmed in 10-kHz segments (or 9-kHz segments, if you live in Europe or Asia). This makes it tough to receive stations from other places in the world. Another problem is that it's tough to DX in a moving car—it's virtually impossible to take notes or even record the program. Nonetheless, I know of a mediumwave DXer from Boston who DXes while commuting home from work. Somehow, he regularly makes astonishing catches of European and African stations without, to the best of my knowledge, getting injured in a car accident. Maybe he uses a voice-activated mini recorder to "take notes."

5-6 The Philips DC-777 AM/FM/SW cassette radio.

If you are strapped for cash and have a decent car stereo available (perhaps it's one that you replaced from a car or maybe the tape deck "died," but you still have the head unit around the house), you can use the radio in your house. There's no rule that says a car stereo can't be used inside. Just find the manual and connect it to a small 12-V bench power supply. A 12-V "wall-wart" power supply won't provide enough current to operate the unit. Some of the inexpensive power supplies are noisy, so you might want to pick up a surplus 12-V gel cell. Gel cells are great, but they are very expensive new, often as much as $50 or $100 for only a few mA-hours of capacity. However, you can often find them at terrific prices at hamfests and other surplus sales.

Modern and tube tabletop receivers

The best radios to use to tune in the mediumwave band are modern communications receivers and 1950s/1960s-era military receivers. Although some people snub their noses at the old heavyweight tube radios, the best models have exceptional records for receiving in the mediumwave bands. Two great examples of ex-U.S. military receivers for the mediumwave bands are the Hammarlund SP-600 and the Collins (although many other companies also produced this radio) R-390A. Both of these receivers were designed for excellent sensitivity and selectivity. However, perhaps their most powerful attribute for mediumwave listening is that they were designed to receive while many-kilowatt transmitters were operating only a few feet away—perfect for DXing if interference from local stations is a problem.

Although the R-390A and the SP-600 have little competition from other tube radios for DXing on the mediumwave bands, one other receiver that falls into this category is the Hammarlund HQ-180 (also available as the 180A, 180C, and 180AX). Like the other two radios, the HQ-180 is still used by a number of DXers for serious mediumwave and shortwave listening.

The big tube radios have three major pitfalls for listeners:

- *Parts availability and repair.* If anything goes wrong with your radio, what will you do?
- *Getting a tip-top radio.* Many of the tube radios on the market aren't quite operating properly—tubes are weak, capacitances have changed, the alignment is off, etc. Just because it's an R-390A, it doesn't mean that it will receive like a new one.
- *Ease of tuning.* You simply cannot quickly tune a tube radio. There are no memories or quickie pushbuttons. In the case of the HQ-180, it is very difficult to even know what frequency you're listening to, unless you have spent years using it.

The best type of receiver for the nontechnical DXer is the modern tabletop. Although the solid-state tabletop receivers were frowned upon by many of the mediumwave DXers through the 1970s and 1980s, some of the modern tabletops are finding a place in the shack of even the die-hard tube radio enthusiast. Some of the most popular modern receivers for mediumwave DXing are the Drake R-8A (and its predecessor, the R-8), the Kenwood R-5000, and the Lowe HF-250.

These receivers are also excellent performers on the shortwave bands. For more information on selecting a shortwave receiver that suits your needs, see Chapter 2.

Antennas

Antennas were covered in Chapter 3, but there is a big difference between those that are effective in the shortwave bands and those that are effective in the mediumwave bands. The difference isn't that the antenna-receiving characteristics suddenly change as you cross 1700 kHz. The differences are caused solely by the stations that are broadcasting and certain physical limitations.

One of the most popular antennas for general shortwave listening is the standard dipole. One of the important characteristics of a dipole is that it has very little gain—it receives fairly well in most directions and doesn't have any strong nulls. Even if you have the real estate to install a 300' long (or longer) dipole antenna, it will not perform well on the mediumwave band. The problem is not that the antenna doesn't work; the problem is that thousands of mediumwave stations are broadcasting from all around the world at the same time. If you use that dipole at nighttime, chances are that the mess you hear on many frequencies will only be a bit louder than if you had used a short random wire antenna. On the mediumwave band, you need an antenna that is extremely directional. The difference isn't getting a louder signal, it's eliminating interference from other stations.

Using conventional shortwave antenna-building wisdom, you just make a longwire longer and longer, until it is very directional at even mediumwave frequencies. To date, the longest antennas that I've used for mediumwave listening have reached somewhere around 1000' (but this winter, I plan to try some *seriously long* antennas!). This approach works; I have caught some nice DX with longwires. Some of the renowned DXpeditions (see Chapter 6) in the state of Washington, Australia, Nova Scotia, and Finland have turned in some truly remarkable results with longwires. These guys crank out antennas that are anywhere from 1500 feet to nearly one mile in length . . . and stations from all over the world come for a visit.

The longwire is the brute-force method of getting directional reception on the mediumwave band. The other common method is pure finesse. The loop antenna is a lousy antenna. Signals that would be strong on just about any antenna are weak on a loop. However, loops have two real advantages: they only receive the magnetic component of a radio signal (noise is significantly reduced) and they are incredibly directional.

These two elements are very important in mediumwave DXing. Of course, noise reduction is important with any radio, but especially so at these frequencies, where nearly everything man-made or natural seems to cause loud buzzing interference. However, the most important feature of the loop is its directionality.

If you are old enough to remember the days before cable TV, then you have an idea of how directional antennas work. My parents live out in the mountains and still don't have cable TV. If you turn the antenna, suddenly some stations will become clear. If you keep turning the antenna, the signal will disappear. That's how a loop antenna works on the mediumwave band. For example, a radio listener might live in Houston, Texas and be listening to 780 kHz, which is a clear channel. If the antenna is turned toward the northwest, KKOH in Reno, Nevada will be received. However, as the antenna is turned slightly north-northeast, KKOH will disappear and gradually be replaced by WBBM in Chicago, Illinois. These are rather elementary examples; a loop antenna can also be used to null out nearby radio stations so that other, more distant stations can be heard.

Loop antennas were very popular in the very early days of radio broadcasting. Many people (not just DXers) sat around with TRF radios and loop antennas, trying to hear news or entertainment from some station. By the 1930s and 1940s, these antennas were seen as an inconvenience, and most were eliminated entirely. About the only loop antennas used during the 1940s were built into the back of console AM/shortwave receivers (Fig. 5-7). With the advent of the truly portable receiver

5-7 A loop antenna from the 1930s.

and the developments of transistor technology, the loop antennas were also miniaturized and wrapped around bars of ferrite. These antennas work well and are directional, but an air-core loop can still outperform a standard loopstick antenna. Some DXers build their own loop antennas to improve their reception.

A few new adjustable air-core loop antennas reached the market in the 1990s. Although these models are expensive, they are excellent for DXing—especially if you don't have a few hundred acres to install an antenna on. Some of the most popular mediumwave loops are available from Kiwa Electronics and Palomar Engineers. If you have the money and interest in mediumwave DXing, then one of these antennas is certainly worth your while.

If you are considering building your own loop antenna (Joe Carr's *Practical Antenna Handbook—2nd Ed.* contains plenty of loop antenna information). You could also buy a home-built antenna from someone else (check the bulletins from the NRC and the IRCA to see if anyone has a loop for sale).

Recent information

Although it is possible to find plenty of information on your own by listening to stations, culling addresses, telephone numbers, fax numbers, and schedules, it is *much* easier to simply have a few good sources of information on hand to guide you in your searches for rare and/or interesting stations.

A few of the best mediumwave references and resources include:
- *World Radio-TV Handbook* is a must-have guide. It includes international listings of shortwave, mediumwave, longwave, FM, and TV stations around the world. The mediumwave, FM, and TV station listings are weak for the United States and Canada, so you might consider another guidebook to supplement your *WRTH*.
- *The NRC AM Radio Log* is a huge listing of mediumwave radio stations in the United States and Canada. It is sorted by frequency and callsign, and it lists addresses, formats, powers (day and night), and more. This "book" (actually a thick stack of photocopies for a three-ring binder) is published by the National Radio Club, but it's listed separately here because it is such an important resource.
- *The M Street Radio Directory* is a complete must-have for the serious United States mediumwave and FM radio DXer. This annual radio industry book contains station listings by state, frequency, radio market, and callsign. It also contains addresses, powers, Arbitron ratings, and more. This one is a bit pricey (more than $30), so international DXers might want to stick with either *The NRC AM Radio Log* or the *WRTH*.
- The *National Radio Club* is one of the best-known mediumwave clubs in the world. They publish a monthly bulletin, produce a radio-style cassette DX program, have annual get-togethers, and sell a wide variety of books and pamphlets about mediumwave DXing. A few of their many book titles include: *Verie Signers List, NRC Night Pattern Book, Beverage Antenna Design and Theory, Loop Antenna Design and Theory*, and *The DXpedition Handbook*. Many of these topics are much too narrow to ever hit the *New York Times* best-seller list, but at least a handful of these books

are must-haves if you are serious about mediumwave DXing. You can contact the NRC Publications Center at: Box 164, Mannsville, NY 13661. The NRC also has a fantastic Web page on the Internet (http://alpha.wcoil.com:80/~gnbc), with plenty of DX logs from their newsletters, book ordering information, and more.

- The *International Radio Club of America* publishes the *DX Monitor* newsletter 34 times per year. The IRCA is one of the oldest radio clubs in North America and it very actively supports mediumwave DXing. You can order the *DX Monitor* from: P.O. Box 1831, Pervis, CA 92572. You can see the IRCA Web page (http://fly.hiwaay.net/~waholler/irca.html), with information on ordering the newsletter, magazines, etc. The IRCA also works with many different radio stations to provide DX tests so that listeners can hear more new stations. For information about DX tests, see the sidebar in this chapter. For information on upcoming DX tests that you can try to log, check the DX test Web page at http://www.intac.com/~mherson/bcbdxtst.html.

- If you want European DX information, particularly from England, check out the *Medium Wave Circle*. This club has been around for more than 40 years and is something like the National Radio Club in terms of the number of publications that it sells. Many, many esoteric titles are listed here, including: *North American DX heard in UK and Eire*, *The full-size, full-performance loop*, *Grey-line DX*, and *Medium wave, a practical approach*. Considering that the prices listed for the books are often less than $2, most of these are probably pamphlets. The *Medium Wave Circle* also has a Web page (http://www.geocities.com/hollywood/5613/mwc.html), but if you want to eavesdrop on what's being heard in the UK, you'll have to subscribe to the newsletter so that you can access the newsletter online.

- Perhaps you want Western European mediumwave DX information from a different point of view. If so, especially if Spanish is your first language, a good group to check out is the *Mediterranean DX Club*, which is stationed in Spain. Currently, the Internet Web site for this club is incomplete. However, several articles are included in both English and Spanish. See http://www.geocities.com/siliconvalley/4847/index-en.html.

- *Monitoring Times* and *Popular Communications* both carry some columns and special features that apply directly to mediumwave DXing. The "Broadcast DXing" column in *Popular Communications* carries many updates of station callsign changes and new additions, which are very helpful for updating your annual resources.

What is a DX Test?

Lynn Hollerman

Many of you, I am sure, have noticed the posting of AM Broadcast Band DX tests here on the Shortwave Echo. Some have asked "What IS a DX test?"

A DX test is a special transmission aired by a station for the specific purpose of giving DXers a chance to log their station—something that, in many cases, might be very difficult or impossible to do under normal circumstances.

DX tests must be aired during what the FCC calls "experimental hours," which is between midnight and 6:00AM local time at the station. Also, most DX tests are scheduled for Monday mornings, because Monday mornings have traditionally been a time for many stations to do routine maintenance on their transmitters. This is good for the station, because many of them will work in a DX test during their normal testing/maintenance. This is good for the DXer, because the band is less congested at this time—thus increasing in many instances the DXers chances of hearing a station.

DX tests programs are as varied as the stations themselves. Most will include Morse code identifications and test tones (two things that really "cut through" the interference better than voice or music), and many will include some special music which will really "stick out" as something out of the ordinary. Examples of this are March music, polkas, TV and movie themes, or Christmas music played at an odd time of the year (such as in October).

There are some DX tests that are scheduled on Saturday or Sunday mornings and, at the request of the stations, on other mornings. However, most are usually scheduled for Monday, Saturday, or Sunday mornings.

The International Radio Club of America has a very active Courtesy Program Committee (CPC), which routinely schedules 50 to 60 such DX tests every DX season (from October to late March or early April when DX conditions on the AM band are usually at their peak). These test listings, along with many other DX loggings, technical articles, radio news, and other information of interest to the AM Broadcast Band DXer, are published in "DX Monitor," the official bulletin of the International Radio Club of America. This bulletin is published 34 times a year. For a sample copy (which includes annual dues and information on how to join on the back cover), send one 32-cent stamp (45 cents to Canada or 2 IRCs overseas) to: The International Radio Club of America, P.O. Box 1831, Perris, CA 92572-1831.

CHAPTER 6

Shortwave listening

By now, you should have realized that this book is about listening to shortwave, so you might wonder why this chapter is simply entitled "Shortwave listening." Tuning shortwave stations is not like flipping through the channels of a TV with cable. You can't just click through and say "Eh, I think I'll listen to the 'movie of the week' from Zaire or 'Guatemalan celebrity wrestling.'"

In one sense, a shortwave receiver is like having a dumb appliance. Smart appliances are those that do things for you to make a particular task easier. Most 35-mm cameras, for example, have autofocus to help beginners take good pictures. It works; most beginners *do* take better photographs with the assistance of this technology. The problem is that the simplification of the photo-shooting process hinders those who are excellent with a camera. The autofocus will not always target what the photographer wants, and the pictures will often be of a much lesser quality than if he or she had been using a camera with no "smart" features. So, the shortwave listener (especially the DXer) will want a "dumb" receiver with plenty of capabilities.

Few shortwave radios have any "smart" functions; most modern receivers have scanning capabilities and a squelch position, and my Kenwood R-5000 does have automatic selectivity. These capabilities are helpful for just general shortwave listening, but they don't help a bit if the listener is DXing. In the best case, your shortwave receiver's are controlled entirely by you.

In another sense, your shortwave radio is a sensitive musical instrument. With your knowledge and artistry, it can perform marvelously, recreating voices and music from around the world. The primary tool or hindrance to what you will hear is simply you. This chapter focuses on how you can improve your shortwave listening, how you fit into the shortwave hobby, how to make the most of the hobby, and how to keep it fun and interesting.

Where to listen

One afternoon, you arrive home from work or school, and you spot a large brown box on the porch. Either someone dumped a box of kittens at your house while you were away or that shortwave receiver finally arrived. You diligently searched through the hobby newsletters and magazines, and called radio dealers for a good deal on a receiver. You chose a rig that seemed like the best deal for your money. It wasn't the most expensive radio you could have bought, but it received good reviews and it seemed like it was a radio that you could grow into.

After calling in the order, you have waited for the past week for it to arrive. Now, it's here. What do you do? Your best bet is to scream "I've got a new toy!" and rip the box open. Look it over, grin a lot, push all of the buttons, and turn the knobs. It looks too cool. Get the giddiness out of your system, and start thinking rationally about what you're going to do with the radio: specifically, how and where are you going to use it?

If you are already a shortwave listener or if you were one years ago, these problems have been mostly or completely worked out. You know where and when you like to listen. However, if this is your first radio, you will probably need to work out your radio location.

Your radio set-up might seem like such a minor detail that it's not worthy of mention. Your listening location is one of the single most important factors for being able to enjoy shortwave radio listening and DXing. It's often even more important than the quality of your antenna or receiver!

If you follow these rules, you will be able to enjoy your radio to the fullest. Place the shortwave radio:
- In a comfortable location, where you like to spend time.
- Where you won't bother anyone and where you can listen any time that you want.
- Where a decent antenna can be installed.
- Where you can access plenty of information.

I know from experience how your radio's location can affect your listening. Over the past 10 years, I have moved a number of times between apartments and houses. One house that we rented had a huge yard, with acres of woods and fields in all directions (perfect for antennas). The house was also large, and I had one room all to myself and my shortwave radios. My shortwave listening prospered here.

However, I changed jobs and we moved to a small apartment with almost no yard. Here, I rarely could listen to the radio. After a few months, we bought a house with a small yard. I spent most of my time remodeling there. Because of the constant movement of furniture and building materials, I had no permanent location within the house to keep my equipment. So, I put my R-390A in the unheated, uninsulated attic. In the summer, I stripped down and dripped sweat while listening; in the winter, I bundled up and shivered a lot.

Fortunately, I also had a portable receiver, so I carried it into rooms where I was sanding, putting down new floor tiles, or nailing up drywall. At other times, I kept it in the kitchen and listened to the radio while I washed dishes or ate. In these cases, my housing dictated how and when I could listen to the radio. Although these situations were much less than optimum, I still did listen from time to time. However, hardcore DXing was mostly impossible.

Finally, we moved to our present location: a slightly larger house with a bit more yard, but still near power lines and other interference. Of course, this was another fixer-upper. I set up a radio room in the basement. It was a perfect place for me to play: I built shelves and moved in desks for all of my radios, books, and magazines. Everything was easy to reach and use. Guess what? I rarely used it. Because it was away from the other two stories of the house, I was cut off from the rest of the family. Because I needed to remodel the rest of the house, I never had totally free time to go down and listen. Even though I had what seemed to be the perfect shortwave-listening location, I could listen less than I had in the previous house. Weeks and months went by without me even turning on a receiver.

Several months ago, we accomplished enough of the work on our house that I could concentrate on installing a home office. The main concern was setting up a location where I could work for hours. I bought and assembled a desk and built shelves above it to store useful information. Then, I added my computer and printer, stereo equipment, shortwave receiver, and data decoder and monitor (Fig. 6-1). For once in my life, I can work while listening to the radio and DXing. For a break, I can check my e-mail or one of the shortwave forums on the Internet, or I can print out a reception report. As an added bonus, the room also has a bed and several shelves of kids' books, so my son can come up and read or play comfortably while I'm working or listening.

In the past, I usually had to choose between working or listening. Now, I can do both. It makes a huge difference. I've probably DXed (not including shortwave program listening) as much in the past few months as I had my entire life. Just because of my location.

6-1 My present shortwave/computer work area.

The bottom line is that your location and work situation greatly affect your shortwave-listening activities. If you spend most of your time in a car or truck, put your shortwave radio there. If you spend lots of time by the computer, like me, put your receiver at your computer desk. If you travel frequently on business, take a portable along.

Using that radio

Now, you have the receiver set up and turned on, an antenna is connected, and you've looked through the manual. What next? For the veteran shortwave listener, using a new receiver is like taking a car for a spin. You aren't concerned with not being able use it, but rather getting a feel for its performance capabilities.

If you are an experienced shortwave listener, skip this section. If you have some experience, skip to the next paragraph. If you are a beginner, keep reading. The first order of business is to tune in a few broadcasts. Even if you do focus on monitoring utility stations or amateurs, the broadcasts are best to hear first because they are the easiest to tune in. Take a look at Table 6-1, which lists the shortwave broadcast bands and their general characteristics. One mistake that beginners often make is to tune through the lower frequencies during the daytime. "I tuned the radio from 2000 to 7000, but I didn't hear a thing, so I gave up!" Don't make this mistake; check the broadcast bands that are active at this time of day.

Tune through these bands and find something interesting to listen to. Try manipulating the different filter (Wide or Narrow on a portable receiver) and mode positions while you are listening. If you have a tabletop receiver, try different settings for the AGC, noise blanker, and notch filter. By listening to different signals with these controls, you will get a feel for what settings work best at different times.

Another mistake that many new listeners make is not realizing that many government stations broadcast hundreds of hours per week in many different languages. I have heard people say "I heard some stations from Mexico." Only a few stations broadcast from Mexico and most are difficult to hear, so I ask how they deduced the country of origin. "Because they were talking in Spanish" is the reply. The chances are much better that it was the Voice of America broadcasting to Latin America than Mexico. In fact, many of the large shortwave operations actively market their country to their target audience, so they recruit native journalists to provide this service. So, when you hear the Voice of America's Urdu service, you will hear native Urdu speakers broadcasting with news, information, sports, and music that would be interesting to listeners in southeast Asia.

Program listening

The most common type of shortwave listening is simply sitting back and enjoying the programming on a radio station. There's a lot of excellent programming on shortwave to enjoy. In-depth news programs from around the world, many different types of folk music, Christianity from 1001 different prospectives, language-learning programs, conspiracy shows, etc.

I've been listening to shortwave and AM-DXing since 1977, off and on. Here are a few of my thoughts; replies are encouraged.

Shortwave listening 87

Table 6-1. General performance characteristics of the various short-wave bands during "normal" years of low sunspot activity

Band	Summer	Spring-Fall	Winter
4750-5050 kHz (60-meter band)	Limited mostly to short- or medium-range reception during the darkness hours. "Local" reception possible only during daylight.	Fair to good for medium-distance reception, with some long-distance possible during maximum darkness.	Best time of the year for long-distance reception from dusk to dawn for darkness paths.
5950-6200 kHz (49-meter band)	Some long-distance reception possible from dusk to dawn. Good medium-range reception possible at the same time.	Good possibilities for long east-west paths, and medium-distance reception good to excellent during darkness.	Generally "open" to any portion of the world also in darkness from dusk to dawn.
7100-7300 kHz (41-meter band)	Usually follows a pattern similar to the 49-meter band above. This segment is "shared" with hams; therefore, reception of weaker distant stations may be more difficult due to some interference situations.		
9500-9775 kHz (31-meter band)	Best long distance possibilities a few hours before midnight to the east, and again near dawn to the west. Normally fair to good during darkness.	Long "openings" good to excellent shortly after sunset until dawn. Medium distances good shortly after dark.	Good to excellent during the darkness period. Many large broadcasters may "beam" into North America during early evening.
11.7-11.975 MHz (25-meter band)	Good to excellent for long-distances from sunset to late evening, and again after sunrise. Broadcasters may "beam" into North America during early evening hours. Some short- and medium-range possibilities during daylight.	Excellent for long-distance during late afternoon and early evening. Some short- and medium-range reception possible throughout most daylight hours.	Long-distance reception reduced to late afternoon and early evening. Medium- to long-range reception fair to good around sunrise. Short range during midmorning.
15.1-15.45 MHz (19-meter band)	Good to excellent for daytime reception and extending into early evening.	Medium-range reception for most of the day. Long distance possible in late afternoon and early evening.	Distant reception during mid-morning and last afternoon. Short to medium during most daylight hours.
17.7-17.9 MHz (16-meter band)	Fair to good for distance to the south during maximum daylight; medium and short range from east or west.	Fair distance possibilities during peak daylight. Short and medium range throughout most of the day.	Fair to good long-range reception during peak daylight. Medium range good for most of the day.
21.45-21.75 MHz (13-meter band)	Short range only during peak daylight.	Some medium range from the south at midday. Short distances most of the afternoon.	Medium and short range during most daylight hours.

"Short" range: 300 to 1800 miles
"Medium" range: 1800 to 3500 miles
"Long" range: Over 3500 miles

Signal paths are measured by the shortest distance from transmitter to receiver, which is the "great circle" route (i.e., over the north pole from central Europe into western North America, etc.).

The daylight-darkness relationship caused by the sun moves from east to west. Therefore, when a particular band "opens" for distant reception, the stations heard will lie to the east of the receiver; this pattern will shift westward as time progresses.

Looking Back at 20 Years of Radio

Charles P. Hobbs

What I miss from back then:
- AFRTS (especially the news and ball games)
- Radio Mayak (from Russia)
- Long-distance telephone calls, in the clear (also, those repeating tone test patterns they used to play between calls)
- Good AM DX (Hearing Oklahoma, Chicago, Atlanta, and Cincinnati from Los Angeles. I once even picked up the VOA station in FL on 1180!)
- Voice of Chile (in English—a rather rare catch)
- Radio Grenada (good jazz)

What I don't miss from back then:
- The Woodpecker. (If you don't know what I'm talking about, consider yourself very fortunate!)
- Communist Propaganda. (They laid it on thick in those days!)

What I like about now:
- Radio New Zealand's new transmitter. Instead of an exotic DX catch, RNZI is now a station you'll want to listen to for hours on end.
- Ireland and Vietnam being relayed on stronger stations.
- Gene Scott. (Regardless of what one might think about him or his religion, you've gotta admire him bouncing back after the FCC took away his radio and TV stations!)

What I don't like about now:
- Right-wing wacko programming. Would someone else please step up to the mike and buy some time?
- Bad propagation (but this too, shall pass. . . .)
- Cutbacks at many stations. (Especially in former hot spots such as Russia, South Africa, and Israel. I'd kind of like to know how well these countries are doing under their new political systems.)

Program information

Unfortunately for the shortwave listener, most stations alternate frequencies throughout the day: some stations only stay on a frequency for about 30 minutes at a time. If you don't have a way to keep everything straight, you will miss many of the programs that you want to hear.

One good way to track your favorite programs and to see what else is on the air is to subscribe to *Monitoring Times* or *Popular Communications* and follow the schedules for English-language shortwave programs. This information can be supplemented by checking through the rec.radio.shortwave group on the Internet for recent schedule updates. Some shortwave broadcasters, such as the Deutsche Welle and the BBC even publish their own magazine/broadcast schedules (Fig. 6-2), complete with information about the upcoming programs and photos that correlate to

6-2 *DW-radio tune in*—a quarterly listening magazine.

the subject matter. Subscribing to these are exceptionally handy if you really enjoy listening to the programs of one particular station.

One of the easiest solutions to the information problem is to use a shortwave schedule and database program. These programs are incorporated in the software of computer radios, such as the WiNRADiO. The receiver-control software also often incorporates shortwave scheduling programs. So, you can search or scan the database and automatically tune to the frequency. Depending on how you program the software, it can even automatically tune the radio to a given frequency at a particular time so that you can hands-off listen to the radio. Because these database programs can be altered, you can update or customize the listings to suit your own personal listening tastes. For more information on computers and shortwave listening, see Chapter 14.

Improving shortwave audio

The motive of light program listening is to hear programming that sounds good. This *is* shortwave, after all, so you can't expect CD-quality audio. However, most shortwave radios sound needlessly bad—especially portables. With portable receivers, the manufacturers' main concern is to squeeze a tiny speaker into an already-tiny radio box. A 49¢ speaker popped into a plastic box the size of a small paperback novel will have audio that sounds only a bit better than if you talked into a kazoo.

Fortunately, the line-out audio from portable receivers is typically much better than the actual speaker output. It was just the small speaker and enclosure that ruined the sound. The easy way around this problem is to listen through headphones. This method works fine while you're working at a desk or computer or while you're in a motel room, but not being able to communicate with others while you listen is a serious drawback. Also, after a while, headphones can become annoying.

For more permanent listening locations, try an external audio system. As I mentioned in the preceding section, I connected my receiver to a stereo system so that the fidelity on strong signals is greatly improved. If you don't have a spare stereo amplifier on hand, you could always get an old combo phono/tape deck/amplifier and use it. I've gotten some of these for free because the tape deck broke, but the amplifier was still in good working order.

Another option is to purchase a Grove SP-200A signal enhancer (Fig. 6-3). This device is a sort of all-in-one output for portable or tabletop receivers. It is a wooden box that contains a speaker, an adjustable notch/peak filter, audio amplifier, bass and treble equalizers, audio squelch, tape recorder activator, and a noise limiter. The SP-200A has the advantage over using a stereo amplifier because of the extra filters, squelch, and noise limiter. At less than $200, the price isn't too steep.

Stereo diversity reception

One interesting possibility to enhance listening and fidelity is a variation of the "Improving shortwave audio" concept. The idea is diversity reception—a concept that goes back to the early days of radio. In fact, diversity-reception radios on the commercial market go back to the ultra-deluxe Hallicrafters DD-1 Skyrider Diversity receiver from 1938.

6-3
The Grove SP-200 sound enhancer.

The DD-1 was a fantastic idea, one that I'm surprised hasn't since been duplicated by shortwave radio manufacturers for the high-end listening market. The technical intricacies of diversity reception could fill a chapter, so here are only a few of the basic points. One of the main problems with listening to shortwave radio is *selective fading* (also simply known as *fading*). Fading is caused by the ionosphere as it refracts signals back to earth. You can learn quite a bit about where a signal has arrived from by how it fades. For example, fluttery signals have often traveled over either the North or South Pole, hence the term *polar flutter*. A signal with no fading might have reached you via the ground wave.

An interesting aspect of fading is that the fades occur over a very small area. For example, you might be listening to a signal on 15345 kHz that is fading in and out at regular intervals. You could move a few wavelengths away (200 feet, for example) and the signal would still be fading at the same intervals, but not in synch with the radio signal fading at the other location.

With the DD-1, Hallicrafters built a receiver that actually contained two separate radios that were tuned with a single VFO (tuning knob). For operation, a separate antenna was connected to each side of the radio. You would tune in a radio station. The signal from each of the two component receivers would pass to the amplifier. There, the strongest signal was selected and amplified. With the DD-1 receiver, the effects of selective fading could be nearly eliminated.

The largest drawback to buying a DD-1 receiver was price. In 1938, a DD-1 cost $750! In an age when a new car cost about the same amount and the typical salary for a worker in the United States was $25 to $50 per week, it's not surprising that few of these radios sold! Today, a typical new car sells for between $12,000 and $20,000. I really doubt that many $15,000 receivers would sell these days. It's not too surprising that only 200 DD-1 receivers were reported to have been manufactured.

One reason that the DD-1 was so expensive was because receivers were expensive in those days. The technology was also still young, the parts were expensive to manufacture, and the high cost of engineering and development hadn't yet been covered. Also, in these essentially pre-TV days, radio was the only form of mass-media entertainment that you could have in your home. Without competition, radio prices had not yet been driven down. The DD-1 was essentially two separate receivers, which contained a total of 22 expensive vacuum tubes, in a luxury cabinet with a very large "hi-fi" speaker.

Although the DD-1 was a miserable failure in the marketplace, those who have used it report a significant reduction in fading. In the November 1993 issue of *Electric Radio*, Joel Levine, one of the few people who has used a DD-1, reported that it really worked—fading was reduced by 10 dB (an impressive amount).

Diversity reception worked well enough that the U.S. Navy spent large amounts of money to equip many of their ships with diversity-receiving systems. The navy diversity systems varied considerably from the DD-1. Instead of using a single radio that had two separate receiving portions with a single amplifier and VFO, the navy system took a modular approach. They stacked two separate receivers in a rack with a separate piece of diversity-reception equipment that would sample and choose the strongest output. The receivers could be connected in master/slave operation so that tuning the one would automatically tune the other. The manual that I picked up for my Hammarlund SP-600 (a huge receiver from the 1950s that was extensively used by the U.S. Navy) contains extensive instructions on using it in a diversity-reception system (Fig. 6-4).

Is there a need or use for diversity reception in the 21st century? I think so! I doubt that diversity reception would add that much to the cost of a receiver these days. A sensitive receiver with synchronous detection and diversity reception would drastically improve reception of fading and weak AM signals. Although no manufacturers are currently selling receivers with diversity reception, I wouldn't be surprised if someone would experiment with the technology again in the next few years. I just hope that this foray into diversity reception for consumers will be more successful than the Hallicrafters DD-1 Skyrider Diversity!

If you want to play around with a bastardized pseudo-diversity system, it is possible if you have two shortwave receivers, two separate antennas, and a stereo system. Connect the output of the one receiver to the left channel of the stereo and the output of the other to the right channel. If you have enough space for antennas, try to keep them a few wavelengths apart. If not, experiment with different antennas. For example, string a longwire in the backyard and a vertical dipole beside the house. In this case, the difference in signal polarization (vertical for the vertical dipole and horizontal for the longwire) will also affect the reception differently for each antenna. However, this is just an experiment, so polarity differences are okay.

The two significant problems with this system are that the receivers can't be tuned at once (in master/slave operation) and the stereo won't choose the best signal and eliminate the other. Instead, tuning each receiver separately with the volume up can be really annoying, and the signal will fade back and forth between the right and left channels. Still, it can be a neat experiment and a good way to comparison test your antennas. Also, this system does sound pretty good when the receivers are precisely tuned to a strong signal—it produces a quasi-stereo effect.

Speaking of stereo, the U.S. pirate station WJLR operated for a while using stereo ISB modulation. ISB stereo is a transmitter that transmits the left-channel audio on the lower sideband and the right-channel audio on the upper sideband. The few people who could figure out the modulation system and had two receivers said that the ISB stereo modulation sounded great. The future of shortwave?

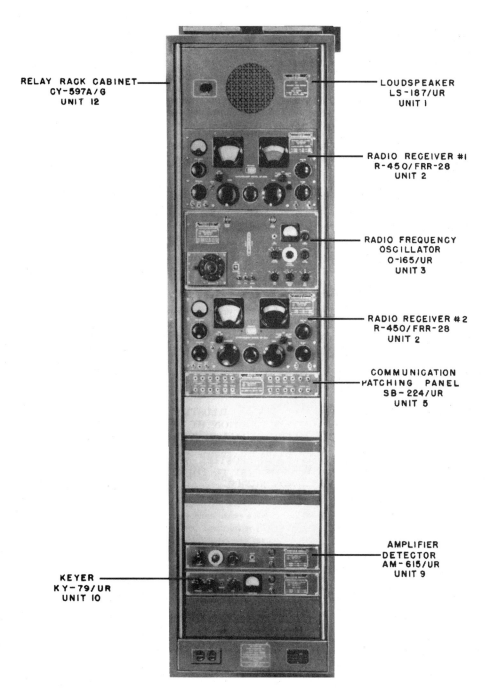

6-4 The U.S. Navy dual-diversity receiving system, which included two Hammarlund SP-600 receivers.

DXing

As mentioned in Chapter 1, *DX* is a Morse code replacement word for *distance*. Over the years, DX has come to mean any difficult-to-hear station, and a DXer is someone who is concerned with hearing these stations.

To the beginner, the differences between a shortwave program listener and a DXer are minor. They both listen to shortwave, right? However, the differences are fantastic. A program listener might catch the John Peel music specials on the BBC, the Newsline Africa show from the Voice of America, or folk music from Swiss Radio International. The goal is an interesting program and a good signal. The total program listener (with no interest in DXing) could care less if the programs were on shortwave or on the local FM radio station. The point is simply the programming.

The opposite is true for the total DXer. The serious DXer is out to hear as many different countries, stations, and transmitter sites as possible. At times, the goal appears to be the medium, not the message. DXers focus on what they haven't yet heard or QSLed so that they can move on to other territory. This alone might sound cold and calculated, but many DXers are fascinated with other lands and cultures. To have a taste of their appreciation of their DX targets, see some of the articles in *The Proceedings of Fine Tuning* (more information in the following section). I think that some of these DXers have a greater knowledge (geography, politics, history, and culture) of distant lands than many people do of their own country!

I have personally learned *so much* from listening to shortwave, especially DXing. I know that this is a cliché, but I feel like DXing has opened the whole world to me. That's really exciting, but a bit scary—seeing the whole world makes you realize how little you know and understand!

Of course, many serious shortwave listeners participate in both general shortwave programming and DXing. I can say from first-hand experience that both are a lot of fun. I listen to a number of programs on shortwave, but I've also QSLed hundreds of stations (including many tiny, "one-lung" operations) from around the world. It might be fun and easy to enjoy a shortwave program, but it's really exciting for me to receive a QSL and letter from a shortwave station that's really happy to have been heard by someone so far away.

Information

Like shortwave program listening, the first order of business for a DXer is to get great information. If you want to hear very difficult stations, you need to know what is broadcasting at different times and on different frequencies. Most of the best DX catches on shortwave are stations with extremely meager budgets that are constantly having technical (and other) problems that are constantly forcing them off the air. A station might break down and not have enough money to fix its shortwave transmitter. A year later, it might return for a few days, only to break down again or go bankrupt. Without fast and accurate information, you would miss this station altogether.

The most basic guides for DX are the *World Radio TV Handbook* and *Passport to World Band Radio*. These guides will provide you with the basic structures to figure out what station is broadcasting on a given frequency. They also provide some scheduling information and plenty of addresses that are essential to anyone hoping to collect QSLs. With these guides as your basic references, you now need frequent news updates to inform you of station changes and propagational characteristics.

Some of the monthly magazines and newsletters—such as *Monitoring Times*, *Popular Communications*, *Shortwave Magazine*, *NASWA*, and *ODXA*—are very helpful for shortwave listeners and DXers, but by far the best information comes from DX logsheets.

Fine Tuning is a long-running hardcopy logsheet that is available from:

Mitch Sams
779 Galilea Ct.
Blue Springs, MO 64014

The *Fine Tuning* group also produces some of the best hardcore DX books available: *The Proceedings of Fine Tuning*. Thus far, several volumes have been published since the first in 1988. In addition to many in-depth articles about shortwave radio equipment and how it relates to DXing, the volumes also contain excellent features on radio and the socio-political-geographical structures of such excellent DX lands as Papua New Guinea, Bolivia, Ecuador, Latin America, and Java. *The Proceedings of Fine Tuning* volumes are a must-have for serious DXers. For more information on these publications, contact:

Fine Tuning Special Publications
c/o John Bryant
Rt. 5, Box 14
Stillwater, OK 74074.

Some of the best logsheets are available online, either posted or via e-mail. One of the best and fastest (hence their slogan "the fastest with the bestest") is *Cumbre DX*. This DX logsheet is only available via e-mail, but it is arrives *fast*! Because it is an e-mail-only publication, overseas subscribers aren't bogged down by long delays or excessive costs. As a result, *Cumbre DX* regularly has loggings from experienced DXers who live in or near some of the world's hot DX regions: Japan, Uraguay, Australia, New Zealand, South Africa, Peru, even Sri Lanka! *Cumbre DX* also has its own radio program, which is often relayed by WHRI. For recent WHRI schedules and postings of *Cumbre DX* back issues, see either of the following Web sites:

```
http://www.grove.net/~cumbre
http://www.gilfer.com/cumbre.html
```

Although North America is beginning to catch up, shortwave radio DXing has traditionally been most popular in Europe. A few excellent sources of online information are also available from European sources. Dansk Shortwave Listener's Club International (DSWCI) has been publishing shortwave DX information for years. You can find information on their e-mail DX logsheets at:

```
http://www.sds.se/org/swl/dswci/dxwin.html
```

With the erratic nature of pirate radio DXing, a number of pirate-only DX logsheets are being published. *SRS-News* (Europe) and *Free Radio Weekly* (North America) are the two most frequently read weekly log sheets. *Pirate Pages* (North America) is available fortnightly via e-mail. To subscribe, contact:

```
jonny@srs.pp.se (SRS-News)
piradio@usa1.com (Free Radio Weekly)
ayoder@cvn.net (Pirate Pages)
```

DX Friends

Don't overlook in any hobby, radio listening or otherwise, the friends and associations that you can make. Shortwave radio listening, although fascinating in itself, often cannot be sustained for lengthy periods of time under solitary conditions. Many DXers that were involved in the hobby for varying lengths of time dropped out because of the lack of friends interested in shortwave.

It is really not important to meet other listeners in person (although doing so is fun). Some of my best friends in the hobby over the past few years have been people I have never met. We write a lot of letters, send e-mail, and talk on the telephone, but the distance between us is too great for us to meet each other. As the years have passed and as more get-togethers have occurred, I have been fortunate enough to meet some of these friends.

Friends in the hobby can improve your DXing considerably, depending on how many stations each listener hears and how well everyone communicates. For example, an article appeared in *FRENDX* a number of years ago entitled "What Communal DX?" that detailed one listener's personal experiences with his friends. The author of the article lived within an hour (driving time) of 10 other shortwave broadcast DXers. The group often staged overnight DX sessions where everyone brought their receivers to a particular house and hunted stations all night. The result of the DX session was a greatly increased number of stations and countries heard for each listener.

> Two of the largest shortwave radio-related get-togethers are the SWL Winterfest (P.O. Box 591, Colmar, PA 18915) in Kulpsville, Pennsylvania; and the annual meeting of the European DX Council (in various European locations). These have been running since the late 1980s, and each attracts hundreds of shortwave listeners from around the world. The meetings have talks by those who are experts in a particular radio subject, so they are a great way to learn about radio, as well as "put a face with a name."

Listening for good DX

After you have an understanding of and general feel for the shortwave bands, you might care to dig into some heavier DX. At this point you probably have heard many of the European broadcasters and a few of those from other continents. Now, it's time to take a dip in the DX pool.

Assuming that English is your primary language and that you know no (or little of) other languages, you should first try looking for local or national broadcasters that broadcast in English. Few countries, mostly from Africa, operate with just regional or national services in English (as opposed to having a large external service that operates in English).

Currently, one of the best signals from Africa in English is also a nice catch. Radio Liberia (also known as the Liberian Communications Network and Radio Liberia International), a clandestine station that is believed to be operated by Charles Taylor's National Patriotic Front of Liberia, has been very well heard on 5100 kHz throughout the summer and autumn of 1996. If you wanted to hear this station, you would need to consider the frequency, times of operation, and the path that it would be received on.

I am located in the Northeastern U.S., and Radio Liberia is located on the northwest coast of Africa. In the autumn, sunset occurs in Liberia at about 1800 UTC. The path from Liberia to Pennsylvania is entirely daylight at 1800 UTC, so 5100 kHz won't work. The path from Liberia to the U.S. must be almost entirely dark for a tropical band signal to be heard across such a great distance. Because November darkness in Pennsylvania runs from approximately 2230 to 1130 UTC and from 1800 to 0500 UTC in Liberia, the primary darkness "window" is the overlap: 2230 to 0500 UTC. Unfortunately, most shortwave radio stations sign off about midnight (Radio Liberia currently signs off at 0200 UTC). So, on the North American East Coast, the best time to hear Radio Liberia is from about 2230 to 0200 UTC.

Because of the instability in the region, Radio Liberia could be captured or destroyed at any time. It could well be off the air by the time that you read this. However, the station is currently heard every night with good signals here (except for some radioteletype interference).

Some of the other strong English-speaking African stations that you might listen for include the Ghana Broadcasting Corporation on 3366 and 4915 kHz, Radio Nigeria on 4770 kHz and the Voice of Nigeria on 7255 kHz (which is supposed to be reactivated soon), and the Namibian Broadcasting Corporation on 3270 kHz (Fig. 6-5) and 3290 kHz.

6-5 NBC Radio, Namibia, a regular performer on 3270 and 3290 kHz.

Something slightly different occurs when you listen for a station from the other direction. For example, the easiest DX targets in English from the South Pacific is the National Broadcasting Corporation of Papua New Guinea on 4890 kHz. Instead of being audible from just before sunset in Eastern North America, like the African stations, the Papua New Guinea stations start fading in after daylight. After about 30 minutes to an hour, they're gone. Because of the correlation between the sunrise and sunset times, the stations from Papua New Guinea are best heard in North America in the autumn months: September through November. The 4890-kHz NBC transmitter is the only one that can be heard regularly throughout the year in Eastern North America (Fig. 6-6).

Now that you've heard most of the easier English-speaking stations, what next? You might care to try some of the Spanish-speaking Latin American stations. If you're like me, you can barely utter any non-English words. Sure, I can say "burrito," "patio," and "armadillo," but do you think I can copy (through the static and fading) some fired up Latin American announcer saying "Desde la provincia de Junin, Cajamarca, Peru, transmite Radio Jaen, en bien de la senal de su programacion del aire, en sus frequencia...." What was that? I think I heard "Peru transmite Radio...."

Not to worry; you can sit back and enjoy some of the Latin American stations without understanding every word. If you are trying to receive a QSL card, some of these

6-6 A QSL featuring all of the NBC shortwave stations in Papua New Guinea. Use the Karai National Service on 4890 kHz to see if reception of the other stations is possible.

stations are strong and clear enough that you can pick out enough details to write a good reception report. A few of the strongest stations are: Radio Quito (Ecuador) on 4919.9 kHz, Radio Nacional del Colombia on 4955 kHz, CARACOL Colombia on 5077 kHz, Faro del Caribe (Costa Rica) on 5055 kHz, Ecos del Torbes (Venezuela) on 4980 kHz, Radio Cultural (Guatemala) on 3300 kHz (Fig. 6-7), Radio Internacional (Honduras) on 4930.6 kHz, and Radio Satelíte (Peru) on 6725.6 kHz (Fig. 6-8).

A somewhat complete discussion of DXing the Latin American stations could take the better part of a week. However, some of the largest stations operate throughout most of the day or even 24 hours per day (such as Faro del Caribe). Other large stations, such as Ecos del Torbes, only broadcast during prime time—in the morning before work and during the eight hours from about the time that work ends until midnight. Many of the lower-budget stations only broadcast for a few hours in the evening (such as Radio Satelíte) or a few in the morning. Because the morning broadcasts are near sunrise and the evening broadcasts are at night, nearly any Latin American station could conceivably be heard in North America. However, depending on the schedule, a station might simply be impossible to hear in Europe or Japan.

The Latin American stations listed in the preceding paragraphs are the easiest to hear and are all fairly distinctive. Things become more difficult with the small, less-established stations for a few different reasons:

- *A few stations might broadcast on the same frequency.* For example, Radio Fides (Peru), Radio Meteorologia (Brazil), Radio K'ekchi (Guatemala), and Radio Lider (Peru) are all listed in the 1996 edition of *Passport* as broadcasting on 4845 kHz. Who are you hearing?

- *Some stations change frequency or have a drifty transmitter.* For example, over the course of a few weeks in October 1996, Radio La Oroya (Peru) was noted as high as 4905.2 kHz and as low as 4904.77 kHz. Given a drifty transmitter on the "right" frequency, the station could easily be confused with another.
- *The commercial Latin American stations are frequently sold.* For example, Radio Continental (Venezuela) was sold and became Radio Amazonas in the mid-1990s. Without a bit of listening and up-to-date references, you could misidentify some of the stations that you are hearing.
- *The Latin American stations sometimes identify themselves as part of a network, with a particular program name, or by the name of a local mediumwave station that they are relaying on shortwave.* For a simple example of the first case, many stations in Colombia identify as part of the CARACOL network, which is something like the ABC radio network in the United States. If you hear "CARACOL" IDs on 5019.8 kHz, you are hearing (as of October 1996) Ecos del Atrato, not a new CARACOL Columbia station. In the latter case, Radio Cristal International (Dominican Republic) relayed local mediumwave station Radio Pueblo on 5012 kHz frequently in 1996.

6-7 The strongest broadcaster from Central America on the 90-meter band: TGNA.

Now you are on your way to some pretty good DX. Once you have gotten the strong stations out of the way, you can focus on hearing some of the tiny and erratic broadcasters, such as those from Peru and Bolivia. Of course, you don't just "forget" the other stations and friends that you've made along your DX journey. This isn't just a formal quest to pad your ego.

As fun as it might be to hear Radio Nacional del Colombia or Faro del Caribe for the first time, these are large regional broadcasters, and they lack the personal

6-8 Radio Satelite is heard surprisingly well in the evenings on 6725.6 kHz, with only 500 watts.

small-town flair that the tiny stations thrive on. On these tiny stations, you will often hear birthday wishes for local residents ("Happy Birthday" in either Spanish or English is commonly aired), ads for local businesses, messages for people from relatives or the town government (telephones aren't common in many locations and radio is faster than mail), and plenty of local folk music.

Here are a few tips to help you identify stations, whether you understand the language or if the signal is just weak. You can't identify a station on just one quality, but given the right circumstances, you might have a good idea who you are hearing.

- *The frequency.* Some stations broadcast on an odd or peculiar frequency. You might have an idea who is on the air just from that frequency. For example, Radio Satelite broadcasts in the middle of nowhere, on 6725.3 kHz. No stations have been reported within 15 kHz of Radio Satelite in years (probably not since they hit the air about 10 years ago).
- *The hour of the day.* Strong signals that are coming from a relatively close location will be heard for much longer than strong signals that travel a greater distance. For example, you are located in the United States, it's a few hours after sunrise, and all of the signals have faded out on 60 meters, except a Spanish-language station on 5025 kHz. Is it Radio Rebelde in Cuba or Radio Quillabamba in Peru? You can bet that it's Radio Rebelde, which is *so* close to the U.S. that the signals will be audible even in the daytime. Most Peruvian stations fade out just after sunrise in the U.S. East Coast.

- *The language of the station.* This can be a bit misleading because many stations broadcast in several different languages. However, Indian languages are typically used in Guatemala, but you wouldn't hear them on a Venezuelean station. Also, if you hear Portuguese, the station is almost certainly coming from Brazil (Angola also uses Portuguese in the tropical bands, but Angolan and Brazilian stations wouldn't often be heard at the same time).
- *The station music.* African stations and some of those from the South Pacific play plenty of Western music; expect almost anything. On Sierra Leone, I have heard everything from U.S. pop and rap to country music. RTV Mali plays everything from pop to African "highlife" music to 1960s British blues bands. Musical selections are typically much less diverse in Latin America. Expect language-adapted Christian hymns from the many Christian evangelical stations in Central America, "lively" and Western music from the Venezuelan and Colombian stations, pop music (mostly from the country) from the Brazilians, and different types of Andean folk music from the stations in Ecuador, Peru, and Bolivia. Peru is well known for its campo folk music and Bolivia for its Andean flute music. Of course, there are many different guidelines and divisions of music on the radio around the world, but this will give you an idea of what to expect.
- *Regional indicator stations.* It's handy to keep an ear on the strong, regular stations from any given region. If the signal is much stronger than usual, chances are good that you are hearing an opening to that area. Then, you can listen for some of the more-difficult-to-hear stations. Also, you can listen for when these stations fade in at their strongest; you will then know the best times to listen for other stations from the region. See Table 6-2 for a list of some of the best indicator stations for different regions of the world.

Table 6-2. Good-opening band-opening stations for various parts of the world

Freq.	Indicator station	To hear
7335 kHz	CHU	Canadian regional stations and North American pirates
15000 kHz	WWV	North American pirates
6725.3 kHz	Radio Satelíte	Evening Peruvian stations
3300 kHz	Radio Cultural	90-meter Central American stations
4753.3 kHz	RRI Ujung Pandang	Indonesian stations
4890 kHz	NBC	Papua New Guinea stations
4835 kHz	KBC (Kenya)	East coast African stations
6245 kHz	Vatican Radio	European regional stations and pirates
5010 kHz	AIR Thiru'puram	Indian regional stations
6937.3 kHz	Yunnan People's Broadcasting Station	Southeast Asian regional stations
2325 kHz	VL8T	Oceania regional stations and pirates

With all of these techniques to hear and identify stations, you can effectively log stations without understanding the language of the broadcast. With some experience listening, though, you should begin to understand a few key words and learn to identify some of the different languages. However, in order to absolutely identify a broadcaster, you need to hear the station identification. Most of the time, this isn't *too* hard, but I have a tough time IDing stations using languages that aren't European-based (Spanish, French, Dutch, German, Italian, etc.).

To catch the identifications and other talk from shortwave stations, be sure to have a tape deck on hand to record what you're listening to. With a tape deck, you can record a broadcast and listen to those IDs over and over and over again. Maybe after a while it will make sense . . . or maybe it will still sound like a bed of noise.

Another way to hear good DX is to be somewhat organized with your informational sources. I keep my *Passport to World Band Radio*, *WRTH*, a year of NASWA's *The Journal*, and a three-ring binder full of *Cumbre DX* printouts all within arm's reach. I check off the stations that I've heard in *Passport* and *WRTH*. I use *Cumbre DX* as my basic DX "hardware." In three different colors, I highlight stations that I haven't yet heard and are listening possibilities at my location: one for local early mornings, one for daytime stations, and one for evening/nighttime.

When DXers reach the advanced stage of listening, some even make up lists of what stations they want to try to hear at different times. This can be really helpful when trying to make the most of a good opening. I'm not quite that serious about my DXing, so I guess that's where I draw the line. You might find the technique to be helpful, though.

Reporting stations to shortwave magazines, bulletins, and logsheets

Aside from writing to stations for QSLs and possibly attending DXpeditions or DX conventions, the only real social aspect to DXing is sending your loggings in to bulletins. When you log stations, everyone sees what you've been hearing and how you report your loggings.

Organized logsheets and logging columns are really beneficial because they are edited by experienced listeners. This makes a big difference in the quality of the loggings that are published. If you would just follow information that is submitted on rec.radio.shortwave on the Internet, you would quickly get lost in all of the incorrect or misleading data. Aside from the honest mistakes, some of the tips are hoaxes. One recent, rather poorly done hoax concerned a supposed new pirate radio ship off the coast of Brazil. One of the best hoaxes in recent years was a posting about the Kirbati army in the U.N. peacekeeping force establishing a relay station in Macedonia. Kirbati is an island in the South Pacific with an estimated population of only 75,000!

Any logging that you send out should contain some program information. Simply stating "music and talk" doesn't cut it—that would sum up 98% of the radio stations on the air. The best loggings mention the ID, type of music played, any advertisements, and program names. Also be sure to follow the general format of the magazine, bulletin, or logsheet that you are sending them to. Log editors are volunteers; they appreciate anything that makes their job a bit easier. A solid logging would look like this:

WREC (U.S. pirate): 6955U, 10/27, 1604-1638* PJ Sparx w/Halloween spcl. Mx by AC/DC, Van Halen, Creedence Clearwater Revival, Ramones, Ozzy Osbourne. Many IDs by Phil Muzik w/NY & BRS adds. Solid S9 sig (Yoder,PA)

A few of the important rules of reporting loggings include:
- *Include all date/time/frequency information.* A report is virtually useless without all of these details.
- *Mention any uncertainties.* If you're not sure about a station's ID or some of the talk, mention it in the report. If you are fairly certain of the station, based on the frequency, programming, time of day, etc., list it as being tentative by placing a "(t)" or "(tent)" after the station name. If you aren't sure what station it is, list it as an unidentified "(UNID)." With UNIDs, it's best to listen for the station for a few days in a row to see if you can identify it before reporting it.
- *Never send fake reports.* There's just no good reason to send in fake reports. This *is* just a hobby, after all.

It does require some effort to get your loggings out on a regular basis, but your information helps other listeners to hear interesting broadcasts, too. Their information will, in turn, help you to hear much more.

Finding and eliminating radio interference

For many shortwave listeners, the greatest hindrance to receiving good DX isn't the expense of a better radio or not being able to install a good antenna. The arch enemy of shortwave listeners is *radio frequency interference (RFI)*.

RFI can result from many different conditions, and many of these will sound different on your radio. You might have a steady buzz (that affects the higher, lower, or all portions of the shortwave bands), steady clicks, a regular splatter of talk and music, or even "glitches" that well up around different frequencies.

If you have any of these problems, take a look around your house before you call the local power company for help. Deep, steady buzzing can be caused by a number of problems, but two could be hiding right in your house. Fluorescent lights will cause interference to the shortwave bands. Most people don't have fluorescent lights throughout their entire house; you can usually just turn them off before you start listening. Another DX plague is touch lamps, lights that you merely touch to turn on or off. These horrible contraptions run electricity through the metal base so that you can turn them on or off; essentially, they are a noise emitter that doubles as a light.

A few household sources of buzzing noises that are intermittent include hair dryers, vacuum cleaners, blenders, refrigerator and freezer motors, and even household heating systems. There isn't a whole lot that you can do about these problems, but I *do* avoid going to the refrigerator for a drink or some ice cream while I'm in the middle of receiving good DX!

For wideband bleeping sounds and noise that wells up on different frequencies, perhaps every 10 kHz or so, look no farther than your personal computer. The wideband bleeping sounds are typically caused by a computer modem. Most of the welling up noise spots are caused by the monitor, so if I am listening to something really good that's being wiped out by interference from my monitor, I just turn off the monitor for a while and do something else while I listen. An article in *Monitoring Times* covered the elimination of computer-caused RFI by ripping apart the monitor and installing components to eliminate the noise. That's a bit too involved to cover here, but just be aware that it is possible to clean up noise from a PC, and it probably won't be long before a book is dedicated to the topic!

Speaking of monitors and welling-up noise spots, TVs also cause these problems. Fortunately, though, most TVs cause fewer of these spots than computer monitors.

It is not at all impossible to determine the origins of outside noise sources, although they can be more difficult. With the help of a trusty portable receiver, you can often locate the RFI source. Just tune in the noise on your radio and walk around outside until the signal peaks. Chances are that you've found it. If the noise leads you to a telephone pole, good. The pole might have a faulty transformer (pole pig), which is causing the noise. Regardless, if the noise is being emitted by a defect in the power system, call the power company and they will probably send someone out to fix the problem. They might even be grateful to you for doing the legwork on their problem.

In the worst case, your receiver will lead you to someone else's house. Most people aren't too friendly when someone with a radio walks up to their house and accuses them of causing radio interference, so there isn't a whole lot that you can do. If their TV is wiping out your favorite frequency, you can't ban them from watching it. About all that you could do is buy them a new TV that doesn't cause this interference (good luck!). One listener that I know lives near a welding shop—she can't exactly make them stop working so that she can listen to her radio. Another listener reported that he DFed a very loud noise source to his neighbor's new automatic security lighting system. The system was very expensive and involves their security, so what can he do?

In most of these cases, there is very little that you can do to listen to the radio interference-free, except move or take a radio DX vacation.

DXpeditions

Of the thousands of listeners who regularly tune across the shortwave bands, probably only a few are satisfied with their reception. As the country becomes more urbanized, the growing population uses more electrical appliances. Electrical interference abounds—especially for shortwave listeners who are crammed into apartment buildings or small suburban lots.

Are you seeking out rare and interesting DX on the mediumwave and shortwave bands? Your quest is doomed if you live in one of these locations. You can move or give up DXing, but neither is an option for me. So, I move temporarily: I take a DXpedition.

A *DXpedition* is a merger between the words *DX* and *expedition*. In amateur radio circles, a DXpedition means going to some tiny uninhabitable island and being one of the first operators to transmit from that "country." In shortwave circles, the meaning is much simpler; just take your equipment on an outing and have a radio adventure.

South Pacific Union of DXers Inc.
SPUD

If you're serious about your radio listening, only one club can provide you with all your needs. South Pacific Union Of Dxers Inc. SPUD is made up of some of the finest DXers in the land and is fast becoming known as "The Club" to be a part of in the region.

Unlike other clubs, SPUD has 99% local membership and does not rely on overseas members contributions. We pride ourselves on local content and local input. Let's face it, a club is only as good as its own members.

The club boasts the best shortwave, tropical band, mediumwave, and HF utility information in the country and our monthly magazine is chocked full of interesting articles for the serious & beginner listener. The club also caters for those who are into scanner, VHF/UHF, and FM/television DXing.

SPUD is also known for its mateship, and we are often referred to as the DX Socialites with regular BBQs, restaurant nights, tours, and monthly meetings.

Regular DXpeditions are held to "The Coorong" in S.A. and to Marlo and Cape Otway in Victoria where members are able to experience exotic and hard-to-hear stations, most of which you are unable to hear from home due to local man-made interference. It's also a good chance to see how other fellow listeners go about the hobby.

You might even get roped into the odd Test match or a kick of the footy. Failing that a few soothing ales around the campfire have also been known to happen. (Heaven forbid.)

SPUD believes that there is more to a club than just putting out a magazine. So if you think that the South Pacific Union Of DXers Inc. is your sort of club, then drop us a line, and check out what we have to offer.

For a sample copy of our magazine, please send two IRCs to:

SPUD Inc.
P.O. Box 293
Coburg, Victoria, 3085

Locations

The first concern is to abide by trespassing laws. Obtain permission from landowners, or utilize state and national parks. Don't overlook land that you don't own, but that you have rights to, such as the back acres of a hunting club that you might belong to. I have tried all of these methods. The difference between a good and a bad setup depends more on the geographic location that you choose than whether it is private, commercial, or public land.

The next consideration in finding a location that will suit your needs is whether the site contains ac power. If the site does not have power, you need to operate from a battery system. Some of the table-top DX receivers only operate with ac input. If you are depending on an ac-only receiver and you are on a site without ac power, you might as well start a fire and practice your smore-making techniques.

Commercial sites include recreational areas and hunt clubs, but they are primarily private campgrounds. They are excellent sources of electrical power. There's nothing finer than reaching out from your warm sleeping bag to tune in a rare broadcast, watching the LEDs glow in the chirping darkness of the forest. If you're interested in this option, though, be sure to call the camp office before you go. Some campgrounds only have 220-V power outlets and a handful have no power at all. The curse of the commercial campgrounds is the flipside of their blessing; if a number of electrical appliances are being used in a nearby RV, your radio might receive as much local electrical noise as rare radio signals.

Don't confuse commercial campgrounds with those in state and national forests. State and national parks are excellent sites because of the extensive acreage, fewer campers (in some situations), and generally lower amounts of electrical interference. However, some parks close at dusk and have no campground provisions. If they do, they are more likely to be *primitive* sites, which have no electricity. Furthermore, some park rangers have a rather dim view of radio listening. During a recent DXpedition, the park ranger didn't believe that my friends and I were planning to listen to the shortwave radio. He even abandoned several dozen fishermen and picnickers to suspiciously keep tabs on us and make sure that we left "his" state park immediately.

Make sure that the owners or keepers of the land that you are using for a DXpedition are aware of what you are doing. Very few people understand what it's all about. It is best to explain to them what shortwave is, what you are doing, why you would want to listen from that particular location, and why you will probably need to string long antennas. In most cases, the rangers or managers will be interested in what you are doing, and they might even drop by to check out your listening post. They might even want to hear what you are listening to.

Another concern when DXing at commercial campgrounds is crowding. It is nearly impossible to DX in crowded campgrounds because there is virtually no direction that you can run the antenna wire. Another problem is audible noise. One beachfront campground that we stayed at in July was full of hundreds of other campers who were so loud that it was difficult to sleep until after midnight.

Fortunately, these problems usually cure themselves. Because of the bad propagation, the worst shortwave listening season is summer, so few people would have a DXpedition then anyway. Otherwise, times that you might encounter crowded campgrounds are Labor Day weekend, the first day of hunting in your state, and during local festivals. Campgrounds within 30 miles or so of popular tourist attractions are crowded all the time "in season." Avoid these until September, or go before the middle of June (the opposite is true for countries south of the Equator). If you want to be sure that few people are staying in the campground, call the manager in advance. Ask how many campsites are currently open and the average number that are open for that time of year.

Antennas

The antennas that you choose are also important aspects of a successful DXpedition. They depend on what you plan to hear and from what direction the signals will emanate. The rarest shortwave broadcast stations operate below 6 MHz, so a longwire antenna (or a derivation, such as the Beverage antenna) is one of the best choices. Longwires are also very simple to install, so they are one of my favorites. However, *don't ever* connect a long longwire to a portable shortwave receiver—most of these receivers have sensitive IF sections. At best, you will overload your receiver and cause images of the stronger stations to appear across the dial. At worst, you could overload the IF section to the point where some components are destroyed.

A longwire antenna (see Chapter 1) is simply one long piece of wire (100' or more in length) that is run from your receiver. The major benefit of the longwire (as far as reception is concerned) is that the longer it is, the more directional it is (with the best points of reception being off the ends of the antenna). For the best directional effect, try to make the wire as many wavelengths long as possible. This is difficult over the mediumwave band, where a wavelength is often over 1000' long. If you can manage to make a 1000' long (or longer) longwire, you will notice some rather directional effects when tuning across the mediumwave region. Because of the directional effects of longwire antennas, you should try to point them at the DX regions that you want to listen to (Fig. 6-9).

6-9 Installing antennas during winter DXpeditions can be cold. Here a longwire droops under the load of an inch of snow.

Pointing antennas in the proper direction to hear what you want can be very tricky business. Instead of just aiming the antenna at the direction that you want to hear, you must check out a *great circle map* (also known as an *azimuthal equidistant map*). These maps are different for every location because they are centered on your location and everything becomes more distorted in shape out from that point.

The great circle map shows the actual direction that you need to point your antenna. Looking at a standard world map, I would assume that because Papua New Guinea is just below the Equator and my location is well North of the Equator, that I should aim my antenna just south of West (about 260°). Wrong! For actual radio paths, according to a great circle map, I should be pointing my antenna at 300°, which I would have assumed to be good for tuning in Alaska! The handy *ARRL Operating Manual* features a number of great circle maps that are centered on various locations around the world. If you're on the Internet, check out the following site:

```
http://www.xray.duke.edu:1080
```

Here, you can make your own specialized great circle map, with many different variations (focus on one part of the world, include gray line information, use grid blocks to show land distortion, etc.). You can even download a free copy of the software from a related FTP site. Neat!

It might be difficult for you to install several antennas that are extremely long. Unless you rent a private cabin or spend a winter weekend in an icy, deserted campground, you will probably have neighbors surrounding your listening area. Don't try to run your wires close to neighboring campsites; your antenna might pick up electrical noise from any appliances that the campers might use over the course of the evening, and depending on how you string the antenna, you could place your neighbors at the risk of injury.

To limit both the cost and the weight of antennas, I purchased a 6000' spool of enameled #26 magnet wire. It's surprisingly strong and very light; it easily strings through trees. I try to keep my antennas strung well above head level so that no one is ever at risk of being injured by them, especially because the #26 enameled wire is very difficult to see. However, unlike some antennas, such as those in the dipole family, it is not necessary for the wire to be strung as high as is humanly possible. The last safety precaution is the most important: *Never cross any wires while stringing antennas. Don't even run your antennas close to power wires.* Not only will you probably pick up noise from the power lines, but you could place yourself in a deadly situation.

Supplies

Aside from your communications receiver, antennas (or wire for longwires), and yourself, what else do you need? Of course, food or a local restaurant is a must. If you take food, should it be ready-to-eat or will you take along everything necessary to prepare that food? I've had everything from excellent homemade chili to ready-to-eat junk food on DXpeditions. Depending on the number of people present, the amount of indoor space, and the weather conditions, either style can work well.

It's handy to have a number of tools available. Such things as straight and Phillips screwdrivers, reels of string or nylon cord, a pocket knife, one or more sets of headphones, solder and a soldering iron, a cassette deck and blank cassettes for recording stations, many audio hookup cables, a hammer, several extension cords and multi-outlet surge protectors, duct tape and electrical tape, and extra antenna insulators are necessary.

For comfort, take such things as sleeping bags, pillows, a folding table and chairs, desk lamps, a comfortably large tent, bug spray (for the summer), and extra weather-appropriate clothing.

As far as DXing supplies go, make sure that you take along a recent copy of the *World-Radio TV Handbook* and/or the *Passport to World Band Radio*, some radio newsletters (such as *Cumbre DX* printouts, *Fine Tuning*, *ODXA*, *NASWA*, etc.) and radio magazines (such as *Radio!*, *Monitoring Times*, and *Popular Communications*). If you have made any charts of stations that qualify as your personal DX targets, this is the time to use them.

Laptop computers can also be extremely useful tools on a DXpedition. Download information from a computer online service that covers radio (such as rec.radio.shortwave on the Internet or the general shortwave section on the Fidonet), or be sure to bring this information with you.

Even if you don't have access to ac power at your location, you can either run the laptop from its built-in battery for a limited time or you can run the computer off an external battery supply (my laptop computer requires 12 V, so it will operate from a standard 12-V gel cell). You can also operate the computer from your car's battery via a laptop supply or inverter. Unfortunately, a number of computers cause radio interference that will hamper your listening. If this is the case, you will be relegated to either using the computer in your car, using a battery supply away from your receiver location, or using it during listening breaks. Test the computer for interference before you go.

Finally, pull out your handy-dandy radio logbook, and take along some ballpoint pens. Be sure that you use ballpoints and not felt tips; I had a few felt-tip logs that were entirely washed away by rain.

Excursion results

On one DXpedition in central Pennsylvania, several veteran DXers reported solid loggings of 14 different Peruvian regional stations and dozens of other countries. On a DXpedition on the coast of Washington state, two DXers netted 12 different Indonesian stations with 700' and 1800' longwires. On another DXpedition at the same location in Washington, the participants logged mediumwave stations from New Zealand, Australia, Tonga, and Fiji.

Australian DXpedition Results

Here are a few loggings from a recent trip to the "Coorong" in South Australia.
- Dave Onley. Coorong, South Australia. (150km Sth/East of Adelaide) Nov 9-15
- Drake-R7 & 400m Bev towards Southern Europe.

Shortwave listening

Freq.	Time	Station	Notes
540	1849	RRI Bandung.	Fair w/MA in Indon. suffering from c/c 7SD.
549	1753	Les Trembles AA w/FA.	1st time in a while at fair level.
549	1854	CNR Fujian. Taiwan-1.	CC program w/traditional folk tunes.
558	1904	DZXL Makati.	Fast talking MA in Tagalog. Fair signal over Aussie.
576	1857	RM Oyash. Mayak.S.	Instrumental mx leading up to 1900 ID & IS.
621	1907	Batra.	Fair under 3RN in AA. MA/FA in conversation.
666	1910	DZRH Navotas.	MA w/telephone call from YL in Tagalog.
666	1922	St Pierre, Reunion.	Fair at times w/FF program of mx & talk.
675	1143	Nei Menggu-CH.	Some great Autonomous CC mx at fair level.
684	1717	CNR Fujian. Taiwan-2.	CC chat & folk mx. Strong level.
783	1052	8AL Alice Springs.	Aussie Rules match between Magpies & ?
819	1228	DYVL Taclobin.	Jingles/laughter/chatter in Tagalog. Fair.
819	1130	Shanxi. 1st prgr.	FA w/CC talk. At times mixing with DYVL.
846	1915	DZNN Quezon City.	Good w/R.Veritas program. mixing w/Aussie.
846	1826	Nyamninia (Kenya) tent:	Mixing at times w/2RN. Male in Swahili w/muffled modulation. Afro mx. Very hard to ID with // SW National svce.
855	1224	RRI Mataram.	Excellent all week w/Indon pops & talk.
855	1230	DZGE Naga City.	FA & MA at poor level with constant chat.

Freq.	Time	Station	Notes
873	1914	RM Novosemeykino.	First time ever heard. Male & female in RR at strong level. Appeared to be chatting about Israel.
909	1238	DYLA Cebu City.	Huge all week with talk/mx in Cebuano.
927	1953	TRT-1 Izmir.	Big over 3UZ w/Burasi ID on hour. Turkish mx & talk.
936	1503	AIR Tiruchirapalli A.	Good all week w/Indian folk mx, talk & nx.
945	1434	CNR-2 Beijing.	Classical mx & FA in CC. OK at times.
945	1448	AIR Sambalpur.	Nightly w/Indian folk mx & chatter. EE at times.
945	1509	Tronoh (Malaysia)	Soft spoken FA then trad mx. Poor.
954	1704	RRI Kendari.	Monster at times w/nx then Indon pops.
954	1815	Al Arish.	AA talk program. MA at fair level. Not noted for a while in Eastern Oz.
999	1630	RRI Jakarta.	Regular spoil sport level. Indon pops & talk.
1008	1922	RRI Median.	The poorer of the Indons. Pop tunes & mix w/Tassie.
1017	1923	TRT-1 Mundanya/Instanbul. //927.	Big w/TT commentary & mx.
1080	1935	Katowice.	Personal 1st for Poland. Took a few mornings to ID. Heard w/Polish & sret x.
1278	1936	Kermanshah (Iran)	Fair w/Middle Eastern mx & Farsi talk.
1287	1934	RFE Melnik/Litomysl.	Fair w/MA in Slovak. Classical mx.
1296	1317	CRI Beijing.	MA & FA in EE F/S broadcast. Fair at times.

Freq.	Time	Station	Notes
1359	1704	BEC33 Hualien.	CC folk tunes & chat.
1359	1842	Batra.	Huge signal. Non stop classical mx & AA talk.
1368	1930	Krakow.	Another highlight w/Polish talk & Instrumental mx.
1386	1901	Nyeri, Kenya.	Surprised at the monster level of this. EE MA/FA w/talk on Kenyan water supplies. Stronger on MW than SW //.
1404	1927	Rasht, Iran.	Approaching the end of Teheran's Turkish pgm. Fair at times over 2PK w/mx & MA. Also mixed w/Romania.
1404	1926	Sighet, Romania.	Romania Actualitai pgm. Fair/poor.
1410	0926	CFUN Vancouver.	Massive signal for the Coorong w/EZL tunes & countless ID's. Hung in there for hours.
1413	1845	BBC Masirah Island.	Local level each morning. Here in Persian.
1431	1941	Kopani. R.Ukraine Int.	Fair w/Instrumental mx & FA.
1440	1941	Dammam.	Usual barnstorming signal in AA. MA & mx.
1458	1943	AIR Bhagalpur.	Here each morning w/EE nx & Indian affairs.
1476	1712	Dubai, UAE.	Strong at times in AA. M/eastern mx & talk.
1494	1925	Al Karanah.	MA in AA. Middle Eastern mx. Here each morning.
1512	1945	Jeddah.	Strong w/AA MA & Mx. Regular.
1521	1903	Duba.	Another strong Middle Eastern w/AA mx & talk.
1530	2001	VOA Sao Tome.	Not as strong as on previous visits. Here mixing with Vatican & 2VM. EE nx.

Freq.	Time	Station	Notes
1539	1947	Sadiyat.	UAE. MA w/AA at fair level. Here each morning.
1544	1912	V.O.Sahara.	Fair w/AA commentary/politics.
1611	1846	Vatican City.	FA in Latvian w/relig pgm. Good each morning.

Dave Onley
SPUD (South Pacific Union Of DXers Inc.)
Melbourne, Australia

Personally, I have usually had much less spectacular results on DXpeditions, in part because I would rather talk about radio and hear some interesting stories than stay glued to my receiver. On one of my most successful DXpeditions, I logged seven different pirates from Europe (running as little as 30 watts) and a number of regional broadcasters from Peru and Bolivia. I was among the only listeners in North America to receive QSLs from some of these stations.

The overall success of DXpeditions has been excellent for me and for many other people. It's amazing what enjoyment you can derive from building superior directional antennas in quiet locations while sacrificing personal comfort and hygiene to listen to the radio. Even when the bands are empty, the presence of friends, stories, and experiences of hiking into the middle of nowhere for a weekend of radio listening are well worth the effort.

7
CHAPTER

International shortwave broadcasts

Just like the Internet, you will be virtually lost if you scan through the shortwave bands without any information to help you find your way. This chapter is composed of countries, frequencies, times, addresses, telephone and fax numbers, and station features to give you a small taste of what's behind that schedule information.

How to use this chapter

Each listing features the following information (where available):
- *Country*   The country specifies the country of origin for the programs, not of the broadcast. Some stations have relay transmitters located around the world.
- *Station name*   Generally, the station name is listed in English, rather than that of the home service language.
- *Areas of the world that the station targets*   Some stations beam their programs to different parts of the world, and others don't. The continent icons should quickly give you a better idea about what you can hear, no matter where in the world you are.
- *Address*   The address is listed without the name of the country. If an address outside of the country is used, the country name is listed in capital letters.
- *Telephone and fax numbers*   Not all numbers were available, but many are listed. For more help on these numbers, try the *World Radio TV Handbook* or the *Passport to World Band Radio*.
- *Computer addresses*   Most of the large international shortwave broadcasters have Internet addresses, but most don't. Some stations, such as Radio Vlaanderen International even have addresses on several different computer networks.

- *Prime frequencies and times* The most common or regular times and frequencies for broadcasts in English are listed.
- *Notes concerning the transmissions* The notes are brief comments about the station's operations—anything from the station being irregular to its future plans to the programming.
- *Non-English-language broadcasting information* Many countries broadcast with no English-language programs. Nearly all of these stations (with the exception of some, like BSKSA from Saudi Arabia) are small operations. Many of these countries, such as Peru and Indonesia, have more than 100 active shortwave stations. For these countries, a brief paragraph describing the countries best-heard operations and frequencies are included; however, to keep this book down to smaller than unabridged dictionary-size, the other schedule and contact information is not included. If you need this, consult the latest edition of *Passport to World Band Radio* or the *World Radio-TV Handbook*.

Albania

Radio Tirana
Radiotelevisione Shqiptar, Rruga Ismail Qemali, Tirana
Prime frequencies: 6120, 7160 kHz 0145–0200 UTC **NA**
6185, 7155 kHz 1715–1730 UTC **EU**
6270, 7270 kHz 1930–2000 UTC **EU**

Notes: Once a powerful communist voice with a large English service, Radio Tirana has been drastically weakened by the changeover to democracy. An interesting development in light of the country's recent politics, the Christian broadcasting organization, Trans World Radio, now utilizes some of the old Radio Tirana transmitters.

Algeria

Radio Algiers International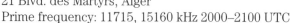
21 Blvd. des Martyrs, Alger
Prime frequency: 11715, 15160 kHz 2000–2100 UTC

Notes: The station doesn't seem to put much emphasis on its English service—the times often change without notice between about 1800–2200 UTC, the frequencies sometimes change, and the broadcasts are often entirely absent.

Angola

After years of civil war, much of Angola has been destroyed. Only a few of their shortwave transmitters are left on the air. The best-heard station from the country is the former clandestine broadcaster, A Voz da Resistencia do Galo Negro (The Voice of the Black Cockerel) with programming in Portuguese on 7100 kHz. New equipment is reported to be forthcoming for Radio Nacional de Angola, so official programming might be easier to hear outside of Central Africa in the coming years.

Antigua

Antigua is a small island in the Caribbean. This is an easy catch on shortwave because of the massive relay transmitters that are used by the BBC and the Deutche Welle. Unfortunately, no local programming is aired by these transmitters.

Argentina

Radiodiffusion Argentina al Exterior
CC 550, 1000 Buenos Aires
Prime frequencies: 11710 kHz 0200–0300 UTC **NA SA**
15345 kHz 1900–2000 UTC **NA EU**

Notes: Although Argentina is a large and important country in the Americas, shortwave is apparently not a top priority. The RAE is reported more regularly in Spanish than in English.

Argentine Antarctica

Argentina maintains a military base in Antarctica, which is fitted with a broadcasting transmitter. The station, Radio Nacional de Arcangel, is occasionally heard on 15476 kHz with Spanish-language programming across the world, but only with a poor signal. Many DXers strive to hear this station because it is the only shortwave broadcaster on the continent and it is often off the air for months at a time.

Armenia

Voice of Armenia
Alek Manukyan St. 5, 375025 Yerevan
Prime frequencies: 4810, 4990, 7480 kHz 1845–1900 UTC **EU**
7480, 9965 kHz 2130–2200 UTC **EU**

Notes: This station was known as Radio Yerevan until 1995. Because of political instability in the region, the Voice of Armenia had been erratically heard in early 1996, but the station operations have stabilized and it is one of the easiest-to-hear stations from the former Soviet Union.

Ascension Island

With a population under 3000 and a location way out in the South Atlantic Ocean, Ascension Island is one of the more exotic and unlikely locations for a shortwave station. Unfortunately, Ascension Island is home to Voice of America and the BBC relay stations and little more. Maybe someday, the island will follow St. Helena and air the little MW/FM combo Ascension Radio on shortwave. Ascension Island Day, anyone?

Australia

Radio Australia
GPO Box 428G, Melbourne 3001, Victoria
613 626 1800 (tel)
613 626 1899 (fax)
raust3@ozemail.com.au (Internet e-mail)
http://www.abc.net.au/ra (Internet Web site)
Prime frequencies: 5995, 9580, 11800 kHz 1230–1700 UTC
6080, 12080 kHz 1430–2100 UTC
7240, 9860 kHz 1800–2130 UTC
9510, 13605, 21725 0800–1130 UTC
9580, 9860 kHz 0800–1230 UTC
9960, 11640, 12080, 15240, 15365, 15415, 17750, 17795, 17780U kHz 0100–0500 UTC
9960, 11640, 12080, 13605, 15240, 15365, 15415, 17715, 17880U kHz 0500–0800 UTC
11695, 11855, 12080, 13755, 15365, 17795, 17860 kHz 2200–2400 UTC
13605, 15510, 17750 kHz 0000–0400 UTC
Also broadcasts in English on 6020, 6060, 6090, 7330, 9615, 9770, 11660, and 11880 kHz.

Notes: Radio Australia is a powerhouse broadcaster in Asia and the South Pacific. In addition to a wide variety of programs available, Radio Australia has got to be *the* sports station on shortwave. The station dedicates many hours to broadcasting Australian sports (cricket, the Australian Football League, and the Australian Rugby League) every day.

Australia (Radio Australia)

Radio Australia

Wednesday, December 20, 1939: five days before Christmas, the world is bracing itself for war. On the fringes of the Arctic, the Finns are locked in battle with the Russians. In South America, a freshly scuttled hulk of a Nazi battleship is settling on the bed of the River Plate after being trapped by British warships. Australia has just dispatched an advance party of servicemen to spearhead the war in the Middle East. At a transmission station just out of Melbourne, a radio service is about to be born. . . .

On the evening of December 20, 1939, the measured tones of the then Prime Minister Robert Menzies were heard inaugurating a permanent Australian overseas broadcasting service with the words: "The time has come to speak for ourselves." "Australia Calling," six years later to take the title "Radio Australia," was launched.

Fifty years later, Radio Australia celebrated its jubilee as the embodiment of the service that Menzies espoused. Acknowledged as having a major share of the shortwave audiences in many Asian and Pacific countries, it continues to speak for Australia and Australians in nine different languages, reaching an audience that numbers more than 50 million regular listeners. Every day, these people are informed and entertained by programs in English, Indonesian, Standard Chinese, Cantonese, Tok Pisin, French, Thai, Japanese, and Vietnamese.

Soundabout
The very best from Australian and international sounds.
Kim Taylor — Producer — Presenter
Nikki Mancer — Production Assistant

Radio Australia in touch with the World

Austria

Radio Austria International (ORF)
A-1136 Wien

431 0222 878 78 3636 (tel)
431 0222 878 78 4404 (fax)
kwp@rai.ping.at (Internet e-mail)
http://www.ping.at/rai (Internet Web site)
Prime frequencies: 5945, 6155, 9495 kHz 1930–2000, 2230–2300 UTC **AF AS EU**
6015 0630–0700 UTC **NA**
6015, 6155, 15410, 17870 kHz 0530–0600 UTC **AS EU NA**
6155, 9655, 11780, 13730 kHz 1530–1600 UTC **AS AUS EU**
6155, 13730 kHz 1330–1400 **EU NA**
6155, 13730, 15240, 17870 kHz 0830–0900 UTC **AS AUS EU**
7325 kHz 0130–0200 UTC **NA**
7325, 9870 kHz 0230–0300 UTC **NA SA**
11780 kHz 1630–1700 UTC **AS**

Notes: Radio Austria International broadcasts news, special musical programs, and features about Austria's culture. The ORF's schedules proudly proclaim "Our editorial independence is guaranteed by law."

Azerbaijan

Radio Dada Gorgud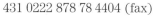
370011 Baku
Prime frequency: 6110 kHz 1800–1830 UTC

Notes: Has also been noted at other times relaying the BBC World Service. This is one of the most difficult countries to hear from the former Soviet Union.

Bahrain

Radio Bahrain
POB 702, Manamah
Prime frequencies: 6010 and 9746 kHz 0300–2100 UTC

Notes: Bahrain was never a shortwave powerhouse, and they were irregularly reported by overseas listeners. However, in 1996, somewhat local listeners reported no trace of Radio Bahrain on either frequency. Apparently, they are currently inactive, but this status could change at any time.

Bangladesh

Radio Bangladesh
POB 2204, Dhaka 1000
Prime frequencies: 7185, 9548 kHz 1230–1300 UTC
7190, 9568 kHz 1745–1815 UTC

Notes: Radio Bangladesh was operated as a clandestine station (known as *Swadhin Bangla Betar Kendra*) until the revolution succeeded in separating Bangladesh from Pakistan in 1971. Since then, Radio Bangladesh has slowly grown and modernized.

Belarus

Radio Belarus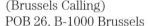

In the days of the Soviet Union, Radio Minsk operated as only an internal broadcaster. The trend continues now that Belarus has separated from Russia and the other former SSRs and now. Now known as Radio Belarus, the station only broadcasts in Belorussian and German. Try 7105 and 7210 kHz around 1930 UTC.

Belgium

Radio Vlaanderen International
(Brussels Calling)
POB 26, B-1000 Brussels
32 2 7328336 (fax)
rvi@brtn.be (Internet e-mail)
http://www.brtn.be/rvi (Internet Web site)
32 3 8253613 (DXA BBS)
Prime frequencies: 5900, 9925 kHz 0030–0055 UTC **NA SA**
5910, 9925 kHz 1900–1925 UTC **AF EU**
5910 kHz 2200–2225 UTC **EU**
5985, 9925, 9940 kHz 0730–0755 UTC **AUS EU**
6035 kHz 1000–1025 UTC **EU**
13685 kHz 1330–1355 UTC **NA**
13685, 13795 kHz 1400–1425 UTC **AS NA**

Notes: Brussels Calling is the name for the English service of Radio Vlaanderen International. Brussels Calling features such programs as Belgium Today, Living in Belgium, The Arts, Press Review, and Music from Flanders.

Benin

Office de Radiodiffusion et Télévision du Benin (ORTB) is one of the more difficult catches in Africa. Benin is a tiny finger-shaped country in the "hook" of West Africa. Unlike its high-powered neighbors, Nigeria and Togo, ORTB's most powerful outlet on 4870 kHz is listed at 30 kW.

Bhutan

> Bhutan Broadcasting Service
> POB 101, Thimphu
> 975 12072 (tel)
> 975 23073 (fax)
> Prime frequency: 5025 kHz 1415–1500 UTC

Notes: Bhutan has been one of the fabled "rare radio countries" for radio listeners in North America and much of the world. Don't expect to hear them unless you are in Asia.

Bolivia

Nestled in the Andean Mountains of South America, Bolivia is one of the hotbeds of commercial shortwave activity in the world. Unless you live in Central or South America, none of the stations from Bolivia are easy catches, and none broadcast in English. A few of the best signals from the country on shortwave include Radio Illimani on 4945 kHz, Radio Santa Cruz on 6135 kHz, Radio Fides on 4845 and 9625 kHz, and Radio Santa Ana on 4649 kHz. Literally dozens of other Bolivian stations are on the air, but most are very difficult to hear.

Bosnia-Hercegovina

One of the most difficult catches in Europe, RTV Bosnia-Hercegovina is also one of the few broadcasters to operate in the SSB mode (on 7108.1 kHz USB). However, if you can't do SSB, just tune down 3 kHz; the other Bosnia-Hercegovinan transmitter is on 7105 kHz AM. Unfortunately, this station is so difficult to hear because the country has been devastated by the Yugoslavian civil war.

Botswana

> Radio Botswana
> Private Bag 0060, Gaberone

Notes: In addition to its national programming, Radio Botswana is well-known for the barnyard animal sounds that it airs as an interval signal before and after its broadcasts. Radio Botswana broadcasts for short periods in English. Try at different times on 4820, 7255, and 9640 kHz.

Brazil

Radio Bras
CP 08840, CEP 70912–790, Brasilia DF
Prime frequencies: 15265 kHz 1800–1920 UTC **EU**
15445 kHz 1200–1320 UTC **NA**

Notes: Radio Bras is one of the only shortwave broadcasters from South America with English-language programming.

Aside from Radio Bras, dozens of other commercial stations operate on shortwave from Brazil. In fact, the 1996 *World Radio-TV Handbook* listed 159 transmitters on shortwave! Unlike the other commercial stations from South America, those from Brazil broadcast primarily in Portuguese with higher power levels (typically 5, 7.5, or 10 kW). Some of the most widely reported Brazilians include: Radio Bandeirantes on 6090 and 9645 kHz, Radio Nacional Amazonas on 11780 kHz, Radio Clube do Para on 4885 kHz, Radio Bare on 4895 kHz, and Radio Anhanguera on 4915 kHz. The programming on Brazilian stations is typically very different from that of other stations in Latin America. Instead of the upbeat Latin American pop of Colombia and Venezuela or the Andean folk music of Peru or Bolivia, many Brazilian stations air pop and rock music, with lyrics in Portuguese.

Bulgaria

Radio Bulgaria
4 Dragan Tsankor Blvd., 1040 Sofia
003592/87 10 60 (fax)
Prime frequencies: 7335, 9700 kHz 2000–2100 UTC **EU**
7375, 9485 kHz 0000–0100, 0500–0600 UTC **NA**
7390, 9700 kHz 2200–2300 UTC **EU**
9440 kHz 1130–1230 UTC **AS**

Notes: Radio Bulgaria was formerly known as Radio Sofia. As it was in the past, Radio Bulgaria is regularly heard with strong signals.

Burkina Faso

Radio Burkina, from the land that was once better known as Upper Volta in west central Africa, is neither a difficult nor an easy catch outside of Africa. Like many of the stations from African countries, Radio Burkina is a government operation that broadcasts in French. The best chance to hear the station is via their 50-kW outlet on 4815 kHz.

Cambodia

National Voice of Cambodia
Address is unreliable
855 23 27319 (fax)
Prime frequency: 11940 kHz 0000–0015, 1200–1215 UTC

Notes: Because of political instability within the country, the National Voice of Cambodia is not regularly heard.

Cameroon

The Cameroon Radio Television (CRTV) national radio system is novel in Africa. Instead of operating with one network, they have several different outlets that each produce their own programs and feature several transmitters. The best-heard CRTV sites are the 100-kW powerhouses from Douala on 4795 and 7150 kHz, and from Garoua on 5010 and 7240 kHz. These stations broadcast in French, although the news is sometimes aired in English.

Canada

CFRX
2 St. Clair Ave. W., Toronto, ON M4V 1L6
416 323 6830 (fax)
Prime frequency: 6070 kHz 24 hrs
Notes: Relays local broadcasts from CFRB-1010 kHz in Toronto, Ontario.

CFVP
POB 2750, Stn. M, Calgary, AB T2P 4P8
403 240 5801 (fax)
Prime frequency: 6030 kHz 24 hrs
Notes: Relays local broadcasts from CFCN-1060 kHz in Calgary, Alberta.

CHNX
POB 400, Halifax, NS B3J 2R2
902 422 5330 (fax)
Prime frequency: 6130 kHz 24 hrs
Notes: Relays local broadcasts from CHNS-960 kHz in Halifax, Nova Scotia.

CKFX
2440 Ash St., Vancouver, BC V5Z 4J6
604 873 0877 (fax)
Prime frequency: 6080 kHz 24 hrs
Notes: Relays local broadcasts from CKWX-1130 kHz from Vancouver, British Columbia. CKFX is currently off the air with transmitter problems, but when active, it uses a mere 10 watts of power!

CKZN
POB 12010, Stn. A, St. John's, NF
709 576 5099 (fax)
Prime frequency: 6160 kHz 24 hrs
Notes: Relays national CBC broadcasts from CBN-640 kHz from St. John's, Newfoundland.

CKZU
POB 4600, Vancouver, BC V6B 4A2
604 662 6350 (fax)
Prime frequency: 6160 kHz 24 hrs

Notes: Relays national CBC broadcasts from CBU-690 kHz from Vancouver, British Columbia. Like the other shortwave broadcaster in Vancouver (CKFX), CKZU is currently off the air.

Radio Canada International
POB 6000, Montreal, PQ H3C 3A8
514 284 0981 (fax)
http://radioworks.cbc.ca/radio/rci/rci.html (Internet Web site)
Prime frequencies: 5925, 5995, 7235, 9805, 11945, 13650, 13690, 15150, 17820
  2100–2159 UTC **AF EU**
5960, 6010, 9535, 9755, 11940 kHz 2300–2329 UTC **NA SA**
5995, 7235, 9805, 11705, 11945, 13690, 15150 kHz 2200–2229 UTC **AF AS EU**
6040, 9535, 11940 kHz, 0000–0029 UTC **NA SA**
6050, 6150, 9740, 9760, 11905 kHz 0600–0629 UTC **AF EU**
6150, 11730 kHz 1200–1229 **AS**
6150, 9505, 9645 kHz 0400–0429 UTC **AS**
6155, 9535, 9755, 11725 kHz 0200–0259 UTC **NA**
6155, 9755 kHz 0300–0329 UTC **NA**
9640, 11855 kHz 1300–1400 UTC **NA**

Notes: Even though Radio Canada International is one of the most popular shortwave stations, it has been threatened with being eliminated in the government cutbacks to the Canadian Broadcasting Corporation (CBC) for years.

Canada (Radio Canada International)
RCI Coalition

RCI RADIO CANADA INTERNATIONAL
50
1945 - 1995

Whether Radio Canada International will continue to exist will be decided in the next few hours. The decision can only be finalized by the Canadian Prime Minister and the federal cabinet and comes, ironically enough, as RCI celebrates its 50th anniversary of service.

The Coalition to Restore Full RCI Funding, with the support of the Canadian International DX Club and the unions representing RCI employees, today called on the government to stop any attempt to shut the service down and to assure a separate, protected annual budget to guarantee the service's existence.

This urgent appeal to the government comes days after the CBC Board of Directors decided to shut the international service down, and CBC human resources was poised to send termination letters to all RCI employees. Although halted temporarily, it appears that the termination letters will in fact be sent this week, and the service closed down, unless the government steps in.

Around the world, 126 countries broadcast internationally, yet Canada with this decision would turn its back on the world and become the only G7 member not to have an international radio service. For this reason, and for the standing and economic benefit of Canada, the Coalition calls on the Prime Minister and the federal cabinet to assure an annual budget of at least 16 million dollars, guaranteed and protected from sudden cuts.

RCI was saved from the axe again.

Central African Republic

Radiodiffusion-Télévision Centrafricaine is another government station that broadcasts in French and is moderately difficult to receive outside of Africa. Listen for them on 5034 kHz.

Chad

Chad, a very large country with a small population, covers a chunk of the central Sahara Desert. For years, Radiodiffusion Nationale Tchadienne was one of the more difficult catches from Africa. Now, Chad is much easier to receive, although still not easy, thanks to a higher-powered transmitter on 4904.5 kHz.

Chile

http://www.chilnet.cl/empresas/radio01.htm (Internet fax to Radio Esperenza & R. Mineria)

Chile, like the other nations in southern South America, have very little shortwave broadcasting. The major station, Radio National de Chile, retired a few years ago. Now, the best chances to hear Chilean radio are via the 10-kW stations: Radio Santa Maria on 6030 kHz and Radio Esperenza on 6090 kHz.

China

Radio China International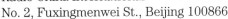
No. 2, Fuxingmenwei St., Beijing 100866
6092274 (tel)
8513135 (tel)
8513174 (fax)
8513175 (fax)
8513176 (fax)
Prime frequencies: 7405 kHz 1400–1600 and 1700–1800 UTC
9710 kHz 0000–0100 and 0300–0400 UTC
9730 kHz 0400–0500
11715 kHz 0000–0100 and 0300–0400 UTC
15440 kHz 1300–1400 and 2100–2300 UTC
Also broadcasts in English on: 3985, 6950, 6955, 7170, 9440, 9535, 9715, 9785, 9920, 11660, 11755, 15110, and 15130.

Notes: China Radio International is one of the last communist shortwave broadcasters in the world (along with Radio Pyongyang, Radio Havana Cuba, and the Voice of Vietnam). CRI has one of the largest relay networks on shortwave, with outlets and/or relay agreements in Mali, Brazil, Canada, Spain, France, French Guiana, Switzerland, and Russia.

Colombia

http://latina.latino.net.co/empresa/caracol/caracol.htm (CARACOL)

Colombia, in northern South America, used to be one of the major shortwave goldmines, with dozens of stations that broadcasted local programming, and often sent out very attractive pennants. Now only about two dozen Colombians are on shortwave. Many of these stations are difficult to receive outside of Central or South America, but Caracol Colombia on 5077 kHz and Radio Nacional de Colombia on 4955 kHz are two of the strongest stations from Latin America.

Congo

Radio-Diffusion Télévision Congolaise is one of the more difficult catches from Africa, despite its listed 100-kW transmitter, which is supposed to operate on 60, 49, 31, and 19 meters. After recently reactivating, it was reported with good signals on 4765 kHz.

Cook Islands

Radio Cook Islands
POB 126, Avarua, Rarotonga
Prime frequency: 11760 kHz 1600–0500 UTC

Notes: Intersperses local programs with relays of news from Radio Australia and Radio New Zealand. Because of the low power, it is rarely heard outside of the South Pacific.

Costa Rica

TIAWR (Adventist World Radio)
Ap. 1177, 4050 Alajuela
Prime frequencies: 6150, 7375, 9725, 13750 kHz 1100–1300, 1900–2000, 2300–2400 UTC

Notes: One of the many branches of Adventist World Radio, TIAWR can be heard with a very strong signal across the Americas.

TIFC (Faro del Caribe)
Ap. 2710, 1000 San Jose
Prime frequencies: 5055 kHz 0300–0400 UTC

Notes: Faro del Caribe ("The Lighthouse of the Caribbean") is a Christian station that broadcasts almost entirely in Spanish, but features a few English programs.

Radio For Peace International
POB 88, Santa Ana
SJO 577, POB 025216, Miami, FL 33102 USA
506 249 1821 (tel)
506 249 1095 (fax)
rfpicr@sol.racsa.co.cr (Internet e-mail)
http://www.clark.net/pub/cwilkins/rfpi/rfpi.html (Internet Web site)
Prime frequencies: 6205, 15050 kHz 1200–0000 UTC

7385 kHz 2100–0800 UTC
9400 kHz USB 24 hours

Notes: Radio For Peace International primarily broadcasts international left-wing programming from peace-oriented, human-rights, and women's-rights groups.

Costa Rica

Radio For Peace International

Radio For Peace International broadcasts a number of peace-oriented radio programs, including FIRE (Feminist International Radio Endeavor).

In January 1995, Paulina Díaz Navas, an indigenous-campesina woman from the southern province of Puntarenas, called FIRE from a public telephone near her home to say she wanted to be on the air live and speak to the FIRE audience about a death threat she had received from a police officer who, at gunpoint, had tried to frighten her into turning over her land. He was apparently working for people who desired her property for their own purposes. She tried to take the case to local court, but according to her testimony on FIRE, the police threatened her neighbors in an attempt to get them to testify against her. However, Paulina received support from women in her town even though her female neighbors had been subjected to sexual harassment to intimidate them.

FIRE helped her by taking the case to the Ombudsperson's office in Costa Rica and called on women's groups and human rights organizations worldwide to send letters of support. The Foundation for a Compassionate Society, based in Austin, Texas, featured her case in their newsletter and led a campaign of support. Paulina told us this effort worked.

"Because of our program," she said, "the Minister of Security, Juan Diego Castro, responded. He sent me the letters that he received from international women's organizations on behalf of my rights. He sent a person to investigate the case, and they realized that I was right to fight for my land. He sent orders to the local police, telling them to protect my land rights." The police officer who threatened her is now in jail.

Cote d'Ivoire

Wedged in the western "hook" of Africa, the small nation of Cote d'Ivoire (formerly known as Ivory Coast) is rarely mentioned in the Western media. Because it only broadcasts in French, Radio Abidjan, the international outlet of Radiodiffusion Télévision Ivoirienne is not even a common name in shortwave broadcasting circles. Despite the relative anonymity, the station has one of the best signals from Africa. Check for Radio Abidjan on 7215 kHz, where it has regularly been heard for more than a decade.

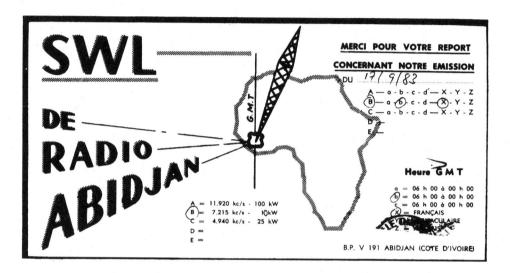

Croatia

Croatian Radio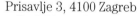
Prisavlje 3, 4100 Zagreb
zelimir.klasan@hrt.com.hr (Internet e-mail)
Prime frequencies: 5895, 7165, 9830, 13830 kHz 0003–0010, 0303–0310, 0703–0710, 0803–0810, 0903–0910, 1003–1010, 1303–1310, 2203–2210 UTC

Notes: Only broadcasts regular programs in Croatian, but features news in English at these times.

Cuba

Radio Havana Cuba
Ap. 6240, Havana
radiohc@tinored.cu (Internet e-mail)
Prime frequencies: 6000, 9830 USB kHz 0100–0300 UTC **NA**
6000, 9820 kHz 0300–0500 UTC **NA**
9820, 9830 USB kHz 0500–0700 UTC **NA**

13715, 13725 USB kHz 2100–2200 UTC **EU**
6180 kHz 2200–2300, 0400-0500 UTC **NA**

Notes: After losing support from the former Soviet Union, the hours of Radio Havana Cuba have been cut back somewhat. Surprisingly, in spite of the strictness of the communist government, the "DXers Unlimited" is very popular among shortwave listeners and the host, Arnie Coro, is somewhat of a shortwave celebrity.

Cyprus

The easiest way to hear Cyprus is also the least interesting: via the 100- and 250-kW transmitters of the BBC. If you are located in Europe or the Middle East, or if you are in the mood for some really difficult DX, try for Bayrak Radio (a tiny, often inactive station) from northern (Turkey-controlled) Cyprus in 6150 kHz.

Czech Republic

Radio Prague
Vinohradska 12, 120 99 Prague
cr@radio.anet.cz (Internet e-mail)
http://www.radio.cz (Internet Web site)
Prime frequencies: Radio Prague operates with 27-minute programs in various languages (including a number of English broadcasts) on 5930, 7345, 9420, 9505, 17485, and 21705 kHz. 5930 and 7345 kHz are most heavily used.

Notes: Some of the featured programs include Letter from Prague, Week in Politics, The Arts, Musical Feature, and Current Affairs.

Denmark

Radio Denmark
Rosenorns Alle 22, DK-1999 Frederikberg C
45 35 20 57 91 (tel)
45 35 20 57 81 (fax)
rdk.ek@login.dknet.dk (Internet e-mail)
http://www.mi.aau.ak/rdk (Internet Web site)

Notes: Radio Denmark had not operated in English for years. In 1996, the station offered an English-language program on alternate Sundays. At the end of 1996, the station returned to all-Danish programming. All programs emanate via the transmitters of Radio Norway.

Dominican Republic

Sharing half an island with Haiti, the Dominican Republic offers a few tiny Spanish-language shortwave broadcasters. The currently most-often heard stations are Radio Cristal International (which has aired several irregular English programs) on 5012 kHz and Radio Amanecer on 6025 kHz.

Ecuador

HCJB
Casilla 17-17-691, Quito
593 2 466 808 (tel)
593 2 447 263 (fax)
POB 39800, Colorado Springs, CO 80494 USA
719 590 9800 (tel)
719 590 9801 (fax)
english@mhs.hcjb.com.ec (Internet e-mail)
22734 HCJB ED (telex)
Prime frequencies: 9445, 9745 kHz 0500–0700 UTC **AUS NA**
9745 kHz 0000–0400 UTC **NA**
11960 kHz 1900–2200 UTC **EU**
12005, 12025 kHz 1100–1600 UTC **NA SA**
21455 USB kHz 0000–1600, 1900–2200 UTC **ALL**

Notes: HCJB is one of the most respected Christian broadcasters because of its projects (hospitals, hydroelectric plants, etc.) to help the people within Ecuador and also because of its interesting programming. In addition to regular Christian shows, the station also features the traditional music of Ecuador, El Mundo Futuro (about the future world of science and technology), and The Computer Corner.

Like Colombia and Venezuela, Ecuador was once a real shortwave DX paradise. In 1970, Ecuador had a whopping 109 stations on shortwave, in addition to the transmitters of HCJB. Now, only 37 stations (still a considerable number) are broadcasting. Some of the Ecuadorans are some of the stronger Latin American stations. For example, Radio Quito on 4919.9 kHz puts a whopping signal into North America. Some of the other better-heard stations include La Voz del Napo on 3280 kHz, La Voz Radio Centinela del Sur on 4770 kHz, and Radio Bahá'í on 4949.9 kHz.

Egypt

Radio Cairo
POB 1000, Cairo
Prime frequencies: 9475 kHz 0200–0330 UTC **NA**
9900 kHz 2115–2245, 2300–0030 UTC **EU NA**
15255 kHz 1630–1830 UTC **AF**
15375 kHz 2030–2200 UTC **AF**
17595 kHz 1215–1330 UTC **AS AUS**

Notes: In spite of the good signals that Radio Cairo delivers to much of the world, the programming is often difficult to understand because of the "rough" sounding audio.

Equatorial Guinea

Radio Africa
POB 851, Malabo
Prime frequency: 15185.7 kHz 1700–2300 UTC

Notes: Unlike the many national-service stations from Africa, Radio Africa is an evangelical Christian broadcaster that features plenty of programs that are produced in the United States. This station can be heard with decent signals across much of the world.

Eritrea

Eritrea is a little strip of land along the northeastern coast of Ethiopia. For years, this land was part of Ethiopia, and civil war raged. At the heart of this civil war was the clandestine station, the Voice of the Broad Masses of Eritrea. Now that the war is finally over and Eritrea is an independent nation, you can still tune in the Voice of the Broad Masses of Eritrea on 4000, 7020, and 7390 kHz. This is one of the more difficult-to-hear stations from Africa.

Estonia

Estonia Radio
Gonsiori 21, EE-100 Tallinn
janc@r2.online.ee (Internet e-mail)
http://www.online.ee/er (Internet Web site)
Prime frequencies: 5925 kHz 1620–1630, 2000–2030 UTC

Notes: Heard irregularly in Europe.

Ethiopia

Radio Ethiopia
POB 1020, Addis Ababa
Prime frequencies: 7165, 9560 kHz 1600–1700 UTC
5990, 7110, 9705, 1030–1100 UTC

Notes: Radio Ethiopia is not one of the more widely reported stations outside of Africa. Still, the country often operates or is afflicted by clandestine stations, which are sought out by many. In 1995, the 9560-kHz transmitter was used for programs of Radio Amahoro, the Voice of Peace for Rwanda. Some stations from Ethiopia that are often heard with better signals than Radio Ethiopia are the Voice of the Revolution of Tigray on 5500 kHz and Radio Fana on 6210 and 6940 kHz.

Finland

YLE Radio Finland
Box 78, 00024 Yleisradio
Box 462, Windsor, CT 06095 USA
800 221 9539 (U.S. toll-free tel)
203 688 5540 (U.S. tel)
3580 14805490 (Finland tel)
rfinland@yle.mailnet.fi (Internet e-mail)
Prime frequencies: 6120, 9730, 11755, 15440 kHz 2130–2200 UTC **AF EU**
9650, 9665, 11845 kHz 2230–2300 UTC **EU**
9730, 15440 kHz 1900–1930 UTC **AF AS**

BUILDINGS AND PLACES OF INTEREST

Finlandia Hall — the concert hall and congress centre where the final document of the CSCE was signed in 1975.

Photo: Pentti Harala

11900, 15400 kHz 1130–1200, 1230-1300, 1330–1400, 1530–1600 UTC **NA**
15115, 17820 kHz 0900–0930 UTC **AUS**
15440 kHz 0430–0500 UTC **AF AS EU**

Notes: With contacts on several continents and an e-mail address, Radio Finland keeps in touch with its listeners—even if it is one of the smaller broadcasters in Europe.

France

Radio France International
BP 9516, F-75016 Paris Cedex 16
rfi.aud@dialup.francenet.fr (Internet e-mail)
http://www-rfi.eunet.fr/rfi1.html (Internet Web site)
Prime frequencies: 6175, 9485, 11615, 12015, 15530 kHz 1600–1700 UTC **AF AS EU**
9805, 11615, 13625, 15325, 17595 kHz 1200–1300 UTC **AF AS AUS EU**
11615, 15210, 15460 kHz 1700–1730 UTC **AF AS**
12030, 15405, 17560 kHz 1400–1500 UTC **AS**

Also broadcasts in English on: 11600, 11700, 12015, 15155, 15210, and 15460.

Notes: RFI is a major force on shortwave, but French-language programming takes top priority.

French Guiana

The easiest way to log French Guiana, the only French-speaking country in South America is via the 500-kW relay transmitters of Radio France International and Swiss Radio International. For a better taste of the country, try RFO Guyane on 5055 kHz. This one often plays non-U.S. reggae music, so it's fun and interesting.

Gabon

Like French Guiana, Gabon is easiest to hear via massive relay transmitters, but its own country's broadcasts are also easy to hear. Africa #1 runs commercial music programs and also relays Radio France International and Radio Japan via its massive 500-kW transmitters. Radiodiffusion-Télévision Gabonaise, the government-operated station, is also sometimes reported on 4777 kHz.

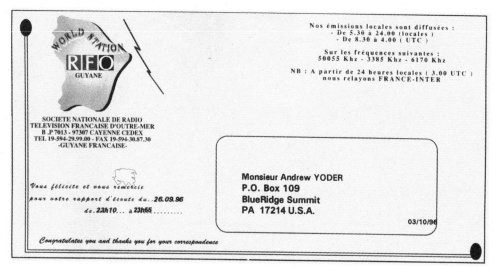

Georgia

Radio Georgia
ul. 68 M. Kostava, 380071 Tbilisi
Prime frequencies: 11805 kHz 0530–0600 UTC
11910 kHz 0700–0730, 1600–1630 UTC

Notes: Radio Georgia is not commonly heard outside of the Middle East.

Germany

http://www.br-online.de (Bayerischer Rundfunk)
http://www.dfn.de/~drln/index.html (Deutschland Rundfunk)

Deutsche Welle
50588 Cologne
POB 50641, Washington, DC 20091
49 221 345 2460 (tel)
49 221 345 2461 (tel)
49 221 389 4599 (fax)
harald.schuetz.@dw.gmd.de (Internet e-mail)
solbach@dw.gmd.de (Internet e-mail)
http://www-dw.gmd.de (Internet Web site)
ftp-dw.gmd.de (Internet program previews)
Prime frequencies: 5960, 6040, 6085, 6145, 9670 kHz 0100–0150 UTC **NA**
5960, 7385 kHz 2000–2050 kHz **EU**
6000, 6160, 7235 2300–2350 UTC **AS**
6015, 6065, 7225, 7265, 9565 kHz 0400–0450 UTC **AF AS**
6020, 6145, 6185, 9650 kHz 0500–0550 UTC **NA**
6035, 7265, 7285, 7355, 9515, 9615, 9815 kHz 0200–0250 **AS**
6045, 6085, 9535, 9650 kHz 0300–0350 UTC **NA**
6160, 7380, 11715, 12055, 17820 kHz 0900–0950 UTC **AS**
6170, 7225, 7305, 9585 kHz 1600–1650 UTC **AS AVS**
7195, 9735, 11965, 13610, 15145 kHz 1600–1700 **AF**
7225, 9565, 11765, 13790, 17820, 21705 kHz 0600–0650 UTC **AF**
9565, 15145, 15410, 17800, 21600 kHz 0900–0950 UTC **AF**
9615, 9690, 11865, 15275 kHz **AF**
9640, 9765, 11785, 11810, 13690, 15135, 15425 kHz 1900–1950 UTC **AF AS**
9670, 9765, 11785 kHz 2100–2150 UTC **AS AUS**
15370, 15410, 17800, 17780 kHz 1100–1150 UTC **AF**

Notes: The Deutsche Welle is regularly one of the most popular shortwave broadcasters across the globe. Like the BBC, the Deutsche Welle produces a *TV Guide* type of program listing every month, complete with descriptions, photographs, and behind-the-scenes features.

Ghana

Ghana Broadcasting Corporation
POB 1633, Accra
Prime frequencies: 3366, 4915 kHz 0525–2300 UTC

Notes: Broadcasts national programming in English, including news, music, editorials, and even the daily lottery numbers.

Greece

Voice of Greece
15432 Messoghion Ave., 432 Athens
01 6397 375 (tel & fax)
01 6396 762 (tel)

dagel@leon.nrcps.ariadne-t.gr (Internet e-mail)
http://alpha.servicenet.ariadne-t.gr/docs/era5_1.html (Internet Web site)
Prime frequencies: 6125, 7450, 9425 kHz 0130–0140, 0340–0350 UTC **NA**
7425, 7450, 15175 kHz 0740–0750 UTC **AUS EU**
9375 kHz 1900–1910 UTC **EU**
9425 kHz 2240–2250 UTC **AUS**
11645, 15650, 17525 kHz 1240–1250 UTC **AF**
11645, 15150 kHz 1840–1850 UTC **AF**
11645, 15175 kHz 1440–1450 UTC **EU NA**

Notes: The Voice of Greece is a relatively minor voice in the crowd of massive European shortwave outlets. The station is primarily intended to keep in touch with Greeks abroad. With all English programs lasting only 10 minutes, you've got to wonder if the station cares if anyone hears these shows.

Guam

KSDA
Prime frequencies: 7395 kHz 1430–1500 UTC **AS**
7400 kHz 1600–1700 UTC **AS**
7455 kHz 1000–1030, 1730–1800 UTC **AS**
9495 kHz 2130–2200 UTC **AS**
9530 kHz 1030–1100 UTC **AS**
9650 kHz 1330–1400 UTC **AS**
13720 kHz 1230–1300 UTC **AS**
15610 kHz 2300–2400 UTC **AS**

Notes: One of the many branches of Adventist World Radio, the broadcasting arm of the Seventh-Day Adventists. KSDA keeps its listeners active by bouncing from frequency to frequency throughout its broadcasting day.

KTWR
1868 Halsey Dr., Agana, Guam 96922
671 477 2838 (fax)
http://www.twr.org (Internet Web page)
Prime frequencies: 9870 kHz 1000–1100 UTC **AS**
11580 kHz 1500–1630 UTC **AS**
11830 kHz 0855–1000 UTC **AS AUS**
15200 kHz 0740–0915 UTC **AS AUS**

Notes: One of the branches of Trans World Radio, a Christian broadcasting organization.

Guatemala

TGNA (Radio Cultural)
Ap. 601, Guatemala City
Prime frequency: 3300 kHz 0300–0600 UTC

Notes: TGNA broadcasts throughout the day in Spanish and local languages but operates in English for several hours every evening.

Aside from TGNA, several other small shortwave stations operate from Guatemala. Most notable are Radio K'ekchi on 4780 kHz, Radio Buenas Nuevas on 4800 kHz, and Radio Maya on 2360 and 3325 kHz. Most of these broadcasters operate with 1000 watts out or less.

Guinea

Guinea's only shortwave offering, Radiodiffusion-Télévision Guinéenne is one of the most difficult-to-hear stations in Africa. The best chances to hear the station are on 4900 kHz and 7124.5 kHz, but unless you are in Africa, don't expect to hear it well.

Guyana

Voice of Guyana
Prime frequency: 3290 kHz 0800–0200 UTC

Notes: For years, the Voice of Guyana was a very difficult catch, buried under signals from WYFR and Radio Sana'a (Yemen) on a frequency in the 48-meter band. The station's recent frequency move has dramatically improved its signals.

Honduras

HRVC (La Voz del Evangelica)
Ap. 3252, Tegucigalpa

POB 828, Wheaton, IL 60187 USA
504 333 933 (fax)
Prime frequency: 4820 kHz 0300–0500 UTC (Monday)

Notes: HRVC is operated by the Conservative Baptist Home Mission with many hours of Spanish-language programming throughout the day. However, the station does feature several hours in English every week.

Aside from HRVC, several other small shortwave stations operate from Honduras. Most notable is Radio Internacional, which booms a big signal into North America on 4830.6 kHz. Most of these broadcasters operate with 1000 watts out or less.

Hungary

Radio Budapest
Brody Sandor u. 5-7, H-1800 Budapest
36 1 1138 8388 (tel)
http://www.eunet.hu:80/radio (Internet Web site)
Prime frequencies: 3975, 5970, 9835 kHz 2200–2230 UTC **EU**
3975, 5970, 9840 kHz 2000–2030 UTC **EU**
5905, 9840 kHz 0200–0230 UTC **NA**
6195, 9840 kHz 0330–0400 UTC **NA**

Notes: Radio Budapest sticks to the well-tested, basic format of news and features.

Iceland

Rikisutvarpid Reykjavik
Efstaleiti 1, 150 Reykjavik
http://www.ruv.is (Internet Web page)

Notes: Broadcasts for local fisherman and other locals abroad. Although Rikisutvarpid can be heard over a significant area, the programming is all in Icelandic, so almost no one can understand it! Try listening on the oddball frequencies of 7749, 9275, 11402, and 13860 kHz.

India

All India Radio
POB 500, New Delhi 110001
http://air.kole.net (Internet Web site)
Prime frequencies: 7140, 7412, 9530, 9700 kHz 1530–1545 UTC **AS**
7150, 7412, 9910, 9950, 11620 kHz 2045–2230 UTC **AUS EU**

7150, 9950, 11620 kHz 2245–0045 UTC **AS**
7412, 9650, 9950, 11620, 13700, 15075 kHz 1745–1945 UTC **AF AS EU**
11620, 13710 kHz 1330–1500 UTC **AS**
15050, 15180, 17387, 17895 kHz 1000–1100 UTC **AS AUS**

Notes: All India Radio provides an interesting look at one of the most populous countries in the world. Although All India Radio does not feature a huge English line-up, plenty of frequencies and programs can be heard worldwide. In addition to the limited English external service, India has many domestic transmitters that are often targeted (especially those from the "radio countries" of Goa and the Andaman Islands) by DXers.

Indonesia

Voice of Indonesia
POB 1157, Jakarta
Prime frequencies: 9525 kHz 0100–0200, 0800–0900, 2000–2100 UTC **AS AUS EU NA SA**

Notes: The Voice of Indonesia announced that their transmitting facilities will be upgraded in the mid to late 1990s for better coverage. Maybe someday?

In the South Pacific, Indonesia is the primary DX paradise for shortwave listeners. The country—a mass of approximately 3000 islands, with no dominant land mass—is ideal for shortwave broadcasting. The 1996 *WRTH* lists 84 different transmitters from Indonesia, mostly different branches of Radio Republik Indonesia (RRI). The reported powers of these stations ranges from 50 kW down to 50 watts! The most commonly reported stations are RRI Ujung Padang on 4753.5 kHz and RRI Sorong on 4874.6 kHz.

For more information on Indonesia, see http://www.hway.net/syahreza. This page isn't specifically radio-related, but it does feature news audio from RRI.

Iran

Voice of the Islamic Republic of Iran
POB 15875/1575, Tehran
98 21 2041051 (tel)
Prime frequencies: 6050, 9022, 9685 kHz 0030–0130 UTC **NA**
7260, 9022 kHz 1930–2030 UTC **EU**
7290, 9635 kHz 1530–1630 UTC **AS**
6165 kHz 2130–2230 UTC **AS**
11875, 11930, 15260 kHz 1130–1230 UTC **AS**

Notes: Less Western-influenced than Radio Kuwait, listening to the Voice of the Islamic Republic of Iran is a great step for understanding Middle Eastern politics and conservative Islam.

Iraq

Radio Iraq International
POB 8145, Baghdad
Prime frequencies: 13680 kHz 1230–1250 UTC **EU**

Notes: The schedule varies considerably, depending on the sanctions against the country and the conditions within. The station had a much larger schedule with better coverage before the Persian Gulf War.

Ireland

Jolly Roger Radio
POB 39, Waterford
Prime frequencies: Most Saturdays and Sundays between 6219 and 6233 kHz at times between 0800 and 1500 UTC

Notes: Jolly Roger Radio is a regular hobby pirate that features country music and relays of other pirate stations from around the world.

Radio Dublin
POB 2077, 4 St. Vincent St. W., Dublin 8
Prime frequency: 6915.5 kHz, irregular

Notes: Radio Dublin is one of the oldest pirate stations in the world, with approximately 30 years of operations on AM, FM, and shortwave. Radio Dublin is sometimes tolerated by the government, but it has been raided and taken off the air many times over the years. Because the only shortwave operations from Ireland are via pirates, Radio Dublin is the unofficial voice of the country.

West Coast Radio
Prime frequencies: 5910 kHz 0100–0200 UTC **NA**
6015 kHz 1500–1600 UTC **EU**
11665 kHz 1800–1900 UTC **AF**

Notes: West Coast Radio began operating in 1996 via relay transmitters in Germany.

Israel

Kol Israel
POB 1082, Jerusalem 91010
972 2302 327 (fax)
Prime frequencies: 7465, 9435, 17545 kHz 0500–0515, 1900–1910 UTC **AS AUS EU NA**
9390, 11605 kHz 1500–1530 **EU NA**
7465, 9365, 9435, 15640 kHz 2000–2100 UTC **AF EU NA**
15640, 15650, 17575 kHz 1000–1030 UTC **NA EU**

Notes: Kol Israel has been fighting off budget cutbacks throughout the early 1990s. Although the entire shortwave operation was threatened to be closed, Kol Israel still continues.

KOL ISRAEL
External Service
P.O.B. 1082
91 010 Jerusalem, Israel

QSL

THANK YOU FOR YOUR RECEPTION REPORT
NOUS VOUS REMERCIONS POUR VOTRE RAPPORT D'ECOUTE
AGRADEMOS SU INFORME DE RECEPCION

____7416____ kHz Date/dia __6·8·83__ / __0100__ GMT

YOUR REPORT HAS BEEN CHECKED AND AGREES WITH OUR LOG
VOTRE RAPPORT A ETE VERIFIE ET TROUVE CONFORME A NOS SPECIFICATIONS
SU INFORME HA SIDO VERIFICADO Y COCUERDA CON NUESTROS DETALLES

Sorry, we're all out of pennants.

Italy

Italian Radio Relay Service
POB 10980, I-20110 Milan
39 2 7063 8151 (fax)
100020.1013@compuserve.com (Internet e-mail)
http://www.nexus.org (Internet Web site)
Prime frequencies: 3985, 7125 kHz 24 hrs
Notes: The IRRS is another "transmitter for hire," as the name describes. In addition to relaying programs from large broadcasters, the IRRS also dedicates several hours every Sunday morning to airing programs from European pirate and hobby broadcasters, such as Rock-It Radio Pamela, Southern Music Radio, International Music Radio, etc.

Radio Europe
c/o Play DX, Via Davanzati 8, I-20158 Milan
Prime frequency: 7294 kHz USB 0830–1200 UTC
Notes: Radio Europe is primarily a relay transmitter for European pirate broadcasters, such as Radio Marabu, Level 48, Radio Joystick, etc.

RAI (Italian Radio and Television Service)
Viale Mazzini 14, 00195 Rome
39 6 322 6070 (fax)
http://www.rai.it (Internet Web site)
Prime frequencies: 5970, 7275 kHz 0425–0440 UTC **AS EU**
6030, 7235 kHz 11905 kHz 1935–1955 UTC **EU**
6005, 9675, 11800 kHz 0050–0110 UTC **NA**
5975, 9710, 11815 kHz 2200–2225 UTC **AS**

International shortwave broadcasts **145**

Notes: Despite the international importance of Italy and its importance with tourists, the country's international radio service (RAI—Radio Rome) is tiny, with broadcasts in English for only about 20 minutes a shot.

Japan

Radio Japan
2-2-1 Jinnan, Shibuya-ku, Tokyo 150-01
http://www.nhk.or.jp/rhnet (Internet Web site)
Prime frequencies: 6110, 6150, 11895, 11910, 11920, 12000 kHz 0500–0600 UTC
AS AUS EU NA
7230, 11740, 11850, 11910, 11920, 15165, 15590, 17810, 17815 kHz 0700–0800
 UTC **AF AS AUS EU**
6035, 9560, 9825, 9860, 11685, 11850 kHz 2100–2200 UTC **AF AS EU**
6035, 7200, 7225, 9535, 11615, 11880, 15210, 15460 kHz 1700–1800 UTC **AS NA**
6180, 9560, 9825, 11850 kHz 2300–2400 UTC **AS EU**
7125, 7200, 7225, 9535, 11705, 11880 kHz 1400–1500 UTC **AS NA**
7200, 7225, 9535, 15355 kHz 1500–1600 UTC **AS NA**
11840, 17810 kHz 0300–0400 UTC **AS**
5960, 11790, 15230 kHz 0300–0400 UTC **NA**
7125, 11815, 11850 kHz 0900–1000 UTC **AS AUS**
11790, 13630 kHz 0100–0200 **NA**
11840, 11860, 15475, 15590, 17685, 18810 kHz 0100–0200 **AS**
11910, 17810 kHz 0600–0700 UTC **AS**
Notes: Tune in to the voice of one of the most powerful economic forces in the world, Radio Japan. The station is heard well throughout the world via relay agreements and transmitter sites in England, Singapore, Canada, Sri Lanka, Gabon, and French Guiana.

Jordan

Radio Jordan
POB 1000, Amman
962 6 788 115 (fax)
http://iconnect.com/jordan/radio.html (Internet Web page)
Prime frequency: 11690 kHz 1200–1730 UTC **EU NA**

Notes: In years past, Radio Jordan was rarely heard outside of Europe and the Middle East, but the 1995-instituted North American service has been heard around the world with news and pop music.

Kazakhstan

Radio Alma Ata
Zheltoksan St. 175A, 480013, Alma Ata
Prime frequencies: 5035, 5260, 5940, 5960, 5970, 9505 kHz 1830–1900 UTC **EU**

Notes: Radio Alma Ata is rarely heard outside of Europe and the Middle East.

Kenya

Voice of Kenya
POB 30456, Harry Thuku Rd., Nairobi
Prime frequencies: 4935 and 6150 kHz 0200–2106 UTC

Notes: The Voice of Kenya broadcasts in English and other languages, such as Swahili. Despite its 100- and 250-kW transmitters, it is not one of the easiest-to-hear English-language stations from Africa.

Kirbati

Radio Kirbati
POB 78, Bairiki, Tarawa
Prime frequency: 3955, 9825 kHz 0500–1000 UTC

Notes: Radio Kiribati, which broadcasts irregularly from a group of tiny islands in the South Pacific, operates with local languages and English.

Kuwait

Radio Kuwait
POB 397, 13004 Safat
Prime frequencies: 11990 kHz 1800–2100 UTC **NA EU**

Notes: Radio Kuwait broadcasts some Islamic and political programming, but an hour or more of the English service, each day, consists of informal announcers playing a wide variety of Western popular music, including: pop, rock, rap, hip-hop, punk, "alternative," and heavy metal.

Laos

Laos is the mostly forgotten nation from southeast Asia that was destabilized during the Vietnam War. Like Cambodia, it's never really stabilized. As you might suspect, Lao National Radio is one of the toughest catches in the region, sometimes heard on 6130 and 7116 kHz. Although Lao National Radio broadcasted in English at one time, it now only operates in Laotian.

Latvia

Radio Latvia International
POB 266, Riga
Prime frequencies: 5935 kHz, 2030–2100 UTC **EU**

Notes: Radio Latvia International primarily broadcasts a national service in Latvian. The half hour in English seems to have been added as an afterthought.

Lebanon

King of Hope
POB 3379, Limassol, Cyprus
Prime frequency: 6280 kHz 0500–2200 UTC

Notes: Operated by High Adventure Broadcasting Ministries (a Christian organization), which also operates the Wings of Hope, Lebanon and KVOH, USA.

> Voice of Lebanon
> POB 165271, Al Ashrafiah, Beirut
> Prime frequency: 6550 kHz 1800–1815, 2325–2336 UTC

Notes: This government station is sometimes off the air as a result of the political turmoil within Lebanon.

> Wings of Hope
> POB 3379, Limassol, Cyprus
> Prime frequency: 11530 kHz 0500–2200 UTC

Notes: Operated by High Adventure Broadcasting Ministries (a Christian organization), which also operates the King of Hope, Lebanon and KVOH, USA.

Lesotho

> Radio Lesotho
> POB 552, Maseru 100
> Prime frequency: 4800 kHz 0300–2100 UTC

Notes: Broadcasts plenty of local music with announcements in English and other regional languages.

Liberia

> Liberia Broadcasting System (ELBC)
> POB 594, 1000 Monrovia 10
> Prime frequency: 7275 kHz 0550–1000, 1355–1730 UTC

Notes: ELBC is the Liberian government station. During the 1990s' revolutions, the transmitter was knocked off the air numerous times, and at one point, two different ELBCs were being operated by rival political factions. ELBC is currently off the air, and the quasi-clandestine Radio Liberia is now best-heard. It's hard to say whether ELBC will return or Radio Liberia will take over as the national voice.

Libya

During the 1980s, when General Khadafy of Libya was in the international news almost every night, Radio Jamahiriya was an easy catch in English. Instead of hardline propaganda, Radio Jamahiriya stuck with a light format of Western pop and rock with Western DJs, and a few editorials about U.S. foreign policy, etc. Today, Libya no longer broadcasts in English, and the shortwave broadcasts mostly consist of Arabic-language programs. A number of old frequencies have reactivated, including 6155, 9655, 9705, 11770, 11815, 15235, and 15415 kHz. Perhaps an English service will return to the line-up.

Lithuania

Radio Vilnius
Konarskio 49, LT-2674 Vilnius
370 2 66 05 26 (fax)
Prime frequency: 6120 kHz 0030–0100 UTC **NA**

Notes: Radio Vilnius, once one of the major outlets of the Soviet Union, has dramatically reduced its output in English.

Madagascar

By far, the easiest way to hear Madagascar, a large island off the southeastern coast of Africa is via the 300-kW transmitters of Radio Netherlands and Deutsche Welle. Otherwise, the only chance is via the rather difficult-to-hear Radio-Television Malagasy on 3358 and 5009 kHz.

Malawi

Malawi Broadcasting Corporation
POB 30133, Chichiri, Blantyre 3
Prime frequency: 3380 kHz 0300–0700 UTC

Notes: The Malawi Broadcasting Corporation is one of the big voices from southern Africa. It broadcasts in English and other local languages with news and entertainment programs.

Malaysia

Voice of Malaysia
POB 11272, 50740 Kuala Lampur
http://www.asiaconnect.com.my/rtm-net (Internet Web site)
Prime frequencies: 6175, 9750, 15295 kHz 0555–0825 UTC **AS AUS**

Notes: The Voice of Malaysia is rarely heard outside of the South Pacific region.

Mali

Mali is an easy target via the China Radio International relay station. However, Radiodiffusion-Télévision Malienne does quite well, too, on 4783 kHz. RTM plays a wide variety of music, including much by Western artists, so it sounds a bit different than the other French-speaking African countries—especially the others in the northern desert.

Malta

Voice of the Mediterranean
POB 143, Valetta
Prime frequencies: 9765 kHz 0600–0800 UTC
11925 kHz 1400–1600 UTC

Notes: The Voice of the Mediterranean is a joint venture between the Maltese and the Libyan governments. For years, the programs were broadcast via the Deutche Welle relay transmitters in Malta. However, now that the DW relay site has been closed, the programs are aired via relay transmitters in Russia.

Mauritania

Office de Radiodiffusion-Télévision de Mauritanie has a bigger name than signal. Located in northwestern Africa, the programming is typical: Koran readings and chanting. In 1996, the station was noted on frequencies ranging from 4825 up to 4845 kHz, always drifting.

Mexico

Radio México International
Apardado Postal No. 19-737, México DF 03900
imer@mpsnet.com.mx (Internet e-mail)
Prime frequencies: 5985, 9705 kHz 0000–0030, 0400–0430, 1400–1430,
    1500–1530, 1900–1930, 2000–2030 UTC

Notes: Radio México International broadcasts irregularly with Spanish- and English-language programs to North America.

Although Mexico has a few shortwave stations on the air, most are small operations that broadcast in the 48- and 31-meter bands, where they are destroyed by other signals. Aside from Radio Mexico International, the best bet is Radio Educación on 6185 kHz.

http://uibero.uia.mx/~jsweeney/rhe_0714.html (Radio Huayacocotla—XEJN)
framos@uibero.uia.mx (Radio Huayacocotla—XEJN)

Moldova

Radio Dniester International
25th October St., 45 Tiraspol Transdniestria, CIS 278000
Prime frequencies: 9620, 11270, 15290 kHz 2030–2100 UTC

Notes: Technically a clandestine station, Radio Dniester International has nonetheless been heard by a number of people outside of the European continent. In fact, the signals are apparently better than the official station, Radio Moldova International.

Radio Moldova International
POB 9972, Chi in u 70, Moldova 277070
Prime frequencies: 7190 kHz 0200–0230 UTC

7235 kHz 1830–1900 UTC
7520 kHz 0330–0355, 0430-0455 UTC **NA**
9620 kHz 2130–2200 UTC
15315 kHz 1430–1900 UTC

Notes: The schedules for Radio Moldova International are erratic—the station is often only heard a few times per month.

Monaco

Trans World Radio
BP 349, M-98007, Monte Carlo
33 92 16 5601 (fax)
http://www.twr.org (Internet Web page)
Prime frequencies: 7115 kHz 0745–0920, 1230–1300 UTC

Notes: This is one of the many Trans World Radio (a Christian evangelical broadcaster) outlets around the world.

Mongolia

Radio Ulaan Baator
POB 365, Ulaan Baator
976 1 323096 (fax)
Prime frequencies: 7290, 12085 kHz 1445–1515 UTC **AS**
7295, 12085 kHz 0330–0400 UTC **AS**
13670, 17900 kHz 1930–2000 UTC **EU**
Also broadcasts in English on: 12015 kHz.

Notes: Mongolia, the land of the Khans, is one of the most mysterious and mythical countries in the world. The country's reputation holds true for its international radio outlet, Radio Ulaan Baator, which is not frequently reported.

Morocco

The easiest way to hear Morocco is via the Voice of America's 500-kW relay transmitters. For transmissions in French and Arabic, try Radiodiffusion-Télévision Marocaine on 15345 kHz and Radio Mediterranée Internationale on 9575 kHz. Both of these are fairly easy to hear.

Mozambique

One of the few former colonies of Portugal, Mozambique (on the southeast coast of the continent) has the only Portuguese-speaking shortwave radio station in Africa. Unfortunately, the transmitters of Radio Mozambique are also very difficult to hear. In 1996, the station was noted on 5896, near 9617, and on 9638 kHz. The 9617-kHz outlet has experienced many problems: On one occasion, it drifted more than two kilohertz in less than an hour!

Myanmar

Radio Myanmar
GPOB 1432, Yangon
95 1 30211 (fax)
Prime frequencies: 4725, 5990, 7185, 9730 kHz 0200–0230, 0700–0730, 1430–1600 UTC

Notes: Radio Myanmar is rarely heard outside of the Indian Subcontinent—especially in English.

Namibia

Namibian Broadcasting Corp.
POB 321, Windhoek 9000
215811 (tel)
62346 (fax)
marietha@nbc_hq.nbc.com.na (Internet e-mail)
Prime frequencies: 3270 kHz 1800–2200 UTC
3290 kHz 0000–0800 UTC
4930 kHz 1000–1800 UTC
4965 kHz 0800–1800 UTC

Notes: Mostly features light pop music in English.

Nepal

Radio Nepal
POB 634, Singha Durbar, Kathmandu
Prime frequencies: 3230, 5005, 7165 kHz 0000–1700 UTC

Notes: Broadcasts in various languages, including English, with the news in English at 0720 UTC and the "Destination Nepal" travel program at 0850 UTC.

Netherlands

Radio Nederland
POB 222, 1200 JG Hilversum
31 035 72 42 11 (tel)
31 035 72 42 52 (fax)
letters@rn-hilversum.nl (Internet e-mail)
letters@rnw.nl (Internet e-mail)
http://www.rnw.nl (Internet Web site)
Prime frequencies: 5905, 6020, 6165, 7305 kHz 0030–0125 UTC **AS NA**
5905, 7305, 9860, 11655 kHz 0130–0225 UTC **AS**
5965, 9830, 13700 kHz 0830–0925 UTC **AUS**
5965, 7260, 9610, 9830 kHz 0930–1025 UTC **AUS**
5995, 6165 kHz 0430–0525 UTC **NA**
6020, 9605, 11655 kHz 1730–1825, 2030–2130 UTC **AF**
6020, 9605, 11655, 15315, 17605 kHz 1830–2025 UTC **AF**

6045, 7190 kHz 1130–1325 UTC **EU**
6165, 9845 kHz 2330–0025 UTC **NA**
7260, 9610 kHz 1030–1125 UTC **AS**
9830, 11895 kHz 0730–0825 UTC **AUS**
9860, 11655 kHz 0230–0325 UTC **AS**
9895, 13700, 15585 kHz 1330–1525 UTC **AS**
9895, 12090 kHz 1530–1625 UTC **AS**

Notes: Although not as large or well staffed as some of the other major international broadcasters, Radio Nederland has won many awards and is one of the favorites of listeners around the world. In addition to its many interesting features, the station is also home to "Media Network," one of the most popular radio-listening (DX) programs on shortwave.

Netherlands

Hello, this is a test from Hilversum, Holland. We think you may be interested in the information below concerning activities at Holland's public external broadcaster.

Details of Media Quiz 1996

The 1996 Media Network Race is on, with a whole bouquet of great prizes if you're first past the finish. The mission is simple. You have to visit 12 international broadcasters but, as usual, the boss is grumbling about travel expenses. So use a map to plan the shortest route anywhere in the world that takes you past any 12 international radio stations. We'll supply you with an imaginary solar-powered car and ferry tickets if you need them. So to win, all you have to do is tell us which route you'd take and work out the total distance in kilometers. You can pick any part of the world, but the 12 radio stations *must* have an international service. So you plan the future, and we'll give you the present.

This time we're giving away 5 Radio Netherlands safari survival kits. Each box contains a copy of the brand-new 1996 *World Radio TV Handbook*, a special radio agenda to keep you on track throughout the year, plus a new exclusive Radio Netherlands coffee mug. The most creative entry will qualify for a mystery bonus.

So grab the map and start puzzling. Send your entries by e-mail to media@rnw.nl, by fax to 31 35 6724239, or by snail mail to Media Race 96, English Department, Radio Netherlands, Box 222, 1200 JG Hilversum in Holland. Entries must be postmarked by January 21st 1996. Bon Voyage!

New Zealand

KIWI Radio
POB 3103, Onekawa, Napier
0064 684 48166 (tel)

Prime frequency: 7475 kHz 0600–0900 kHz (irregular weekends)
Notes: KIWI Radio has been closed several times by the New Zealand government for broadcasting without a license.

Radio New Zealand International
POB 2092, Wellington
64 4 474 1433 (fax)
Prime frequencies: 6070 kHz 1650–1752 UTC **AUS**
6105 kHz 1206–1650 UTC **AUS**
9700 kHz 0816–1206 UTC **AUS**
9810 kHz 1753–1952 **AUS**
11735 kHz 1953–2206 UTC **AUS**
11905 kHz 0459–0758 UTC **AUS**
15115 kHz 2137–0458 UTC **AUS**

Notes: Radio New Zealand International is a small, but popular, broadcaster that lives in the shadow of Radio Australia. In spite of the threats of cutbacks, diehard international RNZI listeners always support the station and help keep it on the air (with letter-writing campaigns, etc.).

ZXLA: Radio for the Print Disabled
First Floor, Levin Shopping Mall, POB 360, Levin 5500
06 368 2229 (tel)
06 368 0151 (fax)
Prime frequencies: 3935 kHz 0600–0900 UTC
3935, 5960, 7290 kHz 1930–0600 UTC

Notes: ZXLA is a national radio station for the blind or vision-impaired. Much of the broadcasts consist of audio books.

Nicaragua

All remaining shortwave stations in Nicaragua are currently inactive.

Nigeria

Voice of Nigeria
PMB 40003, Falomo, Ikoyi, Lagos
234 1 269 1944 (fax)
Prime frequencies: 3326, 4990 kHz 0425–1000, 1700–2305 UTC
4990, 7285 kHz 1000–1700 UTC

Notes: The Voice of Nigeria is heard well in many regions of the world, and most of its broadcast hours are in English.

VOICE OF NIGERIA
THE EXTERNAL SERVICE OF THE FEDERAL RADIO CORPORATION
OF NIGERIA
BROADCASTING HOUSE, LAGOS

14TH MARCH 1983

Thank you for your report of 14TH MARCH 1982 at 0530-0540 hrs G.M.T. I have pleasure in confirming that the transmission was from our V.O.N. transmitter on 7255 KHZ

ANDREW YODER

CHIEF ENGINEER

Printed by Pan African Press (Nig) Ltd.

Rememberance Arcade with Tafawa Balewa Square, Lagos, at the background

North Korea

Radio Pyongyang
Pyongyang
Prime frequencies: 6575, 9345, 9640, 9977 kHz 2000–2050 UTC **AF EU**
6575, 9977, 11335 kHz 1100–1150 UTC **NA**
9325, 9640, 9975, 13785 kHz 1700–1750 UTC **AF EU**
9325, 9640, 9975, 13785 kHz 1500–1550 UTC **AF AS EU**
9345, 9640, 11730, 13760, 15230 1300–1350 UTC **AS EU NA**
11335, 13760, 15130 kHz 0000–0050 UTC **NA SA**
11700, 13650 kHz 2300–2350 UTC **NA SA**
15180, 15230, 17765 kHz 0400–0450, 0600–0650, 0800–0850 UTC **NA EU**
15340, 17765 kHz 0700–0750 UTC **AS**

Notes: Radio Pyongyang, unlike its fellow communist broadcasters in Laos, Vietnam, and Cambodia, is well heard around the world.

Northern Mariana Islands

KFBS
POB 209, Saipan, CM 96950
Prime frequency: 9465 kHz 1830–1845 UTC

Notes: KFBS is a Christian missionary station in the Far East Broadcasting Company network. Nearly all programming is in a variety of Asian languages. See the feature on Seychelles.

KHBI
POB 1387, Saipan, CM 96950
letterbox@wshb.csms.com (Internet e-mail)
http://www.csmonitor.com (Internet Web site)
Prime frequencies: 9355 kHz 1100–1600, 1800–2000 UTC **AS**
9385 kHz 1800–2000 UTC **AF AUS EU**
9385, 11550 kHz 1600–1800 UTC **AF EU**
9430 kHz 1100–1500 UTC **AUS**
9430, 13625 kHz 1000–1100 UTC **AS**
9430, 13840 kHz 0900–1100 UTC **AS AUS**
13625, 15405 kHz 2200–2400 UTC **AS AUS**
13840 kHz 0000–0100, 2000–2300 UTC **AS AUS**
15665 kHz 0800–0900 UTC **EU**

Notes: KHBI is the Pacific branch of Monitor Radio International, the radio voice of the *Christian Science Monitor*.

Norway

Radio Norway International
N-0340 Oslo
47 22 45 7134 (fax)
47 22 45 8008 (tel schedule information)
47 22 45 8009 (tel schedule information)
http://nrk.hiof.no/utenland (Internet Web site)
Prime frequencies: 5905, 7275, 7465 kHz 2300–2330 UTC **AS AUS NA**
5960, 7465, 9590 kHz 1900–1930 UTC **AF EU**
5965, 7180, 9590 kHz 0600–0630 UTC **AS AUS EU**
7440, 7465, 7520 kHz 0200–0230 UTC **NA**
7520 kHz 0400–0430 UTC **NA**
9590, 9945, 15605 kHz 1300–1330 UTC **EU NA**
9590, 11840 kHz 1600–1800 UTC **AF NA**
11730 kHz 1400–1430 UTC **AS**
13800 kHz 0900–0930 UTC **AUS**

Notes: These broadcasts are all only aired on Sundays.

Oman

The easiest way to log Oman is via the 100-kW (soon to be 300-kW) transmitters of the BBC. Otherwise, try Radio Oman on 9540 kHz. Radio Oman is rarely reported, despite its 50- and 100-kW transmitters.

Pakistan

Radio Pakistan
Islamabad 44000

92 51 811861 (fax)
Prime frequencies: 7290, 15485, 17705, 17725, 21730 kHz 0230–0245 UTC **AS**
9400, 11570 kHz 1700–1750 UTC **EU**
9485, 9515, 11570, 11935, 13590, 15555 kHz 1600–1630 UTC **AF AS**
9485, 11570, 13590, 15675 kHz 1400–1410 UTC **AF AS**
15470, 17895 kHz 0800-0850, 1100–1120 UTC **EU**
Notes: Features news, sports, commentaries, and readings from the Koran.

Palau

KHBN
POB 66, Koror, Palau 96940
Prime frequencies: 9730, 9965 kHz 0730–1130, 1400–1430 UTC
Notes: One of the branches of High-Adventure Broadcasting, which also operates the Voice of Hope stations from Lebanon and the United States.

Papua New Guinea

National Broadcasting Commission of Papua New Guinea (NBC)
POB 1539, Boroko
Prime frequency: 4890 kHz 1900–1400 UTC
Notes: Broadcasts in English and various local languages. This is the flagship station for the network of shortwave broadcasters in Papua New Guinea.

Radio Central
POB 1539, Boroko
Prime frequency: 3290 kHz 1900–1400 UTC
Notes: Broadcasts in English and various local languages.

Radio Enga
POB 196, Wabag
Prime frequency: 2410 kHz 1900–1400 UTC
Notes: Broadcasts in English and various local languages.

Radio Milne Bay
POB 111, Alotau
Prime frequency: 3365 kHz 1900–1400 UTC
Notes: Broadcasts in English and various local languages.

Radio Northern
POB 137, Popondetta
Prime frequency: 3345 kHz 1900–1400 UTC
Notes: Broadcasts in English and various local languages.

Radio Western
POB 23, Daru
Prime frequency: 3305 kHz 1900–1400 UTC
Notes: Broadcasts in English and various local languages.

Paraguay

Like the other countries in southern South America, shortwave radio is not much of a priority in Paraguay. The only currently active station from the country is Radio Nacional del Paraguay on 9735 kHz, although it sometimes broadcasts slightly off frequency. Radio Nacional del Paraguay pumps an excellent signal into North America.

Peru

Nestled in the Andean Mountains of South America, Peru is one of the hotbeds of commercial shortwave activity in the world. Unless you live in Central or South America, only a few of the stations from Bolivia are easy catches, and none broadcast in English. A few of the best signals from the country on shortwave include Radio Satelite on 6725.5 kHz, Radio San Ignacio on 6747.1 kHz, Radio Cora on 4914.4 kHz, Radio Atlantida on 4790 kHz, Radio Madre de Dios on 4950 kHz, Radio Chota on 4890.5 kHz, Radio Tropical on 4935 kHz, Radio Altura on 6480 kHz, and Radio Union on 6115 kHz. Literally dozens of other Peruvian stations are on the air (the 1996 *WRTH* lists 154 different transmitters!), but most are very difficult to hear (more than half of those transmitters are listed at 1000 watts or less, and most are probably running at well under the official power).

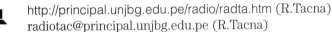
http://principal.unjbg.edu.pe/radio/radta.htm (R.Tacna)
radiotac@principal.unjbg.edu.pe (R.Tacna)

Peru: Musical Transmitters

Don Moore

Because of the difficulties of radio broadcasting in this part of the world, it's not unusual for stations to go off the air for several days, weeks, months, or even years. When an inactive station reappears, it's a good idea to make sure it really is the old station. After all, the owner might have decided to call it quits and sell the transmitter. In rural Peru, you never know for sure just who owns the transmitters, since some change hands frequently.

For example, in the early 1980s, Radio Acunta came on the air from Chota with a 100-watt transmitter on 5800 kHz. Later, the transmitter was moved to Bambamarca and rented to Radio San Francisco, a new station. However, Radio San Francisco didn't make it, and the transmitter was soon back in Chota. By mid-1985, Radio Acunta was having a tough time competing with crosstown rivals Radio Chota and Radio San Juan de Chota, each of which had a 1-kW transmitter, so manager Victor Hoyos called it quits. The transmitter was sold and ended up in San Ignacio, by the Ecuadorian border, where, for several years, it was used by Radio San Miguel Archangel. Radio San Juan de Chota didn't last much longer, and its transmitter was sold to another new Bambamarca station, Radio Onda Popular. In the late 1980s, Radio Nuevo Cajamarca from the town of Nuevo Cajamarca, near Rioja, was heard on 5800 kHz. This was probably the old Radio Acunta transmitter with yet another owner.

Another "musical transmitter" got its start in Moyobamba, on the fringe of the Amazon jungle. In 1982, Radio Moyobamba announcer Miguel Quisipotongo Suxe founded his own station, Estacion C, using a 300-watt Framvel transmitter with a crystal for 6364 kHz. The Moyobamba area is growing fast, and Miguel made enough money to invest in new equipment. He bought a higher powered transmitter, with a crystal for 6324 kHz, he sold the 6364 transmitter to another ex-Radio Moyobamba announcer, Porfirio Centurion. Porfirio called his soon-to-fail station Radio Moderna. A few months later the transmitter ended up in nearby Saposoa, where it was used to broadcast under the name Radio Huallaga. There's no telling who owns the transmitter now.

Philippines

Far East Broadcasting Co. (FEBC)
POB 1, Valenzuela, Metro Manila 0560
Prime frequencies: 11805, 15120, 15270 kHz 0230–0330 UTC **AS**
11995 kHz 1300–1600 UTC **AS**
15450 kHz 0000–0200 UTC **AS**

Notes: The FEBC is a Christian missionary station that broadcasts primarily in the more common languages of Asia. The FEBC also has a sister station in the Seychelles.

Radio Pilipinas
Quezon City, Metro Manila 1103
Prime frequencies: 11815, 11890, 15190 kHz 1730–1800 UTC **AS**
13770, 15330, 17730 kHz 0330–0400 UTC **AS NA SA**

Notes: Radio Pilipinas, the international broadcast station of the Philippines, is by far the smallest of the four shortwave outlets (Radio Pilipinas, Radio Veritas Asia, FEBC Philippines, and the Voice of America relay station in Poro) in the country.

Poland

Polish Radio Warsaw
POB 46, 00-977 Warsaw
48 22 645 93 05 (tel)
48 22 444 123 (fax)
48 22 645 59 17 (fax)
radio1@ikp.atm.com.pl (Internet e-mail)
Prime frequencies: 6035, 6095, 7285 kHz 2030–2155 UTC **EU**
6090, 7270, 7285 kHz 1800–1855 UTC **EU**
6095, 7145, 7270, 9525, 11815 kHz 1300–1355 UTC **EU**

Notes: Formerly known as Radio Polonia, Polish Radio Warsaw is well-heard across much of the world for a few broadcasts in English each day.

Portugal

Radio Portugal
30-1106, Lisbon
Prime frequencies: 6130, 9780, 9815, 15515 kHz 2000–2030 UTC **AF**
6150, 9570 kHz 0330–0400 UTC **NA**
21515 kHz 1430–1500 UTC **AS**

Notes: Radio Portugal broadcasts with a few half-hour programs in English, but the real focus of the station is its Portuguese-language service.

Qatar

Qatar, a peninsula off the main portion of Saudi Arabia, is one of the smallest countries in the Middle East—quite a bit larger than Bahrain, but smaller than even Israel, Cyprus, and Lebanon. For a taste of Qatar's Arabic broadcasts, try 7210, 9570.2, and 11750 kHz.

Romania

Radio Romania International
Str. General Berthelot 60-62, POB 111, Bucharest
401 617 2856 (tel)
401 312 9262 (fax)
Prime frequencies: 5990, 5995, 7105, 7195 2100–2200 UTC **EU**
5990, 6155, 9510, 9570, 11940 kHz 0200–0300 and 0400–0430 UTC **NA**
5995, 7105, 7195, 9690 kHz 1900–2000 UTC **EU**
9550, 9750, 11830, 11940 kHz 1730–1800 UTC **AF**
9690, 11940, 15390, 17745 kHz 1300–1400 UTC **EU**
11740, 11840, 15250, 15270, 17720 kHz 0645–0745 UTC **AUS**
11740, 15335 kHz 1430–1530 UTC **AS**
11940, 15270, 15340, 17745, 17790 kHz 0530–0600 UTC **AF AS**

Notes: Despite the fall of communism in Romania, Radio Romania International (formerly known as Radio Bucharest) continues its large, well-heard English service.

Russia

Voice of Russia
ul. Pyatnitskaya 25, 113326 Moscow
Prime frequencies: 4740, 4975, 9705, 15460, 17860 kHz 1300–1400 UTC **AS**
4920, 5940, 6100, 7180, 9890 kHz 1700–1800, 1900–2100 UTC **EU**
4940, 4975, 6005, 6175, 7210, 9470, 15205 kHz 1700–1800 UTC **AS**
5940, 6110, 7170, 7320, 7440, 9890 kHz 2100–2200 UTC **EU**
5940, 7105, 7180 kHz 2200–0000 UTC **EU NA**
6130, 7180, 7305, 7440, 9890 kHz 1800–1900 **AS EU**
7105, 7125, 7180 kHz 0000–0200 UTC **NA**
7130, 7165, 9470, 9840, 15205 kHz 1400–1600 UTC **AS**
7130, 7210, 7255, 7275, 7305, 9585 kHz 1700–1800 **AS**
5920, 5930, 7105, 7345, 9550, 9580 kHz 0200–0300 UTC **NA**
11655, 13785, 17755, 17860 kHz 1000–1300 UTC **AUS**
12025, 15460, 17570 kHz 0600–0900 UTC **AUS**

Also broadcasts in English on: 5925, 5940, 6150, 6175, 7175, 7220, 9550, 9585, 9635, 9725, 9755, 9820, 9875, 11880, 12030, 12035, 13665, 15120, and 21790 kHz.

Notes: Once one of the very largest shortwave broadcasters in the world, the Voice of Russia (formerly known as Radio Moscow) has cut its number of personnel nearly in half. However, with over 1200 employees, it is still enormous, and the station can be heard around the world.

Rwanda

Radio Rwanda
Kigali
250 7 6185 (fax)
Prime frequencies: 6055 kHz 0515–0525, 1145–1155, 1915–1925 UTC

Notes: Programs are via the facilities of the Deutsche Welle in Rwanda.

Sao Tomé

The only means of hearing Sao Tomé is via the massive 500-kW relay transmitters of the Voice of America.

Saudi Arabia

Saudi Arabia is one of the largest countries in the world that does not broadcast in English (even though its external service operates in 10 different languages, including Indonesian!). Like the other conservative countries in the Middle East and northern Africa, much of Saudi Arabia's programming output consists of Koran readings and chanting. To tune in the Broadcasting Service of the Kingdom of Saudi Arabia (BSKSA), try 6020, 9555, 9715, 9775, 9870, 11708.2, 11818.5, 11870, 15335, 17780, 17795, and 21705 kHz.

Serbia

>Radio Yugoslavia
>Hilanarska 2, 11000 Beograd
>Prime frequencies: 6100, 6185 kHz 2200–2230 UTC **EU**
>6100, 9720 kHz 1930–2000 UTC **AF EU**
>6195, 7115 kHz 0100–0130 UTC **EU NA**
>6195, 7130 kHz 0200–0230 UTC **EU NA**
>11835 kHz 1330–1400 UTC **AUS**

Notes: The United Nations have built up troops in the regions in and around the former country of Yugoslavia in an attempt to prevent Serbia from invading the rest of Croatia and Bosnia. Because most of the world is in either the United Nations or is Islamic, most media outlets are pro-Croatian or Bosnian. For news and commentaries from a Serbian perspective, one of the only outlets is Radio Yugoslavia, which is widely heard.

Seychelles

>FEBA Radio
>POB 234, Mahe
>3418752@mcimail.com (Internet e-mail)
>Prime frequencies: 9810, 11870 kHz 1500–1600 UTC **AS**

Notes: FEBA Radio in Seychelles is a Christian missionary sister station to the FEBC in the Philippines.

Sierra Leone

>Sierra Leone Broadcasting Service
>New England, Freetown

Prime frequencies: 3316, 5980 kHz 0600–2230 UTC

Notes: Sierra Leone is not widely reported outside of Northern Africa and southern Europe.

Singapore

Radio Singapore International
Farrer Rd., POB 5300, Singapore 9128
65 3535300 (tel)
65 2591380 (fax)
Prime frequencies: 6015, 9530 kHz 1100–1400, 2300–2400 UTC

Notes: RSI features a number of programs including Dateline RNI, Chartbeat, and "You Asked For It" with plenty of local musicians from Singapore.

Slovakia

Radio Slovakia International
Mytna 1, 812 90 Bratislava
042 7498 075 (tel)
042 7496 282 (fax)
http://www.xs4all.nl/~xavcom/rozhlas (Internet Web site)
Prime frequencies: 5915, 6055, 7345 kHz 1730–1800, 1930–2000 UTC **EU**
5930, 7300, 9440 kHz 0100–0130 UTC **NA**
11990, 17485, 21705 kHz 0830–0900 UTC **AUS**

Notes: Listening to Radio Slovakia International is like a mini tour of Slovakia. Instead of focusing on the world (like the BBC) or on a region (like Radio Australia), almost everything on RSI pertains to Slovakia. Some of the programs include: Slovak Kitchen, Back Page News, Slovak Personalities, and Listeners' Tribune.

Soloman Islands

Soloman Islands Broadcasting Corp. (SIBC)
POB 654, Honiara
677 23159 (fax)
Prime frequencies: 5020, 9545 kHz 0500–1100, 1900–2200 UTC

Notes: The SIBC is a tiny local broadcast service for the Soloman Islands, broadcasting in English and Pidgin. The station typically features island music, news, and public service announcements, but it is not heard well outside of the South Pacific.

Somalia

Somalia is currently in a state of civil war and shortwave transmitters are controlled by various political factions. A number of different stations have been reported, including several different "Radio Mogadishus." These stations, best described as clandestines, are among the most difficult to log from Africa.

South Africa

Channel Africa
POB 91313, Auckland Park 2006
vorstern@sabc.co.za (Internet e-mail)
http://www.sabc.co.za (Internet Web site)
Prime frequencies: 3220, 5955 kHz 0300–0400 UTC
3220, 7240 kHz 1500–1800 UTC
5955, 9585 kHz 0400–0500 UTC
5955, 11900 kHz 0500–0600 UTC
7155, 9685 kHz 1500–1800 UTC
15240 kHz 1600–1700 UTC
9695 kHz 1600–1700 UTC

Notes: Channel Africa used to be known as the Radio South Africa, and Radio RSA was the largest shortwave broadcaster on the continent. Today, the name is different and the hours on the air have been slashed.

Sentech Shortwave Services
Private Bag X06, Honeydew 2040
475-5112 (fax)
Prime frequencies: 3320, 4810 kHz 2300–0300 UTC

7270 kHz 0500–1430 UTC
9630 kHz 0800–1630 UTC

Notes: These channels all feature programs from different branches of the South African Broadcasting Corporation (Radio 2000, Radio Oranje, Afrikaans Stereo, etc.). However, the transmitters are all operated by the Sentech company.

South Korea

Radio Korea
18 Yoido-dong, Youngdungpro-gu, Seoul 150-790

02 781 3711 (tel)
02 781 3799 (fax)
Prime frequencies: 3970 kHz 1900–2000 UTC **EU**
5975, 7275 kHz 1900–2000 UTC **AS**
5975, 9515, 9870 kHz 1600–1700 UTC **AF AS**
6480, 15575 kHz 2100–2200 UTC **EU**
7285 kHz 1200–1300 UTC **AF AS**
9570, 13670 kHz 0800–0900, 1230–1300 UTC **AUS EU**
9650 kHz 1130–1200 UTC **NA SA**
11725, 11810, 15345 kHz 0200–0300 UTC **NA SA**

Notes: Radio Korea counters the propaganda of Radio Pyongyang with much more laid-back, Western-style programming. The station is heard well around the world because it uses relays in England and Canada, in addition to its own facilities in South Korea.

Spain

Radio Exterior de Espana (Spanish Foreign Radio)
Apartado 156.202, 28080 Madrid
 346 11 49 (tel)
346 18 15 (fax)
Prime frequencies: 6055 kHz 0000–0200, 0500–0600 UTC **NA**
6125, 11775 kHz 2000–2100 UTC **AF EU**
11775 kHz 2200–2300 UTC **AF**

Notes: Like its peninsula "roommate" Portugal, Radio Exterior de Espana has a limited English-language output and instead concentrates its efforts on its home language.

Sri Lanka

Sri Lanka Broadcasting Corporation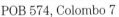
POB 574, Colombo 7
Prime frequencies: 7190, 11800 kHz 0030–0430, 0630–1030, 1330–1730 UTC
9720 kHz, 15425 kHz 0030–0430, 1230–1730 UTC **AS**
9720, 15425 kHz 0445–0515 UTC **NA**
Also broadcasts in English on: 6005, 6075, 15120, and 17850 kHz.
Notes: The SLBC is not widely heard outside of Asia.

St. Helena

Radio St. Helena is a very tiny 100-watt mediumwave station from the extremely remote southwest African island of St. Helena. With some help from a shortwave listener in Sweden, Radio St. Helena is aired for several hours via the St. Helena Cable and Wireless transmitter (utility) on 11092.5 kHz SSB. This special "Radio St. Helena Day" has been one of the more memorable events in DXing for shortwave listeners around the world. It is reported with strong signals across much of the world, despite the 1000-watt power rating.

St. Helena (Radio St. Helena)

Radio St. Helena Day, the one day each year that St. Helena broadcasts on shortwave, is a DXer's dream come true. St. Helena is a speck in the South Atlantic. Its closest neighbors (still, hundreds of miles away) are the tiny islands of Ascension Island, home of massive BBC and Voice of America shortwave relay stations, and Tristan da Cunha, once the location of the 40-watt Tristan Broadcasting Service (known as the all-time greatest catch for shortwave listeners).

Although Ascension Island has an airport, the only access to St. Helena and Tristan da Cunha is via boat. In fact, only one ship, aptly named the RMS St. Helena, travels to St. Helena. This ship travels between Cardiff, England and Cape Town, South Africa. These islands are far from self-sufficient; those who live there are dependent on the RMS St. Helena for virtually everything tangible: food, medical supplies, machinery, mail, and tourists.

> Radio St. Helena Day is a big event for shortwave listeners, but it is also fascinating entertainment for those living on the island. Imagine the excitement for the residents of a very small, very isolated island, hearing greetings from many listeners around the world. "It was really an exciting evening for listeners on the island as well as, from all accounts, listeners in other parts of the world. Being isolated as we are, it was a real thrill to have calls from so many of you, backed up by faxes and letters, stating so genuinely, how happy you were to have received our broadcast," said Tony Leo, station manager. It's not every day on St. Helena that you can hear people from France call you up and say "I wish you very, very well . . . . very strongly."
>
> For more information on the island of St. Helena and on the annual broadcasts, check http://www.sthelena.se on the World Wide Web, or write to: Radio St. Helena, The Castle, Jamestown, St. Helena, South Atlantic Ocean. If you choose the latter, expect to wait a while for a response. After all, the RMS St. Helena can't get to the island every day.

Sudan

Sudan is one of the most difficult countries to hear in Africa. Currently, the external service of the Sudan National Broadcasting Corporation, Radio Omdurman, is heard on 7200 and 9024 kHz. However, a clandestine, the Voice of Sudan, has also been reported on several frequencies, including 9025.1 kHz. Considering that most of the broadcasts are in Arabic (although English programming is sometimes aired), these broadcasts can be very confusing—and they have been confused in some of the radio bulletins.

Suriname

Suriname, the only former Dutch colony of South America, has no external service and is very difficult to hear on shortwave. Only the commercial Radio Apintie operates irregularly with a mere 350 watts on 4990.9 or 5005 kHz in the crowded 60-meter band. Suriname is one of the most difficult countries in the world for most DXers to hear.

Swaziland

Trans World Radio
POB 64, Manzini
268 55333 (fax)

lgunnars+aswz%twrihq@mcimail.com (Internet e-mail)
http://www.twr.org (Internet Web page)
Prime frequencies: 3200 kHz 1745–2045 UTC
3240 kHz 1900–2045 UTC
5055, 6070, 9500 kHz 0430–0835 UTC
9500 kHz 1600–1830 UTC

Notes: This is one of the many Trans World Radio (a Christian evangelical broadcaster) outlets around the world.

Sweden

Radio Sweden
S-105 10 Stockholm
46 8 784 50 00 (tel)
46 8 667 62 83 (fax)
46 8 667 37 01 (schedule by fax)
http://www.sr.se/rs (Internet Web site)
ftp://town.hall.org/radio/Mirrors/RadioSweden/ (To download Media Scan program from Internet)
Prime frequencies: 6065 kHz 0030–0100, 2030–2100 UTC **SA**
6065, 7240, 9655 kHz 1930–2000 UTC **AF AS EU**
6065, 7230 kHz 2130–2200 UTC **AF AS EU**
6065, 7325 kHz 2230–2300 UTC **AF AS EU**
6200 kHz 0230–0300 UTC **NA**
7115 kHz 0330–0400 UTC **NA**
7265, 7290 kHz 0130–0200 UTC **AS AUS SA**
9845, 9885, 11650, 15240 1430–1500 UTC **AS AUS NA**
9895, 11695 kHz 0130–0200 UTC **AS AUS**
11650, 15240 kHz 1230–1300 **NA**
Also broadcasts in English on: 11910 and 13690 kHz.

Notes: The ever-popular Radio Sweden broadcasts such programs as: Upstream, In Touch With Stockholm, Sounds Nordic, Media Scan, and Spectrum.

Sweden

Don't miss Radio Sweden's great holiday program line-up:
- Dec. 23 "Spectrum," our Arts magazine, with a retrospective look at 1995.
- Dec. 24 "Christmas Eve Special" looking at how the Swedes spend Christmas—a festivity based ironically on ancient Nordic pagan rites. We reveal the secrets of gloegg, Advent candles, Lucia, straw goats, and the Christmas gnome.

- Dec. 25 "In Touch With Stockholm" special Christmas edition of our listener program, including famous Swedish sports and music personalities on how they spend Christmas, and their impressions of 1995.
- Dec. 26 "Boxing Day"—A special bumper edition of "SportScan" reviewing all the highlights of 1995.
- Dec. 27 "Money Matters"—A full-length edition of our regular financial, business, and labor market magazine, also reviewing 1995.
- Dec. 28 "Horizon"—A full-length edition of our Science magazine, including a long interview with famed British natural history filmmaker David Attenborough about "The Private Life of Plants."
- Dec. 29 "Sixty Degrees North," looking as usual on a Friday at the Nordic newsweek.
- Dec. 30 "Review of 1995."
- Dec. 31 "Sounds Nordic"—Our music and chat show with plenty of sparkle on New Years Eve.
- Jan. 1 Another chance to hear our review of the 1995 news year here in Sweden.
- Jan. 2 Back to normal . . . . "MediaScan" will be featuring an interview with Swedish Minister of Culture Margot Wallstroem about the future of broadcast media in this country.

Switzerland

Swiss Radio International
Giacomettistrasse 1, 3000 Berne 15
031 350 92 22 (tel)
031 350 95 44 (fax)
Prime frequencies: 5840, 6165 kHz 0615–0630, 0715–0730 UTC **EU**
5850, 7410, 9885, 9905 kHz 1700–1730 UTC **AF AS EU**
6135 kHz 0100–0130, 0400–0430 UTC **AF NA**
6165, 9535 kHz 1100–1130, 1300–1330 UTC **EU**
7155, 13740, 15240 kHz 1330–1400 UTC **AS AUS**
7230, 12075, 13635 kHz 1300–1330 UTC **AS**
9885 kHz 0100–0130, 0400–0430, 0900–0930, 1100–1130, 1500–1530, 2000–2030 UTC **AF NA**
9905 kHz 0100–0130, 0400–0430, 2000–2030 UTC **NA**
12075, 13685 kHz 0900–0930, 1100–1130, 1500–1530 UTC **AS AUS**
13635 kHz 1100–1130, 1700–1730, 2000–2030 UTC **AF AS**

Notes: Switzerland is known for impartiality in politics, so be sure to check out the world features (NewsNet, World Scene, Swiss Scene, and Down To Earth) on Swiss Radio International.

Syria

Radio Damascus
Ommayad Sq., Damascus
Prime frequencies: 12085 kHz 2010–2105 UTC **EU**
12085, 13610 kHz 2110–2210 UTC **AS AUS NA**

Notes: Radio Damascus is a great source of information on the Middle East through such programs as Welcome to Syria, Arab Civilization, Arab Profile, Syria & the World, and Arab Newsweek.

Tahiti

RFO Tahiti has been a long-time favorite of shortwave listeners because of its friendly sound and lively South Pacific music. Hank Bennett, who wrote the first edition of this book in 1973, lamented long and hard that he had never heard the station. It might not be long before many others will have the same experience. RFO Tahiti is still broadcasting on 15167.3 kHz, but the staff has stated that when this transmitter breaks down, it will not be fixed or replaced. Hear it while you can!

Taiwan

Voice of Free China
53 Jen Ai Rd., Sec. 3, Taipei
02 771 0150 (tel)
02 751 9277 (fax)
Prime frequencies: 5810 kHz 2200–2300 UTC
5950 kHz 0200–0400, 0700–0800 UTC **AS AUS NA SA**
7130 kHz 0200–0400, 1200–1300 UTC **AS AUS NA SA**
9680 kHz 0200–0400 UTC **AS AUS NA SA**
11825 kHz 0200–0400 UTC **AS AUS NA SA**
15345 kHz 0200–0400 UTC **AS AUS NA SA**
Also broadcasts in English on: 9610, 9850, 11740, and 11745 kHz.

Notes: The Voice of Free China is one of the major shortwave outlets from Asia. In addition to providing information about the views and culture of Taiwan, the VOFC even informs its listeners with newspapers and station mini magazines.

Tajikistan

Tajik Radio
31 Chapyev St., Dushanbe
Prime frequency: 7245 kHz 0345–0400, 1645–1700 UTC **AS**

Notes: Tajik Radio is rarely heard outside of the Middle East.

Tanzania

Radio Tanzania
POB 9191, Dar es Salaam
Prime frequency: 5050 kHz 0400-0430
Notes: Radio Tanzania is occasionally reported on this frequency or on 5985 kHz with programming in English and Swahili.

Thailand

Radio Thailand
236 Vibhavaadi-Rangsit Rd., Din Dang, Kuay-Khwang, Bankok 10400
2776139 (fax)
Prime frequencies: 7295, 9655, 11905 kHz 1900–2000 UTC **AS EU**
9530, 9655, 11905 kHz 1400–1430 UTC **AF AS**
9655, 9680, 11905 kHz 0000–0100 UTC **AF AS**
9655, 9810, 11905 kHz 1230–1300 UTC **AF AS**

QSL RADIO THAILAND WORLD SERVICE

Broadcast Schedule

GMT	Service
00.00 - 01.00	English
01.00 - 02.00	Thai
03.00 - 03.30	English
03.30 - 04.30	Thai
11.00 - 11.15	Vietnamese
11.15 - 11.30	Khmer
11.30 - 11.45	Lao
11.45 - 12.00	Burmese
12.00 - 12.15	Bahasa Malaysia
12.15 - 12.30	Bahasa Indonesia
13.00 - 13.15	Japanese
13.15 - 13.30	Mandarin
13.30 - 14.00	Thai
18.00 - 19.00	Thai

Graceful Thai doll in dance custume, sporting elk's head.

Photographer : Somchai Nguansa-ngiam

9655, 11890, 11905 kHz 0300–0330 UTC **AS NA**
9655, 11905, 15115 kHz 0530-0600 UTC **AF AS**

Notes: Although Radio Thailand is an international broadcaster, their programs cover regional events (such as concerts, weather, etc.), and it often sounds more like a national service.

Togo

Despite its tiny size, Togo has one of the biggest voices in Africa. Radiodiffusion Togolaise regularly has a good-to-excellent signal around much of the world on 5047 kHz. Unfortunately, Togo broadcasts almost entirely in French, so although the music is fun to listen to, most of the programming is indecipherable for English speakers.

Tunisia

Like many of the Arabic-speaking countries (but unlike nearly all of Africa), most of Radio Tunisia's transmitters are above 10 MHz. As a result of these frequency choices, the best time to listen is during the daytime. For day or nighttime broadcasts, try 7475 kHz. Otherwise, try 11730, 12005, 15450, or 17500 kHz.

Turkey

Voice of Turkey
POB 333, 06.443 Yenesehir, Ankara
490 98 00 (tel)
490 98 11 (tel)
490 98 45 (fax)
490 98 46 (fax)
Prime frequencies: 6000, 9535 kHz 1930–2030 UTC **EU NA SA**
6135, 7280, 9560, 9655 kHz 2300–2350 UTC **AS EU NA**
7300, 17705 kHz 0400–0450 UTC **AF AS AUS NA**
9445, 9630 kHz 1330–1400 UTC **AS EU**
9655, 11805, 11905 kHz 2030–2045 UTC **AS EU**

Notes: With such program names as "A Haven in the East: Turkey" and "Magnificent Istanbul," there is no question that the Voice of Turkey is playing up its role as a tourist lure.

Turkey (Voice of Turkey)

Voice of Turkey

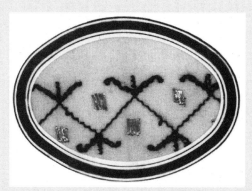

Until the early part of the 20th century, for religious and social reasons, women in Turkey did not take part in artistic activities, such as painting, sculpture, and the performing arts. Their only way of expressing their inner world was through their embroideries, each of which bears a pattern with a particular message. Embriodery was the most effective means of expression for Turkish women.

This card is an example of such work of art created by a Turkish woman. The embroideries convey their messages through individual interpretations of such characteristic themes drawn from nature, a pomegranate signifying abundance, a triolet of figures representing various consecrated subjects, such as God, Nature and Humanity, an eye (symbolizing the evil eye), or an amulet.

They all come from long bygone years.

Uganda

Radio Uganda
POB 7142, Kampala
256 41 254 461 (tel)
256 41 256 888 (fax)
Prime frequencies: 5026 kHz 0300–2100 UTC

Notes: Thankfully, Uganda isn't in the news like it was during the days of Idi Amin. You can catch up on the current events in the country from Radio Uganda on 5026 kHz. Unfortunately for listeners in North and South America, the very strong Radio Rebelde (Cuba) outlet operates on 5025 kHz, eliminating most of your chances to hear Uganda.

Ukraine

Radio Ukraine International
vul Kreshchatik 26, 252001 Kiev
Prime frequencies: 5915, 7150, 7160 kHz 0100–0200, 0400–0500 UTC
5940, 6020, 7180, 7240, 11870 kHz 2200–2300 UTC

Notes: Radio Ukraine International (formerly known as Radio Kiev) is one of the only radio stations from the former Soviet Union that delivers solid signals to much of the world. The broadcasts are quite Western—what a surprise to hear a speed metal song by Metallica during one RUI program!

United Arab Emirates

UAE Radio from Abu Dhabi
POB 63, Abu Dhabi
971 2 451155 (fax)
Prime frequencies: 9605 kHz 2200–2400 UTC
9770 kHz 2200–2400 UTC
11710 kHz 2200–2400 UTC

Notes: UAE Radio from Abu Dhabi is operated by the UAE Ministry of Information and Culture.

UAE Radio Dubai
POB 1695, Dubai
Prime frequencies: 13675, 15395 17630, 21605 kHz, 1030–1110, 1330–1400, 1600–1640 UTC
21605 kHz 1030–1110, 1330–1400 UTC

Notes: UAE Radio from Dubai, like the station from Abu Dhabi, is well heard across most of the world with news, Arabic music and culture, and Koran readings.

UNITED ARAB EMIRATES
Ministry of Information & Culture
Dept. of Broadcasting
P. O. Box : 63 - Abu Dhabi

دولة الإمارات العربية المتحدة
وزارة الإعلام والثقافة
ادارة الاذاعة
ص.ب ٦٣ – ابو ظبي

No. ....
Date January 17, 1995

الرقم : ....
التاريخ : ....

Mr. Andrew Yoder,

Thank you for your letter together with your taped reception report.

We are pleased to confirm that you have listened to our English programmes on 9605 Khz in the 31 metre band. Our frequency schedule is sent herewith. We appreciate your interests in various hobbies and your attention to our broadcasts. Best wishes.

Yours sincerely,

Aida Hamza
Director
Foreign Language Services
UAE Radio from Abu Dhabi.

United Kingdom

British Broadcasting Corp. (BBC)
44 171 257 2211 (tel)
44 171 257 8258 (fax)
44 171 240 4899 (fax)
iac@bbc-ibar.demon.co.uk (Internet e-mail)
http://www.bbcnc.org.uk/worldservice (Internet Web site)
BBC Worldwide program schedule available from: POB 3000, Denville, NJ 07834 USA, 201 627 2427 (tel), 201 627 5827 (fax)
Prime frequencies: 3255 kHz 0300–0600, 1800–2200 UTC **AF**
5965 kHz 0000–0200, 1100–1400 UTC **AS NA**
5970 kHz 0000–0330 UTC **NA SA**
5975 kHz 1400–1830, 2100–0800 UTC **AS NA SA**
5990 kHz 1300–1630 UTC **AS AUS**
6005 kHz 0300–0700, 1900–2200 UTC **AF**
6095 kHz 1100–1400 UTC **AS AUS**
6175 kHz 0330–0800 UTC **NA**

6180 kHz 0300–0730, 1700–2230 UTC **EU**
6190 kHz 0300–2200 UTC **AF**
6195 kHz 1700–0200, 0400–0900, 1000–1600 UTC **EU AS NA**
7110 kHz 2200–0030 UTC **AS**
7145, 21660 kHz 0600–0800 UTC **AS AUS**
7150 kHz 2000–2230 UTC **EU**
7325 kHz 0230–0330, 0600–0900, 2000–2230 UTC **EU NA SA**
9410 kHz 0000–2230 UTC **EU AS**
9510 kHz 1630–1830 UTC **AS**
9515 kHz 1300–1730 UTC **NA**
9580 kHz 1100–1400 UTC **AS AUS**
9590 kHz 2200–0430, 1300–1600 UTC **NA**
9600 kHz 0300–0800 UTC **AF**
9630, 17830 kHz 1830–2100 UTC **AF**
9740 kHz 0500–2000, 2030–2200 UTC **AS AUS**
9915 kHz 2230–0230 UTC **NA SA**
11750 kHz 0000–0200, 0900–1800 UTC **AS SA**
11760 kHz 0300–1400 **EU**
11835 kHz 2000–2300 UTC **AF**
11865 kHz 1100–1630 UTC **NA SA**
11955 kHz 2030–0930, 1100–1300 UTC **AS AUS**
12095 kHz 0330–2100 UTC **AF EU**
15220 kHz 1130–1600 UTC **NA**
15280 kHz 0000–0300, 0330–0500 UTC **AS**
15310 kHz 0300–1500 UTC **AS**
15360 kHz 0000–0330, 0500–0800 UTC **AS AUS**
15400 kHz 0730–1130, 1430–2100 UTC **AF EU**
15420, 17830 kHz 1700–1900 UTC **AF**
15485, 17640 kHz 0700–0830 UTC **EU**
15575 kHz 0400–1800 UTC **EU AS**
17715 kHz 2100–0000 UTC **NA SA**
17830, 21660 kHz 0700–1700 UTC **AF**
17840 kHz 1400–1730 UTC **NA SA**
17885 kHz 0500–1400 UTC **AF**
21660 kHz 0800–1000 UTC **AS**

Also broadcasts in English on: 3915, 3955, 5965, 6135, 7110, 7160, 7250, 9510, 9575, 9580, 9590, 9610, 9760, 9895, 11680, 11730, 11765, 11775, 11840, 11945, 15190, 15380, 15565, 17790, 17880, and 21490 kHz.

Notes: If it's happening in the world, chances are that you'll hear about it on the BBC.

Uruguay

Like the other countries in southern South America, shortwave plays only a very minor role in Uruguay. Only a few stations are noted with weak signals outside of the region. Try Radio Monte Carlo on 9595.1 kHz. A new station is Emissora Cuidad de Montevideo, which has been heard on 9650 kHz. This station hopes to be heard well in the southern U.S. on its second frequency of 15230 kHz.

USA

KAIJ
P.O. Box 270879, Dallas TX 75227
817 277 9929 (fax)
Prime frequencies: 9815 kHz 0100–1300 UTC **NA**
13740 kHz 2300–2400 UTC **NA**
15725 kHz 1400–2200 UTC **NA**

Notes: KAIJ features plenty of the end-times preaching of Dr. Gene Scott.

KJES
Star Rt. Box 300, Mesquite, NM 88048
714 731 4196 (fax)
Prime frequencies: 11715 kHz 1300–1600 UTC **NA**
15385 kHz 1800–1900 and 2000–2100 UTC **AUS NA**

Notes: KJES is a refreshingly simple Christian outreach broadcaster. It features no promotional material, just a straightforward message.

KNLS
Anchor Point, AK 99556
615 371 8791 (fax)
knls@aol.com (Internet e-mail)
http://www.hax.com/wcb/aaaindex.htm (Internet Web site)
Prime frequencies: 7365 kHz 1300–1400 UTC
9615 kHz 0800–0900 UTC

Notes: KNLS (The New Life Station) targets listeners in Russia and the Pacific region. However, they do have several English-language broadcasts and feature an excellent DX program in addition to their primary focus of Christian outreach.

Greetings from **KNLS**, the New Life Station, P.O. Box 473, Anchor Point, Alaska 99556, United States of America

KTBN
POB A, Santa Ana, CA 92711
714 730 0661 (fax)
tbntalk@aol.com (Internet e-mail)
http://www.tbn.org/ktbn.html (Internet Web site)
Prime frequencies: 7510 kHz 0000–1600 UTC **NA**
15590 kHz 1600–2400 UTC **NA**

Notes: Features broadcasts of conservative Christian preachers on the Trinity Broadcasting Network.

KVOH
POB 93937, Los Angeles, CA 90093
Prime frequencies: 9975 kHz 0000–1600 UTC **NA SA**
17775 kHz 1600–0000 UTC **NA**

Notes: Operated by High Adventure Broadcasting Ministries (a Christian organization), which also operates the Wings of Hope, Lebanon and KVOH, USA.

KWHR
Naalehu, HI
Prime frequencies: 6120 kHz 1600–1800 UTC **AUS**
9930 kHz 0800–1600 UTC **AS**
13625 kHz 1800–2000 UTC **AUS**
15405 kHz 2000–2200 UTC **AS**
17510 kHz 2200–0400 UTC **AS**
17780 kHz 0400–0800 UTC **AS**

Notes: KWHR is the Pacific sister station of WHRI (World Harvest Radio).

Voice of America
330 Independence Ave. SW, Washington, DC 20547
202 401 1493 (tel)
202 401 1494 (fax)
letters@voa.gov (Internet from outside the U.S.)
letters-usa@voa.gov (From within the U.S.)
Prime frequencies: 3980, 6040, 9760, 15205, 19379L kHz 1700–2000 UTC **AF EU**
5970, 6035, 6080, 12080 kHz 0500–0630 kHz **AF**
7170 kHz 0400–0600 UTC **AF EU**
6035, 6080, 7340, 7415, 9575 kHz 0300–0500 UTC **AF**
5995, 7170, 11805 kHz 0600–0700 UTC **AF EU**
6035, 11975, 13710, 15410, 15580 kHz 1830–2130 UTC **AF**
6035, 11920, 12040, 13710, 15410, 15445, 17895 kHz 1600–1800 UTC **AF**
6110, 7215, 15395 kHz 1400–1800 UTC **AS**
5995, 6130, 7405, 9455, 9755, 11695, 13740 kHz 0000–0200 UTC **NA SA**
6160, 7215, 9645, 9760 kHz 1530–1700 UTC **AS AUS**
7115, 9740, 11705, 15250, 15370, 17740, 21550 kHz 0100–0300 UTC **AS**
7215, 9770, 11760, 15290, 17820 kHz 2200–2400 UTC **AS AUS**
7215, 9890, 11760, 15185, 15290, 17735, 17820 kHz 0000–0100 UTC **AS AUS**
9645, 9760, 11705, 15425 kHz 1100–1500 UTC **AS AUS**
9760 kHz 1700–2100 UTC **AF AS EU**

11825, 15205 kHz 0500–0700 UTC **AS**
15185, 17735 kHz 2100–0000 UTC **AS AUS**
15205 kHz 1400–1800 UTC **AS**

Notes: The Voice of America is the official voice of the United States Information agency. It represents the views of the U.S. government and airs plenty of popular broadcasting features from locations all over the world.

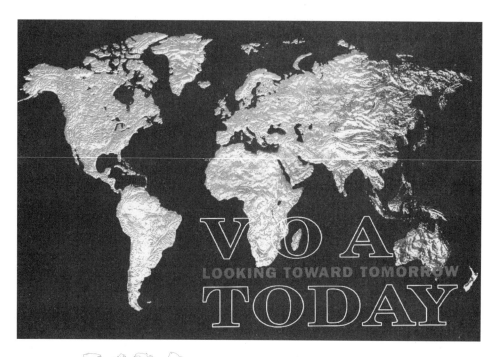

WEWN
POB 100234, Birmingham, AL 35210
205 672 7200 (tel)
70413.40@compuserve.com (Internet e-mail)
Prime frequencies: 5825 kHz 0000–1000 UTC **EU**
6890 kHz 2200–0800 UTC **NA**
7425 kHz 0000–1400, 2000–2200 UTC **NA**
9455, 11875 kHz 1400–1600 UTC **NA**
11820 kHz 2100–0000 UTC **EU**
11875 kHz 1300–1400, 1600–2000 UTC **NA**
13615 kHz 1600–0000 UTC **EU**
15665 kHz 1000–1800 UTC **EU**
15745 kHz 1800–2000 UTC **EU**

Notes: Aside from Vatican Radio, WEWN is one of the only English-language Catholic stations in the world. The station features everything from Catholic preachers to call-in Catholic question talk shows.

WGTG
POB 1131, Copper Hill, TN 37517

315 443 9237 (tel)
5085 kHz 2300–0800 UTC **NA**
9400 kHz 1100–2300 UTC **NA**

Notes: On the air in 1996, WGTG is yet another U.S. private station with paid conservative Christian programming. The station has followed WWCR's lead into the tropical bands.

WHRI
POB 12, South Bend, IN 46624
219 291 8200 (tel)
219 291 9043 (fax)
Prime frequencies: 5760 kHz 0400–1000 UTC **EU**
6040 kHz 1000–1500 UTC
7315, 7355 kHz 0800–1000 UTC
7315 kHz 0000–1000 UTC **SA**
9495 kHz 1000–1300, 1700–0000 UTC **SA**
13760 kHz 1500–2200 UTC **EU**
15105 kHz 1300–1700 UTC **SA**

Notes: Broadcasts various programs in English and Spanish. In addition to the somewhat conservative Christianity of WHRI, the station also relays plenty of programs from anti-Castro clandestines for audiences in Cuba.

WINB
POB 88, Red Lion, PA 17356
717 244 9316 (fax)
Prime frequencies: 11790 kHz 1100-1400 UTC **NA SA**
11915 kHz 2100–0000 UTC **EU**
11950 kHz 0000–1100 UTC **NA SA**
12160 kHz 1900–2100 UTC **EU**
15715 kHz 1400–2400 UTC **AF EU**

Notes: After a political dispute between the owner and the general manager in the summer of 1995, WINB was pulled from the air and was scheduled to return a few months later.

WJCR
POB 91, Upton, KY 42784
Prime frequencies: 7490, 13595 kHz 24 hrs

Notes: Primarily broadcasts Christian country and gospel music.

WMLK
POB C, Bethel, PA 19507
Prime frequencies: 9465 kHz 0400–0900, 1700–2000 UTC **NA**

Notes: WMLK, operated by the Assemblies of Yahweh, is a conservative Christian broadcaster that is not frequently reported.

WRMI
8500 SW 8th St., Suite 252, Miami, FL 33144
305 267 1728 (tel)
305 267 9253 (fax)
71163.1735@compuserve.com (Internet)

Prime frequency: 9955 kHz. Operates with various English- and Spanish-language programs on from 1100–0500 UTC.

Notes: WRMI primarily broadcasts Cuban clandestine and conservative Christian programs. The major programs on the station are La Voz de la Fundacion, Overcomer Ministries, Scream of the Chameleon, and Viva Miami.

WRNO
POB 100, New Orleans, LA 70181
504 889 0602 (fax)
Prime frequencies: 7355 kHz 2300–0400 UTC **NA**
7395 kHz 0400–0700 UTC **NA**
15420 kHz 1500–2300 UTC **EU NA**

Notes: Originally a commercial broadcaster with some paid programming and plenty of relays of the WRNO-FM album rock station, the station has become the relay outlet for many conservative and fascist political voices. The keystone of the programming has been Rush Limbaugh, but it also has featured the extremely right-wing politics of National Vanguard Radio, American Dissident Voices, and Ernst Zundel.

WSHB
POB 860, Boston, MA 02123
letterbox@wshb.csms.com (Internet e-mail)
http://www.csmonitor.com (Internet Web site)
Prime frequencies: 5850 kHz 0200–0400 UTC **NA**
6095 kHz 1000–1400 UTC **NA**
7395 kHz 0900–1200 UTC **SA**
7510 kHz 2000–0000 UTC **EU**
7535 kHz 0000–1000 UTC **AF EU NA**
9430 kHz 0000–0300 UTC **NA SA**
9455 kHz 1200–1400 UTC **NA SA**
9840 kHz 0400–0500 UTC **AF**
11550 kHz 0800–0900 UTC **AUS**
18930 kHz 1700–1900 UTC **AF**
Also broadcasts in English on: 13770 and 17510 kHz.

Notes: WSHB is the North American branch of Monitor Radio International, the radio voice of the *Christian Science Monitor*.

WVHA
POB 1844, Mount Dora, FL 32757
Prime frequencies: 5850 kHz 2200–2300 UTC **AF**
7425 kHz 1300–1500 UTC **AF**
9900 kHz 1800–2200 UTC **AF EU**
13825 kHz 0900–1100 UTC **AF**

Notes: This station was formerly known as WCSN and was one of the Christian Science Monitor stations. The new owners, the Prophesy Countdown, air right-wing Christian end-times programming.

WWCR
1300 WWCR Ave., Nashville, TN 37218
wwcr@aol.com (Internet e-mail)

Prime frequencies: 2390 kHz 0000–1100 UTC **AF AS EU NA**
3210 kHz 0500–1100 UTC **AF AS EU NA**
3215 kHz 2300–0500 UTC **AF EU AS NA**
5070 kHz 2300–1200 UTC **AF EU AS NA**
5935 kHz 0000–1300 UTC **AF AS EU NA**
7435 kHz 1200–1400, 2300–0000 UTC **AF AS EU NA**
9475 kHz 1100–2300 UTC **AF AS EU NA**
12160 kHz 1400–2300 UTC **AF AS EU NA**
13845 kHz 1300–0000 UTC **AF AS EU NA**
15685 kHz 1100–2200 UTC **AF AS EU NA**

Notes: Broadcasts a wide variety of programming, including a country music show, World of Radio, conservative politics (Radio Free America, Hour of the Time, For the People, etc.), Radio Newyork International, etc.

WYFR
290 Hegenberger Rd., Oakland, CA 94621
Prime frequencies: 5950 kHz 1000–1100 UTC **NA**
5950, 11830 kHz 1100–1400 **NA**
5985 kHz 0500–0700 UTC **NA**
6085 kHz 0000–0100 UTC **NA**
6065, 9505 kHz 0100–0500 UTC **NA**
7355, 9985, 13695 kHz 0600–0700 UTC **AF EU**
9885, 11580 kHz 0500–0600 UTC **AF EU**
9985 kHz 0400–0500 UTC **AF EU**
11550 kHz 1300–1500 UTC **AS**
11705, 11830, 17750 kHz 1600–1700 UTC **NA**
11750, 11830 kHz 1400–1600 UTC **NA**
15695, 21745 kHz 1600–1900 UTC **AF EU**
17845, 21525 kHz 2000–2300 UTC **AF EU**

Notes: WYFR is the flagship of the large American Family Radio network, which broadcasts conservative Christian programming on AM, FM, and shortwave.

United States (Voice of America)

Voice of America

In 1991, VOA closely covered the attempted August coup in the former Soviet Union. The local media had switched its programming to fillers, such as classical music and movies, with only an occasional carefully worded statement, depriving the local population of any substantive news. International broadcasts remained the only reliable source of information about what was happening in their own backyard. VOA's Russian Branch and several other language services responded to the need by immediately switching to an all-news format, with the Russian Branch broadcasting 10-minute newscasts on the hour, 24 hours a day.

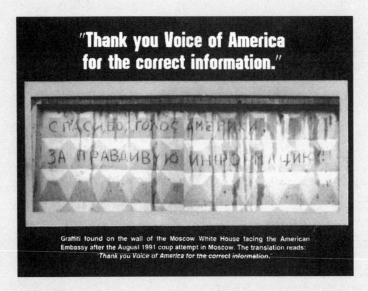

Graffiti found on the wall of the Moscow White House facing the American Embassy after the August 1991 coup attempt in Moscow. The translation reads: "Thank you Voice of America for the correct information."

In the wake of that historic event, VOA learned that its efforts were a virtual life-line for many listeners within the Soviet Union. One woman interviewed on the streets of Moscow said, "We listened to the Voice of America. It was our only link. Please give our thanks to America." At a news conference following the coup, former Soviet President Mikhail Gorbachev praised VOA and two other international broadcasters for their coverage, noting that these broadcasts were his only means of learning what was happening during his detention.

Uzbekistan

Radio Tashkent
72 Weedmore St., London W1H 9L, UK
Room 20, Numpoc Hotel, New Dehli 110001, India
Prime frequencies: 5975, 7285, 9715 kHz 0100–0130, 1330–1355 UTC **AS**
7105, 9540 kHz 2030–2100 UTC **EU**

Notes: Mail theft in Uzbekistan has been so common that several outside maildrops (including the two listed above) have been announced for Radio Tashkent.

Vanuatu

Radio Vanuatu
POB 49, Port Vila
Prime frequencies: 3330, 3945, 4960, 6100, 7260 kHz 0900–1030 UTC

Notes: Radio Vanuatu is a small station from the South Pacific. Its broadcasts are only intended for a local audience. In addition to local programming, Radio Vanuatu also relays programming from some other shortwave stations, such as Radio Australia.

Vatican

Vatican Radio
MC6778@mclink.it
Prime frequencies: 4005, 5880, 7250 kHz 0600–0620, 2050–2110 UTC **EU**
4005, 5880, 7250, 9645 kHz 1720–1730 UTC **AS AUS**
6065, 7305, 9600, 11830 kHz 2245–2305 UTC **AS AUS**
6095, 7305 kHz 0250–0315 UTC **NA SA**
7335, 9650 kHz 0140–0200 UTC **AS AUS**
7360, 9660, 11625 kHz 0500–0530 UTC **AF**
7365, 9645 kHz 2000–2030 UTC **AF**
9660, 11625, 13765 kHz 0630–0700 UTC **AF SA**
9660, 11625, 15570 kHz 1730–1800 UTC **AF**
9940, 11640 kHz 1600–1625 UTC **AF AS**

Notes: Vatican Radio is *the* source of information about the Catholic church. The broadcasts emanate directly from the Vatican, and the Pope regularly addresses his listeners.

Venezuela

Radio Nacional de Venezuela
Ap. 3979, Caracas
rnacional@rapid-systems.com (Internet e-mail)
http://165.247.176.136/radio-nacional/onda_corta.html (Internet Web site)
Prime frequency: 9540 kHz English program one hour per day.

Notes: The English program is rarely reported.

Vietnam

Voice of Vietnam
58 Quan Su St., Hanoi
84 4 255669 (tel)
Prime frequencies: 5940 kHz 0100–0300 UTC **NA**
9840, 12020, 15010 kHz 1000–1025, 1230–1255, 1330–1355, 1600–1625, 1800–1825, 1900–1925, 2330–2355 UTC **AS**
Alternate frequency to 9840 kHz is 12020 kHz.

Notes: The English Service of the Voice of Vietnam had not yielded strong signals to much of the world for years. However, the Voice of Vietnam received a real boost in 1994 after Russia's old Radio Moscow transmitters became available to lease airtime.

Yemen

Another one of the stations from the Middle East that's rarely reported in the Western radio newsletters. Republic of Yemen Radio broadcasts on 9780 kHz. The programming output is entirely in Arabic, so not many listeners outside the Middle East tune in.

Zaire

Although Zaire contains rich mineral deposits, it has been unable to process them and significantly improve the country's standard of living. This is reflected in the external shortwave service La Voix du Zaire, which is often inactive and rarely reported. Recently, though, the station has been reported with better signals on 15244.5 kHz.

Zambia

Radio Christian Voice
Private Bag E-606, Lusaka
260 1 274251 (fax)
Prime frequencies: 4965 kHz 1500–2030 UTC
6065 kHz 1300–1500 UTC

Notes: Radio Christian Voice is a relatively new station on shortwave, having hit the airwaves at the end of 1994.

Zambia National Broadcasting Corporation
POB 50015, Lusaka
260 1 220864 74 (tel)
Prime frequency: 6165, 7234 kHz 0300–2200 UTC

Notes: The ZNBC is well heard in Africa, but the signal pattern is much smaller than that of the new Radio Christian Voice.

Zimbabwe

Zimbabwe Broadcasting Corporation
POB HG 444, Highlands, Harare
263 4 795698 (tel)
263 4 795698 (fax)
Prime frequencies: 3306 kHz 2000–2200 UTC
4828 kHz 0300–0800 UTC

Notes: The ZBC broadcasts a steady schedule on various frequencies. To keep up with the English programs, check the other ZBC frequencies: 3396, 5012, 5975, 6045, and 7175 kHz. The ZBC broadcasts on shortwave are intended for local audiences, and they sound much like a local AM station.

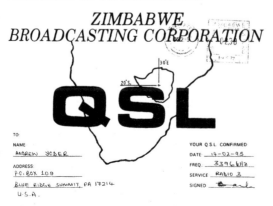

CHAPTER 8

Weird and oddball radio listening

You've enjoyed listening to the BBC, Deutsche Welle, and Radio Netherlands. You've checked out the rhetoric on the communist Radio Pyongyang. You've listened to amateur radio operators, odd sounds on longwave, and distant mediumwave catches. There's more information here than on most cable TV systems.

However, what are you missing? Is there anything out in the ether that "they" don't want you to hear? You won't find it on the cable TV system (although you might find some interesting counter-culture community-access programs). You certainly won't find it on Direct Broadcast Satellite (DSS) systems, where huge networks pay vast fees to rent channels on the satellite.

Occasionally, a subversive form of media will pop up for awhile and then quietly disappear. In the publishing world, it's "zines"—those often-photocopied little newsletters or magazines that you sometimes find in local music stores. The counterpart to this is the independent music scene, which often features plenty of controversial songs and is subversive to the seven or so major companies that dominate the market. Other possibly subversive forms of media could include political newspapers and community-access TV stations. Of course, what is deemed "subversive" varies from country to country and culture to culture. In some European countries, it's common to see women's breasts bared on TV. That's a no-no in the United States. At the other extreme, a woman in Iran had better not even be seen in public without a veil!

No matter what guidelines are used to draw "subversive," the universal medium for this type of activity is on the shortwave radio. It might seem strange that shortwave radio is such a hands-down winner for mystery and subversiveness, but there are a few good reasons:

- *Radio freedom* Many countries have "freedom of the press" to at least some extent. However, because of the finite radio spectrum, anyone who broadcasts without a license is automatically an outlaw, no matter if the programming consists of terrorism how-to talk shows or back-to-back Barry Manilow songs.

- *Shortwave knows no bounds* Unlike other broadcast medias, a small shortwave transmitter with a simple, quick-up antenna can be heard around the world.
- *Shortwave is intangible* Although shortwave equipment can be confiscated or destroyed, the message can't be. So, although copies of a supposedly subversive zine or newspaper could be confiscated at the post office, a radio signal can't be grabbed and tossed out.
- *The equipment is inexpensive* The costs vary around the world, but in North America, thousands of used amateur radio transceivers, capable of worldwide reception, are sold every year for as little as $50 U.S. in working order.

Pirate radio

On a briskly cold hillside, two young men stand, shrouded in darkness. One raises an arm and turns on the backlighting of his LED watch. "Two minutes," he says quietly. The other opens the car door and turns a switch on metal box onto the Standby position. The 25-year-old Heath DX-60B glows softly on the vinyl seat. Both unwind cables, plug them into their respective positions on the Radio Shack mixer. One turns the DX-60B onto the phone position while the other hits Play on a cassette deck. The theme from "Masterpiece Theater" interspersed with the sound of mooing cows booms through a monitor receiver. Meanwhile, part of a hardcore group of radio listeners (from as far as 1500 miles away) sit, glued to their radios, ready to copy down every detail of the program that they are about to hear through the static.

This is a fictional account of a pirate radio station signing on the air in North America, but similar sign-ons have been repeated hundreds of times in the past decade (Fig. 8-1). Just over 15 years ago, only a handful of people operated unli-

8-1 A pirate station broadcasting from out in the woods. A monitor receiver is at the left, the battery is in the center, and two transmitters and a cassette player to the right.

censed broadcast stations, and the term *pirate* was rather nebulous. Through the efforts of such stations as the Voice of the Voyager, Voice of Venus, WFAT, and KPRC, a small underground radio movement began to erupt. By the late 1980s, Radio Newyork International's brief exploits from a vessel were splashed across the front pages of newspapers across the country.

At this point, the pirate radio revolution has not yet swept the nation. For that matter, pirate radio still remains in the underground, but the tremors are growing. The movie "Pump Up The Volume," with star actor Christian Slater, spawned a new interest in community FM pirate radio. This, coupled with new, inexpensive transmitter technology, has given birth to FM "microbroadcasters" across the country. Even some commercial broadcasters have been adopting the Jolly Roger theme with the hopes that it will help draw in new advertising booty. KQLZ, a top-rated FM station in Los Angeles, added the "pirate radio" slogan for a time, and one commercial FM station in Philadelphia cruised the suburbs in a huge black "pirate ship" camper.

Local broadcasters

Literally hundreds of low-power AM and FM broadcast band pirates are active or semiactive across the United States and more are preparing to hit the airwaves. However, many of these broadcasts go unnoticed—heard only by a handful of the operators' friends. The stations often have tiny broadcast ranges. Also, the fact is that many people don't scan the broadcast bands in search of new radio stations. This combination usually dooms a broadcast-band pirate to a sentence of total obscurity.

Still, relative obscurity is fine for most of these community broadcasters. Some, like one FM pirate in western Canada, can receive over 30 phone calls in a broadcast, and yet not be challenged by the Canadian government—the town is just too remote for word of the station to ever get back to the authorities. Others broadcast professional programming that is similar to commercial radio stations. By making their station sound like the others on the broadcast band, they have the benefits of "playing radio" without getting caught. Still others use one of the many available "FM wireless broadcaster" kits that are advertised throughout most electronics magazines. These "Part 15" stations operate at very low power (100 mW) in accordance with FCC rules. Because of the low-power legality loophole, most of these Part 15 stations operate "in the open" with request phone numbers, advertisements, etc. On the other hand, some of the so-called "Part 15" stations are running excessive power or have longer-than-legal antennas, so they are generally somewhat secretive about their operations.

The local broadcast band pirates that have achieved notoriety have done so as the result of either a large publicity campaign or a duel with the FCC. In many cases, both have occurred. Three of the most widely known FM pirates in North America are Free Radio Berkeley from Berkeley, California; Black Liberation Radio from Springfield, Illinois; and KAPW from Phoenix, Arizona.

These three stations are similar in the sense that they all operate with low power (under 40 watts), all have been fined by the FCC, all solicit programming tapes from listeners, and all call for a grass-roots radio movement. KAPW was closed down in 1992 and has apparently run out of funds to pay off the lawyer fees. Black Liberation Radio has been broadcasting regularly for several years and went to a 24-hours, 7-days-per-week format about a year ago. Although Black Liberation Radio was sent a Notice of Apparent Liability, the FCC has been finding it difficult to gain leverage against Mbanna Kantanko—a blind man who has much support from within the city's black community. Black Liberation Radio has a regular newsletter, has produced a video on how to set up your own community radio station, and is currently setting up a Black Liberation Television station.

Currently, Free Radio Berkeley is riding the highest crest of mass media notoriety. The main operator has been an organizer of political protests in the Berkeley area since the late 1960s. As a result, the station's anti-corporate approach is highly organized and powerful. Free Radio Berkeley has not only been winning the media battle of words with the FCC, but they are being successfully represented by lawyers and have even set up transmitter-building workshops for would-be pirates. Free Radio Berkeley is presently the driving force behind the growing California pirate radio scene. Future plans for Free Radio Berkeley? Shortwave broadcasting.

Shortwave pirates

Shortwave radio is prime listening ground for pirates. Here, a small station can broadcast and be easily heard across a large portion of North America. Another enticing feature about shortwave radio for pirates is that the listeners are true radio fanatics—people who love radio and who regularly write letters to the stations that they listen to. Considering that the average AM or FM broadcast band radio listener is lucky to know the callsign of the station that he or she is listening to at any given time, it's no wonder that many pirates choose shortwave.

Although a number of different frequencies have been tested, the present favorite in North America is 6955 kHz (13900 and 15040 kHz are distant runners-up). In general, the best times to listen for pirates are in the 1600–2300 UTC range for 13900 and 15040 kHz and in the 2000–0400 UTC range for 6955 kHz. During the summer months, most of the pirate broadcasting on 43 meters occurs after 0000 UTC. However, during the colder months, the late afternoon propagation on 43 meters improves and signals within 500 miles of the transmitter site are usually solid (Fig. 8-2).

```
VOICE OF THE ROCK
Box 28413, Providence, RI 02908
Dear  ANDY YODER         . Thanks for reporting
   reception of our low powered experimental
station. OUR LOCATION: 42 degrees - 32' North
Latitude.     70 degrees - 47' West Longitude.
You heard us on  6955      kHz at  2344-0130  UTC
on  August 17, 1995   with 10 watts of power,
  using our Radio Animal Grenade Transmitter
              73 - Paul Art          Paul
```

8-2 The Voice of Rock runs occasional test broadcasts with only 10 watts from off the coast of Massachusetts.

A number of people have wondered why certain frequencies, particularly those near the 6955-kHz area, are "pirate frequencies." The 6200- to 6400-kHz area has been used for decades by pirate radio stations in Europe. Their North American counterparts used these frequencies about 15 years ago, until it was found that some ship-to-shore stations operated in this area. Later, stations increasingly began using frequencies in the 7350- to 7500-kHz area. Presumably, this region became a favorite because many amateur radio transmitters will naturally cover this territory, because many licensed American stations also operate here, and because it is near the highest frequency range that is active at night—allowing good coverage and the ability to operate under a cover of darkness. Later, because of increased activity from licensed broadcasters in the out-of-band area of 7300–7500 kHz, the pirates moved to 43 meters—the other side of the 40-meter amateur band.

Unfortunately for listeners in the West and South of the United States, most of the shortwave activity is in the Northeast—in a triangle that's roughly between Baltimore, Maryland; Chicago, Illinois; and Boston, Massachusetts. Although even listeners as far away as Vancouver, Canada can hear most of the North American shortwave pirate activity, the signals are generally weak and bogged down by static there. Still, with the wonders of shortwave propagation, listeners in any part of the country can at least hear fair signals from pirate stations. During the summer of 1996, approximately 15 to 20 different North American pirates were heard by numerous DXers in Europe who were willing to listen past midnight (their local time).

As you might expect from such a potentially diverse group of people, there is no set format for pirate radio stations. The music ranges from old-time country and western to punk rock; the politics range from anarchism to Nazism. The signals, technical standards, and production all range from excellent to poor.

A number of stations with interesting and diverse programming can presently be heard:

- The North American Pirate Radio Relay Service (NAPRS), one of the best-heard U.S. pirates of the 1990s, relays programming from other broadcasters.
- Radio Free Speech has a professional-quality blend of parody songs, fake ads, and Libertarian-type editorials (Fig. 8-3).
- P.J. Sparx of WREC (Radio Free East Coast) plays a variety of music and includes fake advertisements and comedy skits.
- Free Hope Experience combines UFO lore, rock music, and free speech editorials for some interesting and peculiar programming.
- WLIS (We Love Interval Signals) bases its format on interval signals from around the world.
- WRV (Radio Virus) features plenty of new "alternative" music interspersed with editorials and general information about AIDS.
- KDED (The Voice of the Grateful Dead)—all Grateful Dead music, all the time.
- Up Against the Wall Radio mostly airs a variety of music, some comedy material, and plenty of relays of other pirates.
- Radio USA airs home-produced comedy skits, punk rock music, information segments, and commentaries "from a leaky bathtub."

- Captain Squirtlong of WPN (World Parody Network) has been praised and attacked for his comedy skits, many of which have dealt with homosexuality.
- WPRS (Worldwide Pirate Radio Station) began broadcasting in the summer of 1996 with many programs of novelty music and an intercontinental signal.
- Captain Bluebeard of WARR has been heard with plenty of pro-marijuana programming and slogans such as "WARR, the war against the war on drugs."

 # Radio Free Speech News

Volume 2 - Issue 16 1996

Over 220 QSL packages in our 1st Year of broadcasting

In our first year on the air as Radio Free Speech, we have sent out over 220 QSL packages that have included, rulers, bumper stickers, pocket sized Constitutions and our Newsletters. We want to thank all of our listeners for the support, kind words and advise that you have given to us. This really wasn't our first year though. In 1992 we were on 7415khz as "Radio Free America" but we were only able to do about half a dozen shows and then had to stop. So this has been a super year for Radio Free Speech and we've added another transmitter that is a 50-watt unit that lets us broadcast on 6240 , 13900 & 15043khz. Our heartfelt thanks to the Radio Animal for all the technical assistance he has given to us and for the performance of the 10-watt AM "Grenade" transmitter that has allowed us to cover not only the United States but has been heard in Germany, England and Scotland. Special thanks to our friends, Bob the Blade, Farley Q. Fjnork, Beau the Black, Uncle Stevie and the many people who have made Radio Free Speech. - Bill O. Rights

Radio Free Speech broadcasts from the deck of the "Good ship - Defiant" usually anchored 100-miles off the East Coast of the United States of America

Radio Free Speech Relay Service busy this year!

We have been busy not only broadcasting our own programs but helping other pirates with their transmissions also. This year we have relayed programs from CELL Phone Radio, WRAY, KNBS, 6YCAT, Radio Azteca, Radio 3, Infinity Minus One, Voice of the Runaway Maharishi, Radio Free Euphoria, Radio USA and WREC. If you have a program that you'd like relayed, we'd be happy to broadcast one of your programs through our transmitters.

Here's all you do!

Send your program to: Radio Free Speech Relay Service, P.O. Box 1, Belfast, NY 14711 along with sufficient forwarding postage on a cassette tape and then sit by the radio, because we like to get them on the air as quickly as possible.

We have a new Address
P.O. Box 1 - Belfast, NY 14711

Add P.O. Box 1, Belfast, NY 14711 and delete P.O. Box 452, Wellsville, NY to your address book for Radio Free Speech. Ever since they took the porno movie house out of downtown Wellsville, the drop operator has been spending less time their. Actually, the new address is a lot easier to remember and spell. So don't forget Radio Free Speech has two mail-drop addresses. P.O. Box 1, Belfast, NY 14711 & P.O. Box 109, Blue Ridge Summit, PA 17214.
We thank the operator of P.O. Box 452 for his years of help and service to the pirate radio community!

Adolph Hitler was elected in 1923 as the chairman of the Nazi Party by one vote. Your vote does count - Bill O. Rights. 1993

Many thanks to the Mail-drop Operators for all their hard work. If it weren't for the drop-operators, there would be no QSL's or contact with the broadcasters at all.

**Radio Free Speech Rules the Airwaves
Pirate Shortwave**

Radio Free Speech
P.O. Box 1 - Belfast, NY 14711 & P.O. Box 109 - Blue Ridge Summit, PA 17214
6240 - **6955** - 6950 - 7415 - 7420 - 13900 - 15043

8-3 An issue of *Radio Free Speech News* from mid-1996.

Enforcement

As you can see from the list of station formats, the driving force of most of the shortwave pirates is the desire to air alternative radio formats. According to U.S. radio regulations, all American shortwave broadcasters must target foreign audiences. The FCC's minimum power requirements for shortwave transmitters are 50 kW. Because of the high minimum transmitter power regulation, airtime costs are also extremely large. These base costs (typically about $300 per hour) prevent nearly everyone, except religious broadcasters and foreign political coalitions, from buying airtime on American shortwave stations.

In the end, there's no alternative method to broadcasting on shortwave, except to operate without a license. This position doesn't sit well with the FCC, who must try to regulate and maintain control of the radio spectrum. The FCC not only frowns upon the activities of such radio merrymakers, but they actively attempt to close the pirates down and appear to take great pride in doing so. In the early 1990s, however, the FCC faced major budgetary cutbacks. The Commission was also forced to take on new responsibilities, such as regulating the cable television industry, which greatly restricted its power to pursue pirates.

Because they were being stretched so thin, the FCC evidently chose to intimidate would-be violators by hiking the fines. This has worked to some extent, but it has also angered many commercial broadcasters, who also have experienced a huge increase in fine amounts. However, the biggest problem with this restructuring is that the fine for irregular noncommercial unlicensed broadcasting have been upgraded to a level for that of commercial stations. As a result, the fines are currently $10,000 to $20,000 for operating a pirate radio station—higher than the penalties for multiple drunk-driving infractions in most states.

Another inconsistency in FCC decision making is that, when pirate stations in the 7350- to 7500-kHz range are closed, the FCC notes in the media press releases that this is a point-to-point communications band, not a broadcast band. Yet, the FCC has licensed more than 10 American stations to the band. One of the most blatant cases of this type inconsistency occurred with the closure of Texas pirate XERK. When XERK was closed, the FCC announced that the station's 100-watt transmitter on 7435 kHz could have caused interference to a radioteletype station from South America on 7438 kHz. However, within a few months, WWCR (World Wide Christian Radio) was licensed by the FCC to broadcast on 7435 kHz with 100 kW.

Because of these differences, the U.S. having the highest fines for unlicensed broadcasting in the free world (penalties are much higher than those in Russia) and the FCC's handling of such cases (sometimes using procedures of questionable legality), the FCC faces much negative press, lately. The pirates, meanwhile, are often seen as "freedom fighters."

Although the pirate radio community generally operates in a constructive manner and tries to air entertaining and informative programming, there have been some instances of interference. Usually, when interference occurs, two pirates air programming simultaneously on adjacent channels or on the same frequency without realizing that the other station is also on. The worst case of interference occurred in 1992, when a new station, the Voice of the Night, popped up on the air. Lad, the station's 13-year-old operator, enjoyed the attention that he received on shortwave, and he didn't seem to care if that attention was positive or negative. Before long, he was intentionally jamming a number of other pirates on the air and provoking on-air arguments. By late 1992, a combination of threats from other operators and FCC pressure forced the Voice of the Night off the air.

The Voice of the Night serves as a perfect example to the FCC and others who oppose pirate stations that these broadcasters should all be closed down. In spite of the efforts of some radio enthusiasts to set up a low-power shortwave broadcast band for noncommercial stations, it seems likely that the FCC will continue to dismiss the idea. In the meantime, pirate radio stations will continue to risk huge fines to broadcast to the masses. Although it violates federal administrative rules to broadcast, it is certainly legal to listen to and enjoy these pirate stations.

Europe

Although a good case can be made that Americans created pirate radio (the first professional broadcasts are often credited to the Americans, and two of the earliest-known true pirates—WUMS and RXKR—were American), modern-day pirate radio has been inspired by the Europeans.

While commercial radio was booming through the 1930s, 1940s, and 1950s in the United States and Canada, most of the world was subject to tight radio regulations. In many cases, commercial radio was banned; only noncommercial, government radio stations were given control of the airwaves. Thus, each country's station broadcasted the party line with little entertainment variety.

Rock music and the whole "youth culture" concept were opposed by nearly every government in the 1950s and 1960s. However, some people saw the need for this type of radio and the possibility to make money. In 1958, a Danish businessman pieced these possibilities together, bought a fishing boat, and outfitted it with radio transmitters, antennas, and radio studios. Radio Mercur was on the air with rock music for Northern Europe! In 1962, unfortunately, the Danish authorities passed laws barring their country's citizens from advertising on or working for pirate radio stations. This law legally "sank" Radio Mercur.

Just as the pirate activities were coming to a close in Denmark, a golden era of offshore broadcasting was beginning in England. The most famous worldwide offshore station began in the spring of 1964 as Radio Atlanta. This station operated from a 53-year-old German schooner formerly used by the Swedish pirate, Radio Nord. The ship was renamed *Mi Amigo* and continued to serve as a broadcasting vessel for two months until Radio Atlanta merged with another offshore pirate, Radio Caroline (Fig. 8-4). The combined operation, known as Radio Caroline, became the premier rock station in England. It was one of the first broadcasters to play the new sounds in rock music by The Beatles, The Kinks, The Who, and The Rolling Stones. Without Radio Caroline and the other offshore stations, people today might be saying "The Beatles? Who *are* they?"

Before long, about a dozen pirates—such as Radio Veronica, Radio 270, Radio London, Radio City, Britain's Better Music Station, Britain Radio, and Radio England—were operating profitably from off the coast of Western Europe. However, as the offshore pirates proliferated off the coast of a nation, that country would pass anti-broadcasting legislation. By the end of the 1960s, nearly every European nation

8-4 The crew of Radio Caroline aboard their broadcasting vessel.

passed laws against broadcasting, and the offshore pirates virtually disappeared. Radio Caroline was the only station from the 1960s that survived into the next decade—mostly because of onboard U.S. crews and advertisers.

By the beginning of the 1970s, some dedicated radio hobbyists—mostly fans of the then-departed offshore stations—thought "If it could be done offshore, why not on land?" Staffs were assembled, transmitters were built, and some land-based hobby radio operations hit the air with limited schedules. Unlike their ship-based counterparts, these stations were operated noncommercially, often as a hobby and a service to other free-radio fans. By the mid-1970s, mostly small-scale land-based pirates remained.

With station names like Radio Aquarius, Radio Valentine, European Music Radio, Radio Viking, Radio Dublin, Radio Titanic, Westside Radio, and Free Radio Service Holland (Fig. 8-5), the pirates took to the airwaves. These stations broadcasted a few hours every Sunday morning with only about 10 to 300 watts of power in the shortwave bands. In spite of the low power, limited schedule, and shortwave-only operations, the more popular of these stations brought in anywhere as many as 150 letters per broadcast—many more listener letters than most commercial broadcasters receive in a week! The popularity of these stations was so high, in part, because many of the old offshore pirate listeners felt nostalgic for the ship-board stations that had since passed and because the legal broadcasters in Europe provided little rock music.

At first, the radio governing agencies cracked down hard on the land-based pirates. Few were on the air, which made the regulars easy targets for the radio enforcement agents. When and if the stations were raided, the operators were often taken to court and fined. However, as the popularity of the land-based pirates in-

8-5 A multicolor sticker from the Free Radio Service Holland. (FRSH, providing very popular entertainment to Europe for nearly two decades.)

creased, more stations took to the airwaves. By the 1980s, Ireland and England had dozens of regular pirate stations, including a number of 24-hour FM pirates in the cities. The FM pirates were very popular. The radio enforcement agencies focused on raiding these stations, not the shortwave pirates. As a result, very few shortwave pirates have been raided in Europe since the 1980s. Germany seems to be the only country that actively pursues shortwave pirates.

The rest of the world

Unlicensed broadcasting occurs in nearly every corner of the earth. With any large population, you can guess that someone is out there broadcasting without a license. Because of the European-based culture, Australia and New Zealand both have European-style pirates, although not nearly as many of them. The best known of these stations is KIWI Radio from New Zealand, which has been heard around the world by thousands of shortwave listeners. However, the lines are much more difficult to draw between pirate radio and other forms of radio outside of the Western world.

In Central American and African countries, unlicensed left-wing FM community broadcasters have been closed down by their governments and have had their equipment confiscated. These stations fit in a gray area between pirate and clandestine radio (more information of clandestine radio is featured later in this chapter).

Another gray area is between pirate radio and "bootleg" two-way (such as amateur radio and CB radio) communications. These stations are often called "amatirs" in Indonesia (such as Radio Suara Kasih Agung and Radio Arista) and "hooligans" in Russia (such as Radio 10th and Radio Orbita). These stations often talk with other operators over the air and maybe air a few songs. The amatirs and hooligans seem to be somewhat like the North American CB pirates that pop up from time to time.

The last gray area for pirate radio is the area between it and commercial broadcasting. Some countries have almost no broadcasting regulations and virtually no radio enforcement. When "pirates" from these countries go on the air, they are considered "legit" by most everyone who hears them—even in their own country. These stations are most common in South America, where thousands of commercial radio stations are active. Occasionally, some information will leak into the DX community that one of these stations is a pirate (such as Radio Alianza from Ecuador in 1995) or perhaps one station will be closed down by the government and the DXers will then discover that it was unlicensed (such as Radio Católica International from Colombia).

The big difference between most of these stations and what is typically considered "true" pirate radio is information. Most of these station operators know nothing about the pirate radio scene in North America and Europe, so they don't air their broadcasts under the pirate radio banner. Although most of the Western pirates have connections to the radio newsletters, most of the stations in other parts of the world have no such connections. However, with the Internet reaching all parts of the world, some of the information concerning pirate radio is bound to spread and more international pirates can be expected. Already, Radio Piraña International, from South America, has conducted test broadcasts that have been heard in Argentina.

More information

Because pirate radio is booming around the world, many publications are dedicated to the topic. *Pirate Radio: The Incredible Saga of America's Illegal, Unlicensed Broadcasters* by Andrew Yoder (available from HighText, P.O. Box 1489, Solana Beach, CA 92075) is a how-to guide and a historic look at pirate radio broadcasting in North America and around the world. *Pirate Radio Operations* by Andrew Yoder and Earl T. Grey (available from Loompanics Unlimited, P.O. Box 1197, Port Townsend, WA 98368) is an in-depth look at exactly how people put a pirate broadcaster on the air. The annual *Pirate Radio Directory* by George Zeller and Andrew Yoder (available from Tiare Publications, P.O. Box 493, Lake Geneva, WI 53147) is a yearbook of the different pirates that broadcast from North America.

As far as hardcopy newsletters are concerned, *The ACE* is really the only pick in North America. This monthly bulletin contains pirate loggings, QSL information, a column on Part 15 FM broadcasting, and even a clandestine radio column. *The ACE* is available from: P.O. Box 11201, Shawnee Mission, KS 66207.

On the Internet, the unquestioned top pirate Web site is the Free Radio Network (FRN) page. This Web site features links to hundreds of smaller pirate Web sites, articles on how to successfully operate a pirate station, photos of pirates and images of different QSLs, archives of pirate newsletters (such as Free Radio Weekly and SRS News), North American pirate audio clips, and a pirate radio bulletin board so that you can contact others with similar interests. You can find it at:

```
http://www.clandjop.com/~jcruzan/pirate.html
```

Another great source of pirate radio on the Internet is the alt.radio.pirate newsgroup. This group features hundreds of recent posts about pirate radio, although much of it is dedicated to Part 15 and local FM pirate radio. After several pirates without maildrops began requesting reports to alt.radio.pirate in 1996, the newsgroup has also been filling up with recent pirate loggings.

Clandestine radio

Pirate radio and *clandestine radio* are often confused because *pirate* means "to steal" and *clandestine* means "covert." However, the two types of radio are typically very different. *Pirate radio* is hobby broadcasting, pure and simple—some politics might be involved, but they are typically just the personal beliefs of the pirate operators. *Clandestine radio*, on the other hand, is the radio voice of a revolutionary group whose primary intent is to overthrow the government (Fig. 8-6). Table 8-1 lists clandestine radio schedules.

8-6 A QSL from La Voz del CID, a longtime clandestine that broadcasts programming toward Cuba.

Clandestine radio was extensively experimented with during the Spanish Civil War in the late 1930s, but propaganda became an art form during World War II. Most of the true clandestine stations that operated during this time were "black" clandestines—broadcasters that claimed to be something very different from what they really were. For example, one German clandestine station used an ex-Scot as an announcer. He claimed to be a British sailor and would have crass conversations on the air—with the desired effect of turning the British public against their armed forces. This station and many others were featured in articles in *Popular Communications* several years ago.

These days, most clandestine stations aren't quite so sophisticated with their propaganda. Most clandestines will broadcast revolutionary music, slanted news and information, and heated editorials about why the government should be overthrown. Things become interesting when the adversarial government monitors these broad-

Table 8-1. Clandestine radio schedules (Courtesy Harald Kuhl)

UTC	Station	Target area	Language	Frequency
0000–0040	Voice of Free Tajikistan	Tajikistan	Tajik	7080 kHz
0000–0100	Radio Marti	Cuba	Spanish	6010, 9815, 15330 kHz
0030–0100	Voice of National Salvation	South Korea	English	3480, 4400, 4450, 4557 kHz
0030–0130	Thabye Radio	Burma	Vernacular	6355 kHz
0100–0200	Radio Marti	Cuba	Spanish	6010, 7365, 9580, 15330 kHz
0200–0400	Radio Marti	Cuba	Spanish	6030, 7365 kHz
0230–0330	Radio Message of Freedom	Afghanistan	Pushto, Dari	6145, 7090 kHz
0230–0245	Voice of Kashmir Freedom	India	Vernacular	5300, 5750, 6300 kHz
0230–0330	Voice of Kashmir Freedom	India	Urdu	5300, 5750, 6300 kHz
0230–0500	Voice of Rebellious Iraq	Iraq	Arabic, Kurdish	6360, 6600 kHz
0300–0340	Voice of Free Tajikistan	Tajikistan	Tajik	7080 kHz
0300–0400	Voice of the Iraqi People	Iraq	Arabic	3900–3940, 5820–5845, 7025–7045 kHz
0300–0415	Radio Voice of Sharia	Afghanistan	Vernacular	7000 kHz (variable)
0300–0500	Radio Mogadishu, Voice of Somali Pacification	Somalia	Somali	6711, 6722 kHz
0300–0700	Voice of National Salvation	South Korea	Korean	3480, 4400, 4450, 4557 kHz
0330–0425	Voice of Abkhazia	Georgia	Abkhazian, Russian, Georgian	1350, 9505 kHz (not Sundays & Fridays)
0330–0430	Voice of Iraqi Kurdistan	Iraq	Arabic	15660, 1610, 4180 kHz
0330–0500	Voice of Iranian Kordestan	Iran	Kurdish, Persian	4160, 4180–4195, 4280 kHz
0345–0500	Voice of Iraqi Kurdistan	Iraq	Kurdish	4180 kHz
0400–0600	Voice of Sudan	Sudan	Vernacular	8000, 9000, 10000 kHz
0400–0700	Radio Rutomorangingo	Burundi	Vernacular	104.0 MHz & 90m-Band
0400–1000	Radio Marti	Cuba	Spanish	6030 kHz
0420–1000	La Voz del CID	Cuba	Spanish	6305 kHz
0430–0530	Voice of the Iranian Communist Party	Iran	Persian	3870–3910, 6400–6420 kHz (Fridays)
0450–0900	Voice of the Resistance of the Black Cockerel (VORGAN)	Angola	Portugese, Vernacular	9755 kHz

Table 8-1. Continued

UTC	Station	Target area	Language	Frequency
0500–0530	Voice of Abkhazia	Georgia	Abkhasian, Russian, Georgian	1350, 9505 kHz (Fridays)
0500–0600	Voice of Iraqi Kurdistan	Iraq	Kurdish	4180 kHz
0530–0630	Thabye Radio	Burma	Vernacular	6355 kHz
0600–0640	Voice of Free Tajikistan	Tajikistan	Tajikistan	7080 kHz
0730–0830	Radio Message of Freedom	Afghanistan	Pushto, Dari	6145, 7090 kHz
0730–0930	Voice of Rebellious Iraq	Iraq	Arabic, Kurdish	576 kHz
0900–0940	Voice of Free Tajikistan	Tajikistan	Tajik	7080 kHz
0900–1000	Voice of Independent Kurdistan	Kurdistan	Kurdish, Turkish	7020–7030 kHz
0930–1200	Radio Mogadishu, Voice of Somali Pacification	Somalia	Somali	6711, 6722 kHz
0950–1030	Radio Free Bougainville	Bougainville	Vernacular	3850 kHz
0950–1030	Radio United Bougainville	Bougainville	Vernacular	3880 kHz
1000–1200	Radio Marti	Cuba	Spanish	6030 kHz
1000–1215	Radio Free Somalia/Voice of Free Somalia	Somalia	Somali, English	7215 kHz
1000–1700	Voice of National Salvation	South Korea	Korean	3480, 4120, 4400, 4450, 4557, 6020 kHz
1030–1200	Voice of Iraqi Kurdistan	Iraq	Kurdish	4180 kHz
1030–1230	Thabye Radio	Burma	Vernacular	6355 kHz
1050–1430	Voice of the Resistance of the Black Cockerel (VORGAN)	Angola	Portugese, Vernacular	11830 kHz
1100–1130	Democratic Voice of Burma	Myanmar	Burmese	15170 kHz
1100–1140	Voice of Free Tajikistan	Tajikistan	Tajik	7080 kHz
1200–1400	Radio Marti	Cuba	Spanish	9565, 11815 kHz
1200–1430	Voice of Rebellious Iraq	Iraq	Arabic, Kurdish	6030–6090, 6075 kHz
1230–1300	Radio Free Somalia/Voice of Free Somalia	Somalia	Somali	13820 kHz

Weird and oddball radio listening

UTC	Station	Target area	Language	Frequency
1300–	Radio of the Saudi Opposition from Najd and Hijaz	Saudi Arabia	Arabic	11785 kHz
1300–	Holy Medina Radio	Saudi Arabia	Arabic	11785 kHz
1400–1500	Voice of Independent Kurdistan	Kurdistan	Kurdish, Turkish	7020–7030 kHz
1400–1500	Radio Marti	Cuba	Spanish	11930 kHz
1400–1530	Voice of Iranian Kordestan	Iran	Kurdish, Persian	4160, 4180–4195, 4280 kHz
1430–1500	Democratic Voice of Burma	Myanmar	Burmese	11850 kHz
1430–1515	Voice of the Worker	Iran	Persian	4190 kHz
1430–1530	Voice of Abkhazia	Georgia	Abkhasian, Russian, Georgian	1350, 9505 kHz
1500–1700	New Star Broadcasting Station	China PR	Chinese	8300 kHz
1500–1700	Voice of the Struggle of Iranian Kordestan	Iran	Kurdish	4345–4355 kHz
1500–1800	Radio Marti	Cuba	Spanish	11745, 11815, 11930 kHz
1500–1800	Radio Mogadishu, Voice of Somali Pacification	Somalia	Somali	6711, 6722 kHz
1500–1830	Voice of Rebellious Iraq	Iraq	Arabic, Kurdish	6360, 6600 kHz
1530–1730	Voice of Kashmir Freedom	India	Vernacular	5300, 5750, 6300 kHz
1600–1700	Voice of Eritrea	Eritrea	Arabic, Tigrigna	7151.4, 17740 kHz
1600–1715	Radio Free Somalia/Voice of Free Somalia	Somalia	Somali, English	3920 kHz
1600–1700	Voice of Oromo Liberation	Ethiopia	Oromo	5960 kHz (Monday, Wednesday, Saturday)
1600–1800	Voice of Arab Syria	Syrien	Arabic	1377 kHz
1600–1800	Voice of the Ahwaz Revolution	Iran	Arabic	1044 kHz
1600–1900	Mother of Battles Radio	Middle East	Arabic	693, 1377 kHz
1600–2100	Radio Mogadishu (Aydid)	Somalia	Somali, Arabic, English	6870 kHz USB

Table 8-1. (Continued)

UTC	Station	Target area	Language	Frequency
1630–1715	Voice of the Worker	Iran	Persian	4190 kHz
1630–1900	Voice of the Islamic Movement in Iraqi Kurdistan	Iraq	Kurdish, Arabic	4380 kHz
1645–1800	Voice of Iraqi Kurdistan	Iraq	Kurdish	4180 kHz
1650–2100	Voice of the Resistance of the Black Cockerel (VORGAN)	Angola	Portugese, Vernacular, English	7090 kHz
1700–1800	Voice of the Iranian Communist Party	Iran Iran	Persian	3870–3910, 6400–6420 kHz
1715–1915	Voice of Sudan	Sudan	Vernacular	8000, 9000, 10000 kHz
1730–1830	Voice of the Iraqi People	Iraq	Arabic	3900–3940, 5820–5845, 7025–7045 kHz
1700–1900	Iraqi Army Radio	Iraq	Arabic	1530 kHz
1800–1900	Voice of Iraqi Kurdistan	Iraq	Kurdish	4180 kHz
1800–1900	Voice of Palestine	Israel, Arabia	Arabic	1377 kHz
1800–	Republic of Iraq Radio, Voice of the Iraqi People	Iraq	Arabic	9570, 13676, 15133 kHz
1800–0100	National Radio of the Saharan Arab Democratic Republic	Western Sahara	Arabic	1544, 11610 kHz
1800–2300	Radio Marti	Cuba	Spanish	11930 kHz
1900–1930	Voice of Iraqi Kurdistan	Iraq	Arabic	4180 kHz
2000–2300	Mother of Battles Radio	Middle East	Arabic	7151.4, 13650, 15240, 15340 kHz
2000–0030	Voice of National Salvation	South Korea	Korean	3480, 4400, 4450, 4557 kHz
2130–2330	Voice of Rebellious Iraq	Iraq	Arabic, Kurdish	576 kHz
2200–2300	Voice of Free Sahara	Western Sahara	Arabic, Spanish	9640, 15215 kHz
2300–0000	Radio Marti	Cuba	Spanish	9525, 9815, 15330 kHz
2300–0200	Radio of the Provisional Government of National Union and National Salvation of Cambodia	Cambodia	Khmer	5408 kHz

casts and decides to take action. Aside from locating the station and destroying it (which is often impossible because the station is probably either located in an unfriendly, adjoining country or is within the country, behind rebel lines), the most common method of counteracting the broadcasts is to jam them. *Jamming* entails transmitting on the same frequency, with the intention of making the target station difficult or impossible to understand. Interestingly, different types of audio are used on jammer stations—often anything from warbling tones (such as "bubble jammers") to various local or international broadcasts.

Sometimes, the clandestine will jump frequencies to try to avoid the interference. During the early 1980s, it was interesting to listen to Radio Venceremos (El Salvador) broadcasting. They would begin on a particular frequency, then a jammer would wipe them out. Radio Venceremos would move to a nearby frequency to escape the jamming, but within a few minutes, the jammer would move and wipe them out again.

More interesting than jamming for the shortwave listener is when the opposing government sometimes broadcasts a black clandestine that copies the style of the original clandestine station. Typically, the station will try to sound exactly like the real clandestine station, right down to playing the same theme music. However, the editorial might sound arrogant, the announcer might make a few truly tactless comments, or the information presented might be misleading. Black Radio Venceremos stations were occasionally reported in the 1980s; many of these types of stations are reported with broadcasts to North and South Korea.

Depending on where you live, clandestine radio listening can be difficult. Clandestine radio stations are, for obvious reasons, best heard in unstable regions of the world. For decades, the Middle East, Central America, and Southeast Asia have been hotspots. Because of the regions where these problems are occurring, few clandestines broadcast in English. At this writing, one of the only English-language clandestine stations is Radio Liberia, which is operated by one of the powerful military factions. Radio Liberia is well-heard in Europe, Africa, and the eastern Americas on 5100 kHz around 2200 UTC, but the station, time, and frequency could literally change at any time.

For an easy catch of some clandestine radio programs, try listening to some of the U.S. private commercial radio stations, such as WRMI on 9955 kHz and WHRI on 7315 kHz. Both of these broadcasters air programs from various clandestine groups—especially those that broadcast toward Cuba. Many of these programs are produced by revolutionary groups. One, La Voz de Alpha 66, operated via its own transmitter and was raided several times in Florida. Whether these programs are true clandestines is debated within the radio hobby because they just buy time on other radio stations. Regardless of what you think about whether or not these programs are true clandestines, they provide listeners with a good taste of what clandestine radio is all about. In the nature of true clandestine atmosphere, the programs are typically jammed by transmitters from Cuba.

The sidebar in this chapter was written by Don Moore. It was originally printed in the April 1989 issue of *Monitoring Times*. It might seem a bit excessive to focus on a Guatemalan clandestine station that hasn't been on the air in more than 40 years, but it depicts the ultimate in clandestine radio. This story is the ultimate in shoestring radio, and it is a best-case (or worst-case, depending on your point of view) scenario for the effectiveness of radio propaganda. Considering all of the made-for-TV movies and other big-screen productions on the market, it amazes me that no one has produced a movie about the La Voz de Liberacion story; it is more fantastic than just about anything out of Hollywood.

The Clandestine Granddaddy of Central America

Don Moore

To DXers, the 1980s have been the era of the Central American clandestines: Radio Veneremos, Radio Quince de Septiembre, Radio Farabundo Marti, Radio Liberacion, the list seems endless. The political situation never seems to really change, and the stations are there month after month to be logged. Optimally, a political clandestine station gets its job done fast, and then leaves the air, victorious. For that type of success, today's stations have a role model in Central America's first political clandestine. In 1954, the granddaddy of them all came on the air, overthrew a government almost single-handed, and then left the air just two months later. Its story is not well known. However, perhaps at night, on the mountainsides and in the jungles of Central America, the announcers at Radio Venceremos or Radio Quince de Septiembre sit around the fire and talk about La Voz de la Liberacion.

Guatemala, Central America's most important and populous nation, has an unfortunate history of sometimes cruel, sometimes odd, dictators. In 1931, the country was taken over by General Jorge Ubico. One of Ubico's favorite pastimes was to ride around the country on a motorcycle, with a machine gun strapped around his back. In other ways, he was the stereotype of banana republic dictators: anyone who crossed him or violated even the most minor of his laws might just be pushed against an adobe wall and shot. Thousands were. Still, Ubico had his good points: one of his hobbies was shortwave radio, and he preferred using shortwave, instead of the telephone or telegraph, whenever sending messages to officials around the country.

Presumedly it was Ubico's violent one-man rule, not his shortwave hobby that led to his overthrow in 1944. Following massive protests by schoolteachers and students, Ubico was forced to resign and hand over the government to several left-wing army officers, headed by Colonel Jacabo Arbenz. In 1945, elections were held and rule of the country was turned over to a civilian government. During the next elections, in 1950, Arbenz, just 37 years old, ran for the presidency and won handily. His role in the coup of 1944 had not been forgotten.

In the 1950s, most of the countries of Latin America were controlled by right-wing military dictatorships. Many liberal civilian politicians were not allowed to live freely in their own countries. One of Arbenz's first acts was to open Guatemala's doors to political exiles from all over Latin America. However, not only were liberal politicians allowed in, but so were hundreds of exiled Communists and revolutionaries. Although Arbenz said that this was because he believed all men had the right to live freely, regardless of their beliefs, not everyone believed him.

Meanwhile, in the Guatemalan congress, Arbenz was supported by a 51-member coalition that included the four Communist Party representatives. As part of the coalition, Guatemalan communists were given several minor posts in the Arbenz government, mainly in the Agriculture Department. With McCarthyism at its height in the United States, Washington began to keep a watchful eye on Guatemala.

Taking on a Fruit Company

Now Arbenz did something no Guatemalan president before him had ever done; he decided to take on the United Fruit Company. The largest investor in Guatemala, the company was so powerful, that few dared to tangle with it. United Fruit was more than just banana plantations. The only transportation between the interior of the country and the Caribbean coast was United Fruit's railroad line between Guatemala City and Puerto Barrios. The railroad charged the highest rates in the world. United Fruit also owned the only port facilities on Guatemala's Caribbean coast. Arbenz angered United Fruit when he announced that he would give their monopolies some competition, by building a road alongside the railway and constructing a new Caribbean port. Then, in another move, Arbenz forced the company to give severance pay to hundreds of laid-off workers.

Arbenz's disagreements with United Fruit did not stop there. A priority of his government was to give land to Guatemala's hundreds of thousands of landless peasants. There was no question where much of that land would come from: the country's biggest landowner was the United Fruit Company. The company held over half a million acres; 85% of it was uncultivated. In mid-1952, Arbenz issued a decree that all uncultivated land in the country was subject to government seizure so that it could be given to landless peasants. In early 1953, about 200,000 acres of uncultivated United Fruit land was confiscated. Arbenz did plan to pay for the land. Showing that he had a sense of humor, he offered to pay United Fruit exactly what the company said the land was worth—according to the value that the company declared on its tax reports. Arbenz was well aware that the company had been cheating on its taxes for years by declaring the land at only about 4% of its true value. United Fruit was furious.

The U.S. Steps In

The United Fruit Company had its contacts in Washington. John Foster Dulles was Secretary of State, and his brother Allen Dulles was head of the CIA. The Dulles family had extensive business contacts with the United Fruit Company, so the brothers were aware of what was happening in

Guatemala. Assistant Secretary of State for Interamerican Affairs, John Moors Cabot was a stockholder in United Fruit. That watchful eye on Guatemala began to look even closer. In August 1953, the decision was made: Arbenz must go. Allen Dulles brought in some of his best covert action specialists for the task ahead. "Operation Success" had begun.

The CIA had quite a job ahead of it; very few Guatemalans were actually trying to overthrow Arbenz. Because of his land reform program and support for trade unions, the peasants and workers were generally behind him. The middle class, which had neither gained nor lost under Arbenz, was at least willing to tolerate the president until the 1955 elections. Following the 1944 coup, the army had gradually been purged of conservative officers so that those who remained either supported Arbenz or were neutral. Those Guatemalans who did oppose Arbenz were generally free to do so within the established political system. They saw no reason for violence.

Considering all these factors, it's a wonder that "Operation Success" wasn't named "Operation Failure" instead. However, then the CIA had a deep bag of tricks to reach into, and out of it they pulled a World War II propaganda technique called "The Big Lie." Radio would play an important part in this battle.

The key to the plan was psychological warfare. The Guatemalan people had to be convinced that Arbenz no longer controlled the country. This would be accomplished by clandestine radio broadcasts and propaganda leaflet airdrops. Meanwhile, a small military force would be raised to invade Guatemala from a neighboring country. Propaganda would be used to convince the country that this invasion was only a small part of a much larger force of exiled Guatemalans opposed to Arbenz. Other dirty tricks would be used to further confuse and demoralize the population.

It was no secret that the U.S. government was unhappy with Arbenz. For example, the United States Information Agency planted over 200 anti-Arbenz articles in the Latin American press during this time. However, Operation Success had to be done covertly, without any apparent connection to the U.S. government. Not only would such a connection be politically embarrassing to the U.S., but the Guatemalans might realize what was happening and not buy the propaganda. The operation had to take place outside of the United States, and as discreetly as possible.

By early 1954, Operation Success was well underway. Nicaraguan dictator Anastasio Somoza, a staunch enemy of Arbenz, readily agreed to let his country be used as a training base. Guatemalan Colonel Carlos Castillo Armas was brought in to head a rebel "Army of Liberation." Castillo Armas had been exiled after organizing an unsuccessful military coup in 1950. Since then, he had been making a living as a furniture salesman in Tegucigalpa, Honduras. His "army" consisted of about 150 men, a mixture of Guatemalans opposed to Arbenz and Hondurans, Nicaraguans, and American soldiers of fortune, in it for the excitement and the money. Americans and Nationalist Chinese pilots were recruited for the rebel air force.

La Voz de Liberacion

Before any invasion could take place, the country had to be psychologically softened up. Therefore it was important to put the rebel radio station on the air as soon as possible. CIA technicians set up a complete radio base camp on a remote Nicaraguan farm. Additional transmitters were located in Honduras, the Dominican Republic, and even in the U.S. embassy in Guatemala City. Although it was never used, a reserve transmitter was set up on Swan Island (which seven years later would be the site of the CIA's famous anti-Castro clandestine, Radio Swan). Not all of these transmitters were for La Voz de Liberacion. Other uses included fake military command stations and jamming Radio Nacional de Guatemala (TGW) and other Guatemalan radio stations.

Covert action specialist E. Howard Hunt (now well-known for his involvement in the Watergate scandal) was brought in to head the propaganda campaign. David Atlee Philips was appointed his deputy and made head of the radio station. For actual on-air announcing, five Guatemalan men and two Guatemalan women were recruited. The Guatemalans were lead by announcers Mario Lopez Otero and José "Pepe" Toron Barrios.

In early April, 1954, the group was brought together in Florida for technical training at the U.S. military base in Opa Locka. To keep the announcers happy, the men's girlfriends were even flown in for a weekend visit from Guatemala. Their end of training was celebrated with a night on the town in Miami, courtesy of Howard Hunt's expense account. In mid-April, they flew to Managua and a few days later they were out at the radio camp—a barn for the transmitters and studio, and an old "shack" to live in. They had two weeks to finish setting up the station, begin to record programs, and get ready for the hard two month's work ahead of them.

Programs were designed to appeal to patriotism and the base values of the society. The slogan "Trabajo, Pan, y Libertad," or "Work, Bread, and Country" was adopted, to identify with these values. To appeal to all sectors of society, special programs were produced for women, youth, workers, soldiers, army officers, and the elite. The last two groups were especially important. Without ensuring that they would at least be neutral and inactive, the invasion would be doomed to failure.

Advertising Clandestine Radio

The first day of broadcast was scheduled for May 1, International Labor Day. With everybody taking the day off from work, there would be a huge potential audience—if only people knew about the station. Certainly letting its potential listeners know that it is on the air is a problem for any clandestine radio station. After all, an underground radio station can't advertise in the local newspapers. Well, on second thought, maybe it can advertise in the local papers—because Mario and Pepe did just that for La Voz de Liberacion!

A few days before the broadcast, half-page ads were placed in each of Guatemala's daily newspapers. The ads were for a special holiday broadcast from Mexico on shortwave. The program would feature some popular Mexican singers, a famous actress, and well-known Mexican comedian Cantinflas. Of course, the program's time and frequency were included.

When the listeners tuned in, they found the program to be not quite what was advertised. The famous stars were there all right, but on record. Mario and Pepe apologized and explained that the lie was their only way of letting the public know about the initial broadcast. The listeners didn't mind; political intrigue can be a lot more fun than Mexican singers. Here was a station that not only denounced the president, but it claimed that he would soon be overthrown by rebels. Of course, after just one broadcast, very few people took La Voz de Liberacion seriously. Still, the following day Arbenz made a speech on Radio Nacional, TGW, denouncing the station. Any doubts people had as to the seriousness of the rebels were dismissed when the CIA jammers turned on and drowned out Arbenz's speech. Starting day two, La Voz de Liberacion had a regular audience. Even Arbenz, himself, tuned in daily!

The Big Lie Begins

The role of La Voz de Liberacion was quite clear. First, the station had to mobilize those Guatemalans who were opposed to Arbenz into action. Then it had to persuade those who were neutral, that opposing Arbenz would not be such a bad idea, if they wanted to be on the winner's side. When a revolution is in the air, everybody wants to go with the winner. Finally, La Voz de la Liberacion had to persuade those Guatemalans who supported Arbenz that all was already lost and that there was no reason to continue the fight.

To carry this out, La Voz de la Liberacion had to convince the Guatemalan people that Arbenz could not effectively control the country. One way La Voz de Liberacion did this (and also covered up their true identity) was by announcing that the station was broadcasting from the mountains outside Guatemala City. After all, as Mario and Pepe pointed out to the listeners, if Arbenz's army can't find and close down a little clandestine radio station, how can they stop Castillo Armas when he invades the country?

To validate this claim, one night gunshots and screams interrupted the broadcast. The announcers shouted "They've found us," and took off out of the studio, just as soldiers burst through the door yelling "Hands up!" Of course, because the station was in Nicaragua, the Guatemalan army was nowhere near it. However, the ruse worked so well that Guatemalan officials monitoring La Voz de Liberacion believed it. Later that evening, the government radio station, TGW, announced the army had found and closed down La Voz de Liberacion. Now there was no question, either in the eyes of the populace, or the foreign press, that La Voz de Liberacion had really been broadcasting from the Guatemalan mountains. After all, the government radio station itself had said so.

The next day the station returned to the airwaves. Mario and Pepe said thanks to the bungling of Arbenz's soldiers and the bravery of the rebels guarding the station, they had narrowly escaped the trap. Now the station was broadcasting from a new and more secure site. However, because of the imminent danger that they might be caught again, the women announcers would no longer be working at the station.

Radio Grounds the Air Force

Although air support is the key to most modern military operations, the CIA could only supply a few obsolete bombers to the "Army of Liberation." For them to have anything more modern would be like putting a "Made in the USA" banner on the invasion. Yet, there was no way these planes could face up in combat against the Guatemalan Air Force's up-to-date fighters. The Guatemalan Air Force was the biggest factor standing in the way of a successful invasion, because it would control the skies. Not only would government planes be able to freely bomb and strafe the rebels, but, more important, by simply flying over them, the air force could report back on how small and insignificant the invasion really was. If modern planes could not be sent to take care of the Guatemalan air force, something else would have to do it. That something was La Voz de Liberacion.

The station started airing programs praising and telling about courageous Soviet pilots who defected by flying their planes to the west. No direct appeals were made to Guatemalan pilots, but it worked. On June 5, Air Force Colonel Rodolfo Mendoza Azurdia defected, flying his plane to nearby Nicaragua.

Soon after, Mendoza was brought out to the station for a visit. He was asked to do a special broadcast and call for his fellow pilots to defect. Not wanting to cause any hardships to his family, which was still in Guatemala, he refused. Mario and Pepe told him that was okay, they understood and invited him to share dinner and a bottle of scotch with them that evening.

Mario and Pepe made sure that Mendoza drank more than his share of the scotch. Soon the pilot was drunk. Praising his bravery, the two announcers said it was a shame he couldn't give a speech on their station. However, if he did, what would he say, how would he say it? With the persuasion of the bottle to support him, the intoxicated aviator launched into a impassioned speech, putting Arbenz down and telling his fellow pilots how and why they should defect. Each time he started to falter and lose interest, Mario and Pepe asked him more questions, so that he continued in his heated discourse. Finally though, Mendoza was talked out. The scotch took over, and he began snoozing on the floor. Mario and Pepe went over to an old sofa and took out the tape recorder that they had hidden under the cushions. Back in the studio, it just took a little work to cut out their questions and splice the pilot's comments into a coherent, but lively, speech, ready for broadcast the next morning.

It worked perfectly. Arbenz was convinced that, given the chance, more of his pilots would defect with their planes. He ordered the Air Force grounded—and not a single Air Force plane was permitted to take off for the duration of the crisis.

The Air War Starts

Now the skies were safe, and Castillo Armas' air force could go to work. From Tegucigalpa, Honduras, cargo planes took off regularly to drop propaganda leaflets over the capital and principal towns. La Voz de Liberacion played its part in the air war, each night airing announcements instructing the planes where to drop supplies for nonexistent rebels in the mountains. Pleas were made for listeners to help the rebels by locating potential drop sites. Occasional drops were even made so that local people would find the supplies and report them to the government. This created still more uncertainty as to Arbenz's ability to control the countryside.

Even more tension was created when Arbenz decreed a nightly blackout in Guatemala City. The official reason for the blackout was to prevent rebels from bombing the city, as had been threatened on La Voz de la Liberacion. Some thought Arbenz was really trying to make it harder for people to listen to La Voz de Liberacion. If so, it wasn't a very well thought-out plan, because many Guatemalans had either battery radios or their own electrical generators.

Regardless of Arbenz's reasoning, Mario and Pepe found ways to use the blackout to their advantage. Listeners were requested to place lighted candles on their patios, to help the rebel air force find Guatemala City at night. It was explained that this was necessary if the pilots were to be able to orient themselves in their supply drops to the rebels in the hills. Many listeners believed this, and thousands of candles were placed on patios.

The following day, the Arbenz government announced that lighting candles was prohibited. Mario and Pepe still weren't finished, however. The next night, they were on the air, thanking listeners for helping the rebels by lighting candles. This would make the pilots' job very easy, they explained, when the rebels decided to bomb the military bases. Because their supporters were everywhere, the military bases were the only places without candles. All the pilots would have to do would be to look for the dark areas and bomb those. The next night candles blazed all over the city, including the army camps!

Taking Care of the Army

Even with the air force grounded, the CIA's little rebel force was no match for the 6000-man Guatemalan army. Something had to be done to make sure it never came down to a real battle. The break came when CIA agents learned that Arbenz was considering arming the peasants and trade unions who supported him. Arbenz did not totally trust his army, and he wasn't sure how many rebels he was facing. The extra troops could be useful.

However, what might have been a good idea to start with, turned into a disaster when Howard Hunt and David Atlee Philips found out. The rebel air force was called on to drop leaflets over Guatemala City and other large towns, saying that this was an insult to the army and that it was just the first step of Arbenz's plan to destroy the army and replace it with a civilian militia. Fearing for their future, army officers began to wonder what Arbenz was really planning, and Arbenz started to distrust his officers even more. He would keep the army in the barracks until it was all over.

The Invasion

On June 18, 1954, Castillo Armas and his rebel army crossed the border between Honduras and Guatemala, right on schedule. Castillo Armas lead the invasion, riding in an old station wagon, while his 150 soldiers followed behind in several rundown cattle trucks. They drove a few miles to the border town of Esquilpulas, then set up camp. No one opposed them. That night, La Voz de Liberacion announced that the vanguard of Castillo Armas' army had crossed the border and captured Esquipulas after a fierce battle. Mario and Pepe went on to say that, from their location near Guatemala City, they were unable to confirm the rumor that Castillo Armas had 5000 men.

Now the CIA began launching occasional bombing and strafing raids from Puerto Cabezas, Nicaragua. Bombs were dropped on military bases around the country and on the port at Puerto Barrios, but none yet on the capital city. Sometimes, when bombs ran low, the pilots would drop empty soda bottles. The noise they made when hitting the pavement sounded just like a bomb going off. Guatemalans began referring to the bombings as *sulfatos* (laxatives) because of the effect that they supposedly had on government officials. Actually the bombings probably had that effect on anyone nearby!

The war was at a standstill. Castillo Armas and his men settled down in Esquipulas; they were too few to continue the invasion and, for the moment, their work was done. Meanwhile the Arbenz government was confused. There was no reliable communication with the border area, and Arbenz refused to let the army go to fight the rebels. Sometimes it seemed the only real news the government could get was from the rebel radio station—and none of it was good. Arbenz sat tight and kept his army in Guatemala City.

Mario and Pepe continued their tricks. One favorite ploy was to use disinformation to start rumors, such as announcing that there was no truth to the rumor that the water of Lake Atitlan had been poisoned. Other times, they would go on the air on a frequency very close to that of the government station, TGW, and mimic the station and put out false announcements to confuse the listeners. La Voz de la Liberacion also broadcast messages to fake rebel camps and reports of fierce battles that never happened.

For weeks, the CIA had been monitoring and noting frequencies used for Guatemalan army radio communications. Now they put this knowledge to use by broadcasting false commands and announcements on these frequencies, thoroughly confusing the army and government. Even the U.S. Embassy helped in starting rumors, as embassy staff called up Guatemalan friends and asked them questions such as "Is it true that Zacapa has fallen to the rebels?" Still, though, the stalemate continued.

The Final Countdown

Now it was time to get serious. On Friday, June 25, for the first time, bombs were dropped on the army base outside Guatemala City. The noise and smoke convinced inhabitants of the nearby city that the end was near. Thousands began to flee, blocking all the roads leaving town. On Sunday morning, June 27, La Voz de la Liberacion was on the air, announcing that two large columns of rebels were approaching Guatemala City. Appeals were broadcast, asking the refugees to get off the roads and let the rebel trucks pass. Mario and Pepe spent the day broadcasting news of troop movements, redeploying hundreds of fictitious rebel soldiers. Guatemala City was totally in panic. Meanwhile, Castillo Armas and his 150 rebels were still relaxing in Esquipulas. Their only chance for success was if La Voz de Liberacion's propaganda broadcasts over the past two months had done their job so that everyone would believe this final big lie.

Sunday night, at 9:15 pm, Arbenz went on Radio Nacional, TGW, to address the country. More Guatemalans were probably listening to La Voz de la Liberacion than to TGW, and those who were listening to TGW had to put up with the jamming. Arbenz summed up the situation the country was in and blamed the United States for backing the rebels who had invaded the country. He then said that he had decided the only way to restore peace to the country was for him to resign from the presidency. He was going into exile in Mexico and would turn the government over to his friend, and Army chief of staff, Colonel Carlos Enrique Diaz.

For the next few days, the scene of action was Guatemala City. Diaz and other officers formed and dissolved juntas daily, trying to find one that would suit the U.S. ambassador and be recognized by the United States. The only solution was to allow Castillo Armas a position in the government. Castillo Armas and his troops flew into Guatemala City. After seeing how insignificant the rebel army really was and realizing how easily he could have defeated it, Diaz went home and cried for several days. Meanwhile, with a few more days of political maneuvering, guided by the U.S. ambassador, Castillo Armas became sole president of Guatemala.

The war was over, La Voz de la Liberacion had won. It was much easier than anyone had believed possible. David Atlee Philips, the CIA head of the clandestine station was listening when Arbenz made his speech. Philips said he fully expected Arbenz to tell the people about how the invasion was a farce and to announce that everything was under control. That's all he would have had to do, and the invasion would have been crushed. Philips couldn't believe that Arbenz (and all the Guatemalan government) had been so taken in by the station's propaganda, and he was shocked by Arbenz's resignation. This was the man who ran the radio station that had brought the resignation about.

Aftermath

It's work a success, La Voz de la Liberacion shut off its transmitters forever. The transmitters probably found their way to other battlegrounds around the world. However, for most of the people involved, there was no happy ending.

Arbenz spent the next 10 years moving around Europe and Latin America, before being granted permanent residency in Mexico in 1965. He died there in 1970, by drowning in his bathtub. Howard Hunt, of course, went on to become a household name in the United States, after Watergate. David Atlee Philips stayed with the CIA until 1974, when he resigned, critical of the agency's workings. Since then he has written books on the CIA. Castillo Armas proved to be a corrupt ruler and, in 1957, was assassinated by one of his own bodyguards. His was the first in a long string of military governments in Guatemala, finally ending in 1986. Mario and Pepe became victims of the political violence that began in Guatemala in the 1960s and that continues to today. Going to work one morning, Pepe was shot down in front of his family. Not long afterwards, Mario was machine gunned in a supermarket parking lot.

For the CIA and the U.S. government, success in Guatemala probably came too easy. Seven years later, David Atlee Philips was brought in to run Radio Swan, in preparation for the Bay of Pigs invasion of Cuba. Many other agents who had worked with the Guatemala operation also were also brought in to help out. The Bays of Pigs, though, was as big a failure as Guatemala was a success.

There are numerous theories as to why the Bay of Pigs was a disaster. Perhaps part of the reason was one of the exiled Latin American communists living in Guatemala in 1954, a young Argentine doctor named Che Guevara. He watched what happened, learned, and when the end came, took off for Mexico. There he met and became friends with Fidel Castro. A few years later, Castro was the leader of Cuba, and Guevara his second in command. When Radio Swan came on the air, Guevara knew what was happening. He had been through it all before.

More information

For more information on clandestine stations of the 1980s and 1990s, check out *Shortwave Clandestine Confidential* by Gerry Dexter and *The Clandestine Broadcasting Directory* by Mathias Kropf (available from Tiare Publications, P.O. Box 493,

Lake Geneva, WI 53147), both of which take a focused look at the many clandestine radio organizations around the world.

For recent clandestine schedules on the Internet, see The Clandestine Radio Broadcasting Web page at:

`http://up4c03.gwdg.de/~kuhl/cla`

Also be sure to check The Clandestine Black Book at:

`http://nickel.ucs.indiana.edu/~lpinhey/undgr.html`

Numbers stations

The last, and certainly least interesting, of the mystery stations on shortwave are often known as *spy numbers stations*. If you tune around the shortwave bands for any length of time, you will cross some stations that only broadcast numbers in groups of four or five over and over again. After 20 or so minutes of these numbers ad nauseum, the announcer will say "finalé, finalé" and the transmission will end.

What did you just hear? Was it a spy communicating back to her government in a code? Was it a drug dealer? Was it someone communicating with alien spacecraft? Or was it simply a transmitter test? No one in the shortwave hobby really knows what these transmissions are used for, but everyone suspects that governments use these stations to send secret coded messages. A number of books about spying and espionage have somewhat similar accounts of how the secret agents receive their plans in the field. They pull out a shortwave radio and their code book, tune up the proper frequency, and decode the numbers that they receive over the air.

Decoding

Of course, many different codes can be, and are probably, used to decipher these broadcasts. This is where the books differ in their information on numbers stations. The code blocks could have a number that triggers whether or not the block should be decoded. For example, perhaps all number blocks are discarded, except those that begin with the number 2. From that point, every number block could represent a different letter.

Another possibility is that each number block could represent a word or phrase. For example, 18364 28375 96234 74215 23481 might be sent in a message. Maybe 28375 is the standard code for "meet at McDonald's." 74215 could represent the location "Greencastle, Pennsylvania." 18364 might represent Agent 36. Series 96234 and 23481 might just be filler nonwords. So, the final message could be: "Meet Agent 36 at the McDonald's in Greencastle, Pennsylvania and learn the secrets of the playground area—especially the twisting sliding board tubes and the ball pit."

Still another possibility could involve the use of a key book. In this case, the numbers would refer to words on the page. For example, the first two digits might refer to the page number and the last two (I'm using a four-digit code in this case) could refer to its word number on the page. This code would be nearly impossible to break because almost any very common book could be used. Using this code and the seemingly passive *Build Your Own Shortwave Antennas—2nd Ed.* by Andrew Yoder as the key, the code: 0461 1317 1318 9710 8107 8108 1319 1320 actually means "Buy ice cream for Joe Carr at Baskin-Robbins!" Now that's sinister!

Several years ago, an article was published that proposed to disprove the spy numbers theory. The author stated that these transmissions are merely tests, and the 4- and 5-digit codes are too archaic to be used by a government anymore. I didn't believe it for a minute. Dozens or hundreds of numbers broadcasts are transmitted every day with very high power levels. There is absolutely no way that the governments of the world would waste equipment, manpower, and millions of watts of power per day—for the sole purpose of testing a frequency or training young agents how to decode number groups!

Tuning in

Numbers stations primarily operate on frequencies outside of the standard shortwave broadcast bands. However, even though these are supposedly spy stations, the broadcasts are right out in the open: you can hear big signals for a fairly long period of time (usually about 10 to 40 minutes) on a regular schedule. So, if you can hear a station at 1800 UTC on a given frequency, chances are very good that you can tune in the next day at the same time and frequency and hear the station again.

One of the more interesting aspects of the numbers stations, which adds some intrigue to monitoring them (otherwise, repetitions of numbers are simply boring) are the voice types that you can hear. The voices on these stations are not "live;" the audio consists of digitally recorded numbers, which sound very much like those used by the telephone company if a number has been changed. The delivery is very stilted. Because of the lack of personality on most of these stations, one of the most interesting numbers broadcasts is from an operation known to hobbyists as "the Lincolnshire Poacher." This station features the same type of digital announcer, but the voice speaks with a British accent (most stations use female voices) and the transmissions are often preceded with a digital musical interval.

The numbers broadcasts are aired in many different languages, but the most common are in Spanish, German, and English. Some of the other languages reported include Russian, Korean, French, and Czech. As you can expect from the international world of espionage, these different languages have nothing to do with the location of the broadcast or what governmental or military group is using them.

More information

For a much more in-depth look at numbers transmissions, see *The Underground Frequency Guide—3rd Ed.* by Don Schimmel (available from HighText, P.O. Box 1489, Solana Beach, CA 92075). This book covers everything that you need to know about numbers: where they come from, who transmits them, how they are broadcasted, etc. Some other excellent books on the subject include: *Secret Signals: The Euronumbers Mystery* by Simon Mason; *Uno, Dos, Quatro: A Guide to Numbers Stations* by Havana Moon; and *Los Numeros* by Havana Moon (available from Tiare Publications, P.O. Box 493, Lake Geneva, WI 53147).

9
CHAPTER

Utilities monitoring

When the subject of shortwave radio comes up in a conversation, a person might think of amateur radio stations, or perhaps even of the shortwave broadcasters, such as the Voice of America or the BBC. However, one oft-forgotten territory in shortwave radio listening is utilities monitoring.

The reason for this ignorance lies in the definition of a utility station. DXers and program listeners alike are often puzzled about what fits into this category. Actually, it's quite easy. In the shortwave-listening hobby, every station on the air is a broadcast station, an amateur radio station, or a utility station. Utility stations are everywhere through the radio spectrum—from as low as 10 kHz to as high as 30 GHz. Like the amateur and international broadcast bands, utility stations are also often placed into their own segments of the radio spectrum.

Utilities stations transmit information that is useful to small audiences in the military, government, scientific community, maritime and aviation groups, etc. None of these services are intended for the general public.

You might wonder why anyone would want to monitor such specialized, private radio traffic. One reason that utilities are of interest is that these stations sometimes provide the news behind the news. For example, when the large jetliner (Flight 800) crashed off the coast of Long Island, New York, in July 1996, reports of the body and wreckage search were broadcast over nearly all radio and TV stations in the United States for several weeks. However, if you were really interested in the events, lived within a few hundred miles of the crash site, and owned a shortwave radio, you could tune in to the Coast Guard search-and-rescue frequencies and actually hear the news as it happened. Not only would you immediately have the information, but you could hear the events straight from those who were performing the work, without the extra coloration of the topic from the news media. Of course, these events alone also make listening to utilities seem secretive and mysterious—another reason why someone would care to tune in.

The lack of high power and regular schedules of many utility stations requires extra patience and effort in tuning, but don't let that deter you. The fun that you receive from monitoring these stations will more than make up for the effort expended. Remember that there are many more utility stations than broadcasters on shortwave, and some transmit from such exotic places as Antarctica and the Fiji Islands.

What can you hear on the utilities?

What radio hobbyists call *utilities* or *utes* comprise a vast number of stations, operations, and spectrum space on the shortwave bands. The following list doesn't cover all of these services, but you'll get a good taste of what's out there:

- *Traveler's Information Stations (TIS)*   These are the odd roadside radio stations that broadcast loop tapes of traveler's or tourist information in the United States (Fig. 9-1). Most of these "broadcasters" transmit with only 10 watts or less on the mediumwave band, typically on 530 or 1610 kHz (although the 1610-kHz frequency could be changed if the expanded mediumwave band proposals pass through). In addition to the low power, many TISs connect their transmitters only into a small antenna or even just the roadside guard rails, so they are extremely difficult DX catches! TIS stations fit solidly in the gray area between utilities and broadcasting, and they are often listed as either.
- *Spy numbers stations*   All through the shortwave bands, you can tune in coded blocks of numbers being repeated, probably for the benefit of international intelligence agencies and their operatives in the field. For more information on these stations, see the previous chapter, where the topic was covered in much better detail.

9-1 The control boards at the TIS station for the Philadelphia International Airport.

- *Feeder broadcasts*   At one time, government broadcast stations used utility transmitters to feed programming to studios or transmitters in remote locations. Now, most feeders are on satellites, but a number still remain on shortwave. In the United States, a few commercial stations are licensed to run remote transmissions in a 25-MHz utility band in the FM mode and just above the old mediumwave band (1600 kHz). This is another area where broadcasting and utilities overlap.
- *Beacons*   Beacon stations transmit a single letter or several-letter callsigns in Morse code, over and over. These stations are typically used for navigational purposes of some sort. At times, they are owned and operated by a federal aviation agency or by a local airport. At other times, they are installed and used by an army or air force for air or land-based tracking; likewise, a navy might also use them for air or ship tracking or navigation. For more information, see Chapter 4, which features a section on airport navigational beacons in the longwave band.
- *Ship-to-shore communications*   This very broad description covers a number of different modes and services. Voice (SSB), RTTY, FEC, and AMTOR modes are often used for these communications. For most routine communications, high-speed data transmissions are used. Very often, you will hear routine communications, such as information about sick passengers, requests for supplies, etc. One FEC transmission that I monitored must have been from a luxury cruise ship; they put in a large order for minced mutton!
    - *Emergency communications*   If you are the type of person who listens to the scanner for ambulances and slows down to catch a glimpse of car accidents, you should enjoy listening to the emergency marine channel of 2182 kHz. This frequency is reserved for ships that are in a state of emergency: on fire, sinking, dead engines, etc. If you stay tuned here long enough (especially at night because the frequency is too low to carry far during the daylight hours), you will hear some real distress calls. SSB is used here because, in an emergency, you wouldn't want to slowly type digital information at keyboard!
    - *Ship telephone calls*   Few people seem to realize that, if they are on a ship and are making a telephone call, they are communicating via radio, not telephone lines or fiberoptic cables. A few times per year, if I get bored with DXing and it doesn't seem like anything good is on, I'll tune through some of the shipping frequencies and listen for telephone calls. For the most emotional calls, listen during the holidays. It's amazing what the families of sailors will say over the radio during this time of year after their loved one has been at sea for a few months.
- *Shore communications*   Shore communications are much less interesting than ship communications. Of course, sometimes ship and shore stations are in communication with each other, but the shore stations are frequently active and transmitting even when they aren't communicating with a ship. Some of the most common stations to hear worldwide are those from Globe Wireless (Fig. 9-2), which operates a large worldwide network of stations

9-2 One of the major ship-to-shore communications networks, Globe Wireless.

that are heard on schedule with transmissions in FEC mode. Globe Wireless even has an Internet Web site (http://www.GlobeWireless.com), and for a fee, you can send a message to a ship via Internet e-mail!

- *Military communications*   Although almost every mode can be used for military communications, the one that is most interesting (and apparently most common) is voice SSB. Many of these transmissions seem to emanate from air forces, but they can be used anytime that voice communications are best (fast, immediate communications are necessary; hands are too busy to type; can't carry digital equipment into the field; etc.). These transmissions are either scrambled or use obviously contrived code names for different people, objects, and positions. You might hear something like "Gatekeeper 12, this is Milkshake. What is your position?"

    One gray area in military communications is U.S. MARS transmissions. *MARS (military amateur radio service)* transmissions are simply amateur radio two-way conversations that are conducted by amateur radio operators who are or were in the military and have received special MARS licensing to operate radio nets and utilize frequencies outside of the amateur radio bands. Most of these communications are simple conversations, only used to keep a schedule, but in times of warfare, such as during the Vietnam War, these operators often transferred messages from soldiers to their families.

- *Aviation transmissions*   Many of the plane-to-control-tower communications are low-power transmissions in the VHF range (just above the FM broadcast band). However, VHF frequencies can't be used for worldwide coverage, so shortwave frequencies are often utilized. Most of these transmissions are routine, but you never know what to expect. Robert Hall of South Africa reported in *The RTTY Listener* (Fig. 9-3) how he heard air-to-ground communications from a jet liner that had lost two engines and was in danger of crashing in the Atlantic. Hall lost signal and wondered what happened. Later, he read an article about the event in the *New York Times*. The jetliner did land safely, but the situation had been quite tense for a while!
    - ~ *VOLMET*   VOLMET transmissions sound pretty strange at first, but, like numbers stations, quickly get old. VOLMET transmissions also aren't nearly as cryptic or mysterious. They are merely aeronautical reports of weather, wind speed, etc. at airports around the world. At my location, New York Radio and Shannon Aeradio (Ireland) are the best-heard VOLMET stations.

Utilities monitoring **223**

The RTTY Listener

Issue 35 — **December 1994**

Monitoring from Sweden

Jan Larsson *SM6CWA*, of Ed, Sweden engages in his monitoring and amateur activities from this picturesque location at 59.8° North 11.52° East. Kenwood equipment shown: two TS-940S', TS-440S and TL-922A. Jan recently added a Universal M-7000 decoder, not shown.

Texas Monitoring Station

This complete station belongs to Eric Ronning *KE9YE*, of Texas. Key equipment includes: Kenwood TS-950S transceiver, SM-220 Monitor Scope, 2 Universal M-7000s, 2 AEA PK-232s, and a Dovetron Terminal Unit.

HF Aero Frequency Changes
by Robert Evans

November 27th brought forth some the WARC-92 frequency changes in the military aeronautical mobile service. The implementation of a 3 kHz channelization scheme saw many frequencies change, but usually only by ± 1 or 2 kHz. Many Coast Guard and Canadian Forces frequencies were noted on their slightly different frequencies. December 22nd saw the United States Air Force adopting its new frequencies. The Navy is also making changes slated to be complete by December 31st, 1994.

When tuning shortwave aeronautical you may need to tune up or down 1 or 2 kHz from previously known frequencies. Please see pages 30 and 31 for preliminary information on this important change.

M-8000 Upgrade Available

The Universal M-8000 version 1 owners may now purchase an upgrade kit to add the following modes:

◆ **ARQ6-90** is an increasingly popular French diplomatic mode. N. American day time reception is common. ARQ6-90 is 200 baud and usually 400 Hz shift. Watch on: 15801, 17455, 18033, 18744, 19932, 20095, 20153 and 20550 kHz.

◆ **PACTOR** is the fastest growing digital H.F. amateur mode. Pactor combines the best features of Packet and SITOR. Tune from 14060 to 14090 kHz.

◆ **ACARS** (Aircraft Communications Addressing Reporting System) is hearable virtually anywhere in the U.S.A. on any scanner with air band. It is used to pass technical information and messages between airports and commercial aircraft. Tune to <u>131.550</u>, 130.025 and 129.125 MHz.

This upgrade kit consists of a supplementary manual, a plug-in EPROM chip and two resistors. The installation of the resistors is required *only* for ACARS. The provided resistors are to be mounted "piggy back" on to two existing resistors and can be soldered in from above the board.

The **M-8000v2 Upgrade Kit** is order #2613 and is available for $69.95 (+$3 UPS). Upgrade kits for other Universal products are expected in 1995.

9-3 If you are interested in monitoring data stations, try *The RTTY Listener.*

~ *U.S. Presidential transmissions* Perhaps you'll be fortunate enough to have your radio on and within range when Air Force 1 (which carries the President of the U.S.) and Air Force 2 (carries the Vice President of the U.S.) can be heard in flight. These transmissions are rather uncommon. Although you probably won't hear any near-crash panic, secret messages, or jogging tips, these catches are still really neat to hear.
- *Commercial transmissions* Most commercial transmissions are two-way conversations in one of the VHF bands. However, in some circumstances, such as in offshore oil exploration and drilling, it is necessary to use long-distance communications. In the case of the oil companies, you could hear communications from airplanes, helicopters, oil platforms, or ships. The communications could range from two-way talk about current conditions to RTTY information about the products and procedures.
- *Press services* Press agencies send the news via radio to newspapers, radio stations, and other subscribers around the world. At one time, all press agencies around the world sent their news via shortwave RTTY, and it was by far the best way to receive the best and most up-to-date news (Fig. 9-4). Now, most of the agencies are sending their information via higher data rates over satellites. However, a number of smaller press agencies, such as those from North Korea, Morocco, and Senegal still operate with RTTY on shortwave. These stations are a great source of information from some of the countries and entertainment from others (such as from KCNA, North Korea).
- *Fax* Fax transmissions are sent by many different services, such as by commercial, military, shore stations, etc. However, I pulled them into this

9-4 KUP, a data information station from the San Francisco *Examiner* in 1930.

separate section because nearly all fax transmissions on shortwave are the same: weather map faxes. They might be used for different purposes, but all essentially look the same. According to some of the old-time radio columns in *Popular Communications*, many radio faxes were sent in the early days of radio—everything from newspaper photos to the newspapers themselves. The *New York Times* was even transmitted after midnight via radio fax on at least one of the New York mediumwave stations in the 1940s! Receiving radio faxes is not for those with little time; most weather faxes require about 15 minutes (or longer!) to download.

The legality of listening to utilities

In some countries, such as the United States, it is perfectly legal to listen to any type of utility station, right down to the most personal communications. However, you break the law if you divulge specific information concerning these transmissions. Depending on the situation, you could even wind up with a fine and jail term (for a jail term, you would probably have to receive some sensitive information and use it maliciously).

Although you are free to listen to anything on the shortwave bands in the U.S. and Canada, many countries don't provide these freedoms. Many nations ban the monitoring of government utility stations, for fear of risking national security. Because of the instability of many countries, this ban could be in response to a legitimate fear.

If you plan to listen to utility stations and you live outside of the U.S. or Canada, be sure to check with your government radio regulatory agency to see what you are legally allowed to listen to and what penalties apply for disobeying these laws.

Equipment

Utility stations operate in nearly any mode imaginable on shortwave. A few of the modes include:

- AM for most numbers stations (see Chapter 8 for more information on these transmissions). For AM transmissions, no special equipment is required, just a shortwave receiver (Fig. 9-5).
- FM (narrowband FM, in particular) to feed audio from one remote location to a broadcast transmitter. These transmissions are very uncommon on the shortwave bands. For years, the FM mode was almost never included on shortwave receivers. Now, the pricier ($900 and up) communications receivers include the FM mode).
- SSB for nearly all voice communications. All amateur radio transceivers and communications receivers, and about ⅓ of the portable shortwave radios on the market have either a switch for LSB/USB operation or have a BFO so that these signals can be copied (Fig. 9-6).
- RTTY (radioteletype), AMTOR, ASCII, SITOR, radiofax, and Morse code (CW) for a wide variety of data communications. To copy Morse code, the

P.O BOX 500
RICHLAND, PA 17087

* * EXPERIMENTAL STATION * *

KA2XAU

A. Yoder This will confirm your reception

Transmitter power output 12 (twelve) Watts
Frequency 1620 KHz   Antenna Dipole H & V
Modulation A-3 (A.M.) Date 11 01 92

Subject to the provisions of the Communications Act of 1934, subsequent acts, and treaties, and all regulations heretofore or hereafter made by this Commission, and further subject to the conditions and requirements set forth in this license, the licensee hereof is hereby authorized to use and operate the radio transmitting facilities hereinafter described for radio communications.

Frequency kHz	Authorized Power (watts)	Emission Designator
540	250 (ERP)	10KA2E
1620	250 (ERP)	10KA2E
9715	250 (ERP)	10KA2E
11705	250 (ERP)	10KA2E
15300	250 (ERP)	10KA2E
17745	250 (ERP)	10KA2E
89.7 MHz	10 (ERP)	75KF3E
100.5 MHz	10 (ERP)	75KF3E

Operation: In accordance with Sec. 5.202(a) of the Commission's Rules.

Mr Yoder,
    Thank you for your report. It is appreciated, especially in view of the low power we were operating with during this antenna test. Certainly you are welcome to reprint the above comfirmation. Best wishes with your book.

Craig Baker, Owner C.M. Baker Electronics and KA2XAU

9-5 A QSL from KA2XAU, an experimental broadcast station.

9-6 The Web page for the Baltimore District of the Army Corps of Engineers, which operates WUB4 and a number of other shortwave utility stations.

requirements are the same as that for receiving SSB signals. The other modes, however, require a very stable receiver with either the SSB or RTTY mode. Even though a digital receiver will display one set frequency, the radio will actually slightly drift up or down in frequency. For these modes, a good, stable communications receiver is a necessity. Also, a digital decoder is necessary to decode the data as it passes through the receiver. These decoders are either separate boxes or plug-in computer boards that rely on separate software to help translate the signals.

Choosing a high-speed data decoder

Data decoders are very specialized, sophisticated pieces of radio equipment. As a result, getting into monitoring most of the utility communications (more than half of the utility communications on shortwave are via digital data modes) requires a substantial investment from the listener.

Three different general types of data decoders are currently available: the deluxe stand-alone unit (video monitor required), the mini stand-alone unit with display, and the plug-in computer board. Like most anything, each has advantages and disadvantages.

The plug-in computer board will most efficiently use your resources. Rather than spend lots of money on a video monitor, display, and microprocessing system, you can let your investment in a computer pay for all of those features. Although these models don't usually have quite as many features as a deluxe stand-alone unit, they will fulfill the general needs of most shortwave hobbyists. The trendsetter in this market, the Universal M-400 costs approximately $400. In my case, the plug-in computer board isn't appropriate because most of my shortwave-listening time occurs while I am sitting at my desk, "writing" on my computer.

The mini stand-alone unit with a display is best for hobbyists who don't have a computer and don't have enough money to buy a better model. This type of decoder tries to make the cheapest unit possible that can decode different radio modes and yet still show the characters on a tiny, built-in display. The end result is that the manufacturers try to cram as many features as they can into a tiny, relatively inexpensive box. These units are made by several companies, including Universal and MFJ.

The last option is to purchase a deluxe stand-alone unit. These models are microprocessor-controlled boxes that can decode a number of different data modes and have many different settings so that you can receive nonstandard data transmissions. One of these communications terminals has many more buttons, knobs, and switches than an expensive shortwave receiver. Universal Radio's models even have a card of codes so that you can remember all of the different functions of the keypads! The problem is that one of these units will probably cost as much or more than your shortwave receiver. Most people aren't willing to spend $2500 to $3000 (for a receiver/decoder combination) to receive radioteletype.

If you are seriously into data reception and you can afford to spend a few hundred dollars for a decoder, your best bet is looking into the used market. A unit such as the Universal M-7000 is a great way to get into the hobby at a better price—this model is an excellent stand-alone unit without a monitor. It was recently replaced by the Universal M-8000. Now, many M-7000s are available used for about half of their original price, which is quite a bit cheaper than a new M-8000. Many M-7000s are available via classified ads in *Popular Communications*, *Monitoring Times*, other radio newsletters, and the Internet. If you want to be cautious about getting one that is in good working order, check with the source; Universal Radio buys, sells, and trades. They typically have a number of used M-7000s on hand (Fig. 9-7).

Setting up a data-receiving station

For my personal monitoring of data transmissions, I got a used Universal M-7000 communications terminal. I already owned a Kenwood R-5000 receiver, so I was set to receive some data . . . or so I thought.

The first problem was that the M-7000 doesn't have a video display; it requires output from a monochrome monitor only. No problem, I thought. I go to hamfests all of the time, so I felt that getting a computer monitor would be a simple project. Unfortunately, the M-7000 will not work with a video input from a standard digital monochrome monitor, only from a composite computer monitor. Composite monitors were used in the early days of personal computing, when the Radio Shack Color Computer was all the rage. After searching through a few hamfests, I discovered that composite monitors are also very difficult to locate these days. I tried connecting

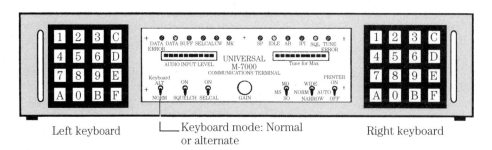

9-7 The front panel of the M-7000 communications terminal.

into the video input of a TV and the input of a VCR, but nothing worked. Next, I placed some *Wanted:* ads on Internet newsgroups. Nothing. So, I broke down and purchased an old composite monitor from Universal. Fortunately, they seem to keep a few on hand.

The output of the M-7000 and the input of the monitor each require ¼" phono plugs, so it is easy enough to run a standard audio cable between the two. To connect the receiver to the terminal, you only need to run an audio cable from the record jack of the receiver (a ⅛" plug on the R-5000) to the input of the M-7000 (a ¼" phono plug). This is really simple, but for the best results, you should use shielded cables to prevent them from receiving electrical interference and noise. The M-7000 that I own (and I suspect that others are like it) emits a fair amount of RF noise, so it would be best to use an antenna with a shielded transmission line (such as a dipole), not just a random-wire antenna.

You will probably want to place the equipment in a location where you can frequently use it and where it is easily accessible. This is very difficult if you have a modern radio shack—with a computer receiver, cassette deck or VCR, printer, etc. It is essential to have the monitor at eye level and the controls of the equipment within easy reach. With this much equipment, it is best to have a standard operating position, such as at a desk. If you are already a shortwave listener and you have a setup that is too cramped to easily access everything, try adding a desk or table to come up with an additional operating position. If you have a poor working arrangement, chances are that you won't use your system very often.

This is a common-sense radio-listening rule, but be sure that you also have some shelves within reach so that you can keep books and recent magazines/newsletters as a constant reference source. So many different data stations operate in so many different modes, speeds, shifts, etc., that you *must* have references nearby or you will be lost.

Tuning in high-speed data transmissions

If you don't have a data decoder or if you are already proficient at using it, skip this section. Decoding digital data streams via radio is much more difficult than tuning in broadcasting stations or SSB voice stations. The receiver tuning must be precise, or the data will be garbled. For that matter, even if you are perfectly tuned in to the station, some of the data will probably be garbled from static crashes, fades, or errors in the data itself.

Because the M-7000 (and the very-similar-in-function M-8000) is one of the most popular data decoders in the world, it is used as the reference here. Here are a few steps for tuning in your first station:
1. Turn on the radio, the terminal, and the monitor.
2. Tune the radio (in either the LSB, USB, or CW modes) to a strong Morse code station in the amateur radio bands (3500 to 3600 kHz and 7000 to 7100 kHz are two of the most commonly used areas). Different random characters will appear on the monitor as you tune around; ignore them.
3. Set the decoder to the medium-speed CW mode.
4. At this point, the Audio Input Level LEDs on the decoder should be "bouncing" in time with the CW that you are hearing on the receiver. Adjust the Gain control so that the signal peaks high into the LEDs, yet drops to no

or almost no LEDs between dots and dashes. If no LEDs light, be sure that the decoder is turned on, then check the audio cables.
5. Fine-tune the radio so that the MK (mark) and SP (space) LEDs light alternately.
6. Fine-tune the decoder so that the Data LED lights. At this point (or earlier), you should notice that some real words (not just garbage) are appearing on the monitor. Remember that with CW transmissions, plenty of ham radio lingo is used (such as "de" for "from," Q signal codes, "CQ," strings of "VVVVV," etc.), so don't mistake actual decoded transmissions for garbage.

Morse code is best to test the equipment with because it is much easier to tune these signals and because the data runs at a much slower character-per-second rate. You can easily match up the sound of the CW with the flickering light from the MK and SP LEDs. Things will be even easier if you can copy Morse code just by listening to the signals. You can verify that the decoder is operating properly that way (i.e., that you have tuned the signal in properly with the receiver and that the settings on the decoder are correct).

Once you play around with receiving CW, then it is time to move on to some more difficult modes. The easiest high-speed data modes to receive are FEC (SITOR A) and AMTOR (SITOR B). To receive these modes, it helps to know what these signals sound like and what frequency ranges are most commonly used. Identifying these two modes by ear is easy because they each have a very distinctive sound. The easiest of the two is AMTOR, which sounds somewhat like a cricket: "chirp . . . chirp . . . chirp . . ." If you have been around the bands long enough to know what radioteletype sounds like, you will have an idea of the sound of FEC. Imagine an enchanted RTTY transmitter that sings, and you will have a good idea of how FEC sounds.

When you tune across one of these signals and you want to decode it, keep an eye on the LEDs of the decoder (of course, you must already have the decoder set in either the FEC or AMTOR mode). Like decoding Morse code, you want a good balance between the lighting of the MK and SP LEDs. The most important LEDs to watch, however, are the Data Error, Tune Error, and Data LEDs. Typically, the signal will be tuned in when the Tune Error and Data Error LEDs flicker and the Data LED lights and remains lit. The Error LEDs are to give you an idea of whether the signal is properly tuned in, but when the data LED is lit, that means that it is receiving and decoding data.

Very often, you will tune in a very strong FEC or AMTOR signal and prepare yourself for an easy reception. According to the LEDs on the decoder, you should have the signal tuned in, but nothing (or maybe only an occasional character) is appearing on the screen. Check to see if the Idle LED is also lit. As the name suggests, *idle* is when the data station is on the air and transmitting, but it is merely sending a "blank" signal that puts your decoder into idle mode. If the station is running in idle mode, it probably means that it will soon be transmitting some actual data. Otherwise, it would just go off the air. Sometimes when I have something better to do, I will just leave the radio, decoder, and radio all turned on and tuned to that radio station. In the meantime, I continue working on other projects and occasionally check back to see if the station has sent any real traffic. Chances are that after an hour or so, some real data will be on the computer screen.

Another problem that you might encounter with monitoring FEC and AMTOR stations is that they aren't all in English. This might not seem like a big deal at first. However, consider that many languages, such as Russian and Chinese, do not use a Latin alphabet. It will appear that these stations are not tuned in properly. No matter how you tune the signal or change the settings, only garbage will appear on the monitor. To save time, it's best just to tune across the bands and pass over the stations that you can't quickly decode. FEC and AMTOR are sent at standard shifts and speeds, so it is easy to find a few signals that you can tune in.

The final major frontier for data monitors is radioteletype. The name is the flipside of *radiotelephone*: *radiotelephone* meant using the radio as a telephone. With *radioteletype*, you type instead. *Radioteletype* is an ancient mode of digital communications—dating back to the early days of radio communications. In those days, a large mechanical device was used to send and receive data. If you wanted to send data, you would type on what looked like an old mechanical or electric typewriter. Hitting each button would key the transmitter for the high-speed clicks that represented each letter. The device could also decode data into letters. Once you had the RTTY signal tuned in and the decoder turned on and set to the proper speed/shift, the decoder would "read" the impulses and activate the correct letter hammer on the printer. The decoder would clicky-clack through a message and print it out on a big roll of paper. The paper looks like a big roll of fax paper (but of lesser quality) and the printing system operated somewhat like an archaic version of a daisy-wheel computer printer.

After years of these radioteletype systems being developed by independent companies, governments, and the militaries of countries around the world, you can imagine just how many different variations of these systems mutated.

So, many different speeds and shifts evolved over time. Now you tune in an RTTY signal. What settings should you use to copy it? Better yet, few transmissions are sent in English, and many are encrypted. The result is that RTTY stations are difficult for the beginner to tune in. Chances are good that you won't understand it, and even if you could, it might be encrypted.

For radioteletype monitoring, you really must have some up-to-date sources of utilities information. Without this information, you will be totally lost on the shortwave bands, and you will spend most of your time trying to figure out which stations you can copy. See the "For more information" section at the end of this chapter.

Receiving utility QSLs

As you might expect, receiving QSLs from utility stations is not easy. For the most part, few people know who is in charge of the radio communications because radio is generally not at the heart of whatever operation is being performed. For example, if the station is located at an oil-drilling rig, the primary concern is oil; if it is on a warship, war; on a sinking ship, safety; or if you are trying to QSL a numbers station, the primary concern is presumably intelligence or espionage.

Because of the difficulty in obtaining utility QSLs (Fig. 9-8), you must take the extra measures if you want to attain success:

- *Be courteous and tactful* Many utility stations don't care to be heard by the general public. Chances are that, if you telephone for information on contact personnel or mailing addresses, you will encounter many people who know nothing about radio listening or QSLing. Any telephone calls or QSL reports are probably an annoyance, at best. If you are rude or pushy, you will be lucky if you even receive information on where you can send your report (although you might be told what you can do with it . . .).
- *Don't include too much information* This point sounds really strange, but utility stations are supposed to provide secure communications. If you hear the radio station from a warship in a secret maneuver, for example, and you disclose too much information about their position, not only might you not receive a QSL, but you could get in trouble. Just how much trouble depends on the station, the country of the station, the country that you live in, and the nature of the information that you heard. The tough call is sorting out how much information is too much and how little is too little for a QSL card. It's a judgment call, so use your best!
- *Be persistent* This might seem like a contradiction of *be courteous and tactful*, but it doesn't have to be. Just keep trying different telephone and fax numbers, or mailing and e-mail addresses; don't get frustrated.
- *Make the QSLing effort easy* Be sure to explain what it is that you want (just saying "I want a QSL card" might not mean anything to the radio operator or the secretary that opens the mail)—a letter or card that states that you heard their station on such and such, then list the details. Also, let them know that you are a radio hobbyist and that you listen to shortwave radio. Tell them that you enjoy receiving these verification letters or cards (something like collecting stamps). To really make it easy for the radio operator, include a *PFC (prepared form card)*. A PFC can be a letter or card that is basically a QSL with nearly all of the information placed in; all the operator needs to do is fill in a few details, stamp the card (usually in the case of shipboard radio stations), and sign his or her name.
- *Make the QSLing effort worthwhile* It's always best to include return postage, $1 U.S., or an IRC (International Reply Coupon) when writing for QSLs. Especially in cases of stations in developing nations, station funds are usually exclusively used to maintain radio equipment, not for postage. If I was the radio operator, I know where I'd spend my money! Include plenty of details about yourself, your hobbies, your family, and your area. Although you might not reap the fruits of your labor with some responses, some radio operators will enjoy verifying your report just as much as you will appreciate getting the QSL back.

The Rideshare Company 108 Charter Oak Avenue, Hartford, CT 06106 (203) 525-8267 Fax (203) 522-8445

February 27, 1995

Mr. Andrew R. Yoder
P. O. Box 109
Blue Ridge Summit, PA 17214

Dear Mr. Yoder:

Thanks for your reception report on the Connecticut Department of Transportation's travelers information station broadcasting on 1620 kHz.

This letter is to verify that you heard the station, based on your description of the message we were broadcasting. The transmitter and antenna are located at the DOT garage on Bloomfield Avenue (Route 305) in Windsor, which is adjacent to Exit 37 on I-91, about 7 miles north of Downtown Hartford.

The station's call letters are KPD-581, and its power output is 10 watts. We are using a 15-foot loaded vertical antenna, mounted on a 35-foot telephone pole. Broadcasts began September 17, 1993.

Although the station is licensed to the DOT, The Rideshare Company worked on the Department's behalf to put the station on the air, as part of a contract to help educate commuters about the use of High Occupancy Vehicle (HOV) lanes, special lanes for vehicles carrying with two or more passengers on Interstate 91 between Hartford and Windsor.

Since signing on, we have received reception reports from various parts of Massachusetts and from Pennsylvania, Washington, D.C. and North and South Carolina. Although the broadcasts are primarily intended for motorists traveling through the I-91 corridor, we are pleased to know that the station is putting out a good signal.

The DOT has had a similar station broadcasting for several years on 530 kHz, originally from Manchester and now from East Hartford, both east of Hartford on I-84; we've never received any out-of-state reception reports, no doubt because of the many other stations on that frequency.

Thank you again for your letter.

Sincerely,

Doug Maine
Manager of Communications

Printed on recycled paper

9-8 An interesting QSL from KPD-581, a TIS station from the Connecticut Department of Transportation on 1620 kHz.

For more information

Some of the best sources of utility information include the "Communications Confidential" column in *Popular Communications*, the "Utility World" column in *Monitoring Times*, *The RTTY Listener* newsletter (published by Universal Radio, 6830 Americana Pkwy., Reynoldsburg, OH 43068), and *The Worldwide UTE News Club (WUN)* (available from Tim Braun, 15915 Smithey Dr., Haymarket, VA 22069 or via the Internet by sending the message "subscribe wun" to majordomo@phoque.info.uqam.ca).

Popular Communications and *Monitoring Times* are handy for the shortwave hobbyist because they contain information on all sorts of radio monitoring. With one or two magazines, you can dabble in all types of listening. *The RTTY Listener* is published by Universal Radio; they have plenty of advice for owners of their decoding equipment. WUN is a vast supply of utility information—addresses, frequencies, pages of loggings, etc. If you are serious about utilities monitoring, WUN is a must-have.

Some of the available books on the subject include: *The Essential Radioteletype Frequency List* and *Radioteletype Monitoring: The Complete Guide* by Dallas W. Williams (available from Tiare Publications, P.O. Box 493, Lake Geneva, WI 53147 and http://www.tiare.com). *RTTY Today*, *RTTY Press—2nd Ed.*, *Worldwide Marine Facsimile*, and *SW FAX Frequency Guide* are all available from Universal Radio (6830 Americana Pkwy., Reynoldsburg, OH 43068 and http://www.universal-radio.com). Some excellent utility books are: *The Worldwide Weatherfax Guide*, *Radio Data Code Manual*, and *The Guide to Utility Stations* by Joerg Klingenfuss (available from Klingenfuss Publications, Hagenloer Str. 14, D-72070 Tuebingen, Germany and http://ourworld.compuserve.com/homepages/klingenfuss). Some other guidebooks include: *US Maritime Frequency Directory* by Robert Gad and Robert A. Coburn, *The Worldwide Aeronautical Communications Frequency Directory—2nd Ed.* by Robert E. Evans, *International Callsign Directory* by Gayle Van Horn, *World Press Services Frequencies* by Thomas Harrington, *The Shortwave Directory—8th Ed.* by Bob Grove, and *Confidential Frequency List* by Geoff Halligey (available from Grove Enterprises, P.O. Box 98, 300 S. Hwy., 64 W., Brasstown, NC 28902 and http://www.grove.net).

CHAPTER 10

CB radio

Faster than you could say "10-4, good buddy," the CB radio craze swept through the United States in the late 1970s. Novelty CB songs were on the radio, and CB theme movies were being produced by Hollywood. Suddenly, movie stars and celebrities were posing with CB radios—both for paid advertisements and to update their image. It was even funnier than the current crowd of celebrities posing as Internet geeks and cyberpunks. However, despite the hullabaloo, the CB fad was gone before you could say "I lost $50,000 making these @#$%& '10-4' belt buckles."

What is CB radio?

The CB radio band is a chunk of the radio spectrum, where anyone can have two-way communications without a license. A few of the technical stipulations are:
- Only transceivers with an output power of 5 watts (AM) or 12 watts (SSB).
- Only FCC-approved commercially manufactured equipment.
- Only operation on set channels (no VFOs).
- Only SSB- and AM-mode voice transmissions.
- Only an antenna 20 feet higher than what it is mounted on or 60 feet above the ground.

CB operations are permitted by the FCC, just like those on the amateur radio bands:
- No commercial transmissions.
- No broadcasting (only two-way transmissions).
- No profanity, jamming, or other malicious conduct.
- No coded or cryptic transmissions.
- No illegal activities.
- No stations beyond 155.3 miles away may be contacted.

Of course, this is just a list. If you are interested in CB activities, don't use this as your guide. Check out FCC rules, Part 95 Subpart D for the exact dos and don'ts.

The FCC used to license CB operators. However, they became swamped with applications and the associated paperwork, so they dropped all licensing requirements. Now, in the United States, you only need to find a CB and an antenna and go on the air. CB rules and regulations vary from country to country. In some countries, the frequencies are even different. So, be sure to check with your local authorities or national communications regulatory body for information.

Roots

CB (Citizen's Band) radio stretches back much further than the 1970s. To trace the history of CB, you must go back to the 1940s. After World War II, representatives from countries around the world met to realign the radio spectrum. The 27-MHz band, which had been used for a great amount of German and U.S. military communications during the War, was now reallocated for amateur radio in much of the world, and also for "industrial, scientific, and medical applications." According to Tom Kneitel's article "The Skeletons In CB's Closet" in the July 1979 issue of *CB Radio/S9*, the amateur radio signals had to compete with a barrage of interference from diathermy and heart-treatment machines.

Of course, the commercial and government interests knew that the band was not reliable for regular communications, so they allowed the amateurs to use it. These organizations never just "give" chunks of the radio spectrum away to noncommercial groups (re, how amateur radio received the shortwave bands in Chapter 1).

By the mid-1950s, few manufacturers of amateur radio equipment included the 11-meter amateur radio band on their transmitters. Chances are that it was because the band wasn't popular enough to warrant adding the electronics components necessary to add the extra position (this was before the days of transceivers with digitally synthesized VFOs). As a result, most of the amateurs active on the 11-meter band were experimenters who picked up military surplus transmitters and either used them as-is or modified them.

According to Kneitel, the FCC had also licensed another radio band for a "citizens radio service" on 460 MHz. Because of the extremely high frequency, transmissions on this range would "get out" more like a baby monitor than anything that you could actually communicate to the outside world with. As a result, the FCC decided to make the 11-meter amateur band into a new Citizens Band, but the amateurs who had been using 11 meters received nothing in return. The amateur operators were intelligent enough to realize that all of the radio bands that they were given could just as easily and quickly be given to someone else. So, they fought the FCC's ruling with petitions and articles in the amateur radio magazines of the day. Of course, the FCC won; for decades they did virtually whatever they wanted with the radio spectrum. In September 1958, the FCC began issuing the first licenses for CB radio operators.

Even though the amateur radio operators lost on paper, they won in the long run. It was true that the amateurs no longer controlled the 11-meter band, but they could receive a license to operate there. In addition, millions of other people could get a taste of two-way radio on the new citizens band. In this sense, the CB band was a hands-on primer course for amateur radio. In September 1958, the FCC began issuing the first licenses for CB radio operators.

In the 1960s, some companies began to develop the first true CB radio designs, although some truly strange radios turned up during the first decade of the citizens band revolution. One really peculiar model was the Hammarlund (a long-time manufacturer of amateur radio equipment and shortwave communications receivers) HQ-105TR. This ugly duckling was a low-end shortwave receiver with a built-in transmitter for the CB band. This concept might have worked, except that Hammarlund had a long, fine reputation for building very large, top-quality receivers. The HQ-105TR combined all of these concepts into one receiver: it was too well-built to be inexpensive, yet not good enough to be bought by serious hobbyists. Even though it contained a CB transmitter, it was much too large to use as anything but a base station. It bombed on the market. A few other amateur radio companies, such as Heath and Johnson, more successfully made the transition to CB radio manufacturing.

Although the CB radio band plugged along for more than a decade and a half, it did not catch the public's eye until the 1970s. Apparently, at that time, the truck drivers began to use CB radios for standard communications on the road. By the mid-1970s, they developed their own little culture that anyone could listen to on the CB band. In 1976, the phenomenon was interpreted into a song that became a hit: "Convoy" by B.J. McCall. The song was so popular that a movie version of "Convoy" was released in 1978, and CB radio became the biggest fad since "Woodstock."

However, the era of the CB fad, like the Dallas Cowboys' Super Bowls and the popularity of country music 8-tracks, ended by 1980. All of the CB radio magazines folded (such as *CB Times*, *Personal Communications*, and *CB Magazine*), as did a number of CB radio companies. The oldest surviving CB magazine *CB Radio/S9* (which had been around for about 20 years) became *S9 Hobby Radio*. It incorporated a number of general-interest shortwave topics in an attempt to save itself. It didn't work. However, within a year, the concept, editor, and several columnists resurfaced in a new magazine, *Popular Communications*.

The final collapse of CB radio was the loss of all of the CB hobby magazines. These magazines really transformed the communications into a hobby—with QSL columns, information on CB from around the world, etc. It appeared in 1996 that the hobby might make a real turn-around when Ross Communications (publishers of *CQ* and *Popular Communications*) began publishing a new magazine called *CB Radio*. This excellent magazine started with a bang by employing an experienced staff, including former *S9* and *Popular Communications* editor Tom Kneitel and current *Popular Communications* editor Harold Ort. However, after about six months of publishing, *CB Radio* folded, once again setting the hobby back.

CB radio today

Like the end of *S9*, the CB fad died out, but the radio communications continue. Unfortunately, much of the hobby element has disappeared from the CB bands. These days, CB radio is primarily used only for communications in the transport industry, emergency calls, or illegal activities. However, country music is popular again and the Dallas Cowboys have been in the Super Bowl recently. Is CB radio next to return?

CB radio losses

In addition to the CB radio fad ending in the late 1970s, CB operators and hobbyists were also lost to the amateur radio no-code tech license and to cellular telephone communications. Because CB transceivers have been available to almost everyone for very little or even for free (used) since the 1980s, and because the FCC has not enforced its regulations over the CB band, it has been utterly chaotic. After some hobbyists experienced abuse, jamming, and on-air arguments, they were more than happy to jump over to amateur radio when the FCC passed the no-code license. Higher costs, more-difficult-to-obtain equipment, and self-policing of amateur radio operators reduces (although they certainly don't eliminate) many of the problems that tend to occur on the CB frequencies.

The sudden boom of cellular telephones is a recent detraction from CB radio. In the 1970s and 1980s, the only way to contact people while traveling in your car—either for personal or emergency communications—was via amateur or CB radio. Many of the people who would have contacted their family via CB to tell them that the car broke down or that they would be late for dinner, now use the cellular telephone.

CB radio: Still a viable communications medium

Because other communications mediums have stripped away many former supporters of CB radio, the hobby is not, and will never be, the same. However, despite these changes, CB radio is still a usable communications medium, and it can be an interesting hobby.

Plenty of users still remain for a variety of reasons. In many locations, amateur radio repeaters are as busy as the CB frequencies were in the 1970s. With 40 channels to choose from (see Table 10-1) and fewer people using them, sometimes it's just easier and calmer to dust off the old CB and talk there. Also, the startup cost is much lower. If you don't have $200 or $300 in the budget for an HT transceiver, try picking up a $40 CB. The low cost is nice for teenagers who are starting into the radio hobby (or who just like to talk) and have a rather limited income.

Although CB just can't compete with the precision, reliability, and relative privacy of cellular telephones, the high expense of cell phones is enough to drive anyone back to CB.

The one area where CB radio surpasses all other modes of communications is in cases where a number of people who aren't radio hobbyists need to be in contact with each other. For example, a group of my friends traveled around the United States in two cars. After 15,000 miles of driving and trying to stick together, you can bet those CBs came in handy! Another perfect use for CB is for community-watch programs. In many large cities, whole communities have been virtually ensieged by violent crime. To fight back, the concerned citizens have set up neighborhood watch groups to help police the area. Aside from your eyes, the next most effective weapon against crime is your voice. With CB radio, crime-watch networks can be coordinated and mobilized, making neighborhoods a safer place.

Table 10-1. CB channels and frequencies

Channel	Frequency (in kHz)	Common use
1	26965	
2	26975	
3	26985	Unofficial marine channel
4	27005	
5	27015	
6	27025	
7	27035	
8	27055	
9	27065	Emergency/traveler assistance
10	27075	
11	27085	
12	27105	
13	27115	Unofficial marine channel
14	27125	Common walkie-talkie channel
15	27135	
16	27155	
17	27165	
18	27175	
19	27185	Truckers/highway channel
20	27205	
21	27215	
22	27225	
23	27255	Shared with remote-control devices
24	27235	
25	27245	
26	27265	
27	27275	
28	27285	
29	27295	
30	27305	
31	27315	
32	27325	
33	27335	
34	27345	
35	27355	Regional SSB channel*
36	27365	Regional SSB channel*
37	27375	Regional SSB channel*
38	27385	Regional SSB channel*
39	27395	Regional SSB channel*
40	27405	Regional SSB channel*

* Common SSB channels by informal agreement. AM and SSB are permitted on any frequency.

Community Radio Network (CRN)
Rob Bellville

Community Radio Network (CRN) was created in response to some of our social problems. We all know that, even in the best neighborhoods, there are rising crime rates and a lack of social interaction between residents. Technology, in some cases, creates more problems than it solves.

CRN takes technology, specifically CB radio, and opens a conduit through which local interaction takes place. By creating a "channel" through which the residents of a community can communicate, stronger bonds will develop and this will give rise to improved community situations.

CRN can be molded and customized to fit your community's needs. One popular application of a CRN is the formation of crime watch groups. Community residents can communicate via radio to other residents when they spot criminal or suspicious activity to alert the police. In conjunction with crime watch activities, a resident with a scanner can monitor local law enforcement communications and alert other residents to area activity. Having extra sets of eyes are always a benefit to police. In this way, a community can be *pro*-active rather than *re*-active and help to solve problems before they occur or get out of hand.

Another idea for a CRN is to utilize radio communications between kids, the elderly, house-bound, or anyone else who might need assistance. Sometimes all someone needs is to have a small job done for them especially if they are unable to do it themselves. Establishing a social relationship via radio helps to cement a personal relationship with others. It is much easier to help out a friend than a stranger.

By monitoring channel 9 (the CB radio assistance frequency), a community can assist visitors to their area with traveler's assistance, helping tourists get to their destination and emergency help. Alerting others to severe weather conditions can be life-saving and a valuable service.

Starting a CRN is easier than you think. If you are a radio enthusiast, you already possess enough knowledge to start one. Contact your local community newspapers, social organizations, and civic groups and inform them of CRN. Many of these groups could help start a CRN either by supplying financial help or media involvement to get the word out. Many philanthropic groups would find this project ideally suited to their charter.

Radio groups, such as REACT, that routine monitor CB radio for emergency traffic are in an ideal position to start a CRN because they are in effect already supporting a specialized CRN.

While there is no nationwide body that regulates CRNs, we do ask that you register with us so that we can promote this project more effectively. Participation in this project is totally voluntary and without cost. Even if you do not call your group a CRN, you are still welcomed to register with us and be a part of the CRN.

Outbanders

CB managed to retain a hobby element (conducting QSOs, writing for QSLs, etc.), but typically only with the *outbanders* (a.k.a. *freebanders*). Outbanders are people who operate like hams but are unlicensed. They only operate on self-assigned channels just outside of the CB bands. Most of these operators originally challenged the FCC's rule against talking to someone more than 155.3 miles from their locations. They apparently figured "If I'm gonna risk a fine for talking to the wrong person, why not go the whole way?" So, these operators moved out of band. They often use amateur radio equipment and linear amplifiers to conduct cross-country and intercontinental communications.

The outbanders are typically fairly organized, right down to QSL bureaus, multicolor QSL cards (Fig. 10-1), and callsigns. You can even discover announcements about rare countries being on the air, DXpeditions, etc. Even the communications are organized and "proper." You won't hear a bunch of "potty-mouthed drug dealers" or "people too stupid" to get their amateur radio licenses here.

So, why don't these people go ahead and get their amateur radio licenses? I don't really know, but I suppose that the reasons vary from person to person. However, most of these station operators probably have so many radio friends and are so immersed in the CB outbander hobby that it would be tough for them to leave (Fig. 10-2).

Contacting international CB radio groups

The following is a list of international CB radio groups. Many of these might be outbander organizations. Other outbander groups exist, but those included here are

10-1 An outbander QSL from 2-AT-246, Kenny—a member of the Alfa Tango DX group.

10-2 Several CB transceivers (top left and right) and a Silitronix 1011D amateur radio transceiver (bottom left), which is commonly used for outbanding.

among the most famous. Although just the club letters are listed here, most club names are the phonetic pronunciations of these letters. For example, if you were talking about AT or BB, you would say "Alfa Tango" or "Bravo Bravo." One interesting point about these groups is that they give callsigns to their members. Usually, these callsigns consist of a number to designate their area of the world, then the club letters, then the number of the club registration. So, if you heard a callsign of 32AT016, you could guess that this outbander was the 16th "licensed" operator from country or sector 32 of the Alfa Tango group.

AC HQ
P.O. Box 2107
5700DA
Helmond, Holland

AT HQ
http://mbox.nau.it/alfatango

BB International DX Radio Group
P.O. Box 2923
67617, Kaiserslautern, Germany

BC HQ
P.O. Box 123
22321, Hamburg, Germany

BR HQ
P.O. Box 20676
Maun, Botswana

BRC QSL Bureau
P.O. Box 33 3271
Zichem, Belgium

CB Group International
P.O. Box 1893
Elkhart, IN 45615

Club CB Costa Verde
P.O. Box 4403
4007 Porto, Portugal

DT International DX Group
P.O. Box 174
Preston, PR1 OBS UK

Eagles Lads DX-Group
P.O. Box 1240
46356 Heiden, Germany

PW DX Group
P.O. Box 62
65570 Lempdes, France

RC HQ
P.O. Box 2032
18026 Bourges Cedex, France

WH International DX Group
P.O. Box 7123
69 302 Lyon, Cedex 7, France

Zulu Bravo International DX Group
http://www2.softec.es/~andoni

Internet CB resources

If you are truly interested in CB radio (whether for community service, general communications, or outband operations), the first Internet page to visit is CB Resources on the Internet at:

```
http://www.ultranet.com/~bellvill/cb
```

This listing is by no means complete, but it contains the best set of CB links that I've seen. From this point, you can check out Web pages from CB manufacturers and dealers, outbander groups, community CB clubs, individual CBer hobby pages, and even the *WoodyWorld CB Gazette*.

If you use Internet phone software and have a microphone connected to your computer, you can use a CB remotely in either Port Alberni, British Columbia, Canada or in Phoenix, Arizona. I guess that CBers in Port Alberni and Phoenix must be getting used to talking to people all over the world via the Internet by now! You can find these at:

http://freenet.alberni.net/cbradio.html (Port Alberni)
http://www.primenet.com/~n7ory/cb_links.html (Phoenix)

If you want to communicate with other CBers out there, find out what the latest FCC (often not-so-affectionately referred to as *Uncle Charlie*) news is, how to make better antennas, find the perfect transceiver, etc., investigate CB radio on the Usenet at rec.radio.CB.

For more background on the topic, see also the rec.radio.cb faq at:

http://www.lth.se/cgi-bin/gfaq?cb-radio-faq%2fpart1.gz

11
CHAPTER

Amateur radio

Amateur radio is an exciting medium; you can communicate with others around the world via telegraphy (Morse code), telephony (voice), a variety of digital modes, and even via television (SSTV) transmissions. Of all the different topics covered in this book, amateur and CB radio alone allow hobby transmitting; the others involve only monitoring to whatever is being transmitted.

Unlike the other areas of shortwave radio, amateur radio (also known as *ham radio*) has a much higher profile in the United States. Radio amateurs are often in the news for helping with emergency communications and when promoting the hobby to their local community. People often say "Oh, you're a ham radio operator. Do you talk to people around the world?" That's one element of amateur radio, but it's just the beginning.

Unlike FM and TV DXing, and CB radio, you can't write a book on amateur radio. You can write *dozens* of books on the subject. Therefore, this chapter is by no means a be-all, end-all text on amateur radio. It's something to get you interested in the hobby and to get you started with either getting an amateur license or monitoring/QSLing ham operators.

For more information on amateur radio, read *The Beginner's Handbook of Amateur Radio—3rd Ed.* by Clay Laster, W5ZPV. Some other excellent sources of information are published by the Amateur Radio Relay League. The ARRL, like its monthly magazine, *QST*, has been in existence since the earliest days of amateur radio. Their annual *The ARRL Handbook for Radio Amateurs* (this title has changed over the years, but if you've seen a few, you'll recognize them despite the varying names) is an excellent project book and technical reference. However, my favorite amateur radio book to date is *The ARRL Operating Manual*. This massive reference (more than 600 pages in a large, 2-column format) is not only interesting, it is fun!

One other note about this chapter: It is written specifically about amateur radio operations in the United States. This book will be read by people from other nations, but especially by people in the United States. Because amateur radio rules vary from country to country, the U.S. was chosen as the focal point. The rules concerning amateur radio stations are essentially the same in Canada, however, so most of the specifications in this chapter apply to both countries. If you live elsewhere, many of the general topics of this chapter will apply, but check with your local amateur radio group for information on the specifics of amateur radio frequency allotments, license classes, power output, etc.

Functions of amateur radio

When you get right down to it, the most basic function of amateur radio is communications and friendship that can transcend geographical and political barriers.

An issue of *QST* from 1967 featured an article based on an American amateur's experience when he fought in the Pacific during World War II. The man took a special observation assignment to a tiny island that was occupied by the Japanese. He discovered an outpost on the island that was manned by only one person—a radio officer. He hid outside the shack and waited to kill him. When the man came out, he knocked him to the ground and was about to end his life. Then he noticed that the man was holding a copy of *QST*. The American slowly put away his weapon, and the two talked about amateur radio for an hour or so. As it turned out, they had communicated and exchanged QSL cards before the War. By mutual agreement, they went on their separate ways, without bloodshed.

I have no idea if this story is truth or fiction (it was billed as a true story), but events of this nature (usually less dramatic) have occurred as a result of amateur radio communications. In the harshest of times, such as during a war, when humanity is reduced to sheer numbers, communications are the most important humanizing factor. Amateur radio is also one of the most reliable means of short- or long-distance communications.

Aside from the content of the communications, another important aspect of amateur radio is the means by which it is transmitted. Amateur radio operators must be conscious of the technical aspects of their equipment and its operation. Because of this love of the technical, amateurs often experiment. Some hams are into "fox" and balloon hunting. Others bounce their signals off the moon to talk with someone or chat with the crew of the Space Shuttle. Vintage radio enthusiasts often build or repair equipment so that it will operate just as it would have decades ago. Still others labor for hours building their own equipment—not out of monetary necessity, but just because they want to.

This pioneering aspect of amateur radio has drawn some of the technically best and brightest into the hobby. In fact, many electronics engineers got started in their field after first experimenting with amateur radio. Also, amateur radio electronics are some of the best and most reliable in the field because of the work and experimentation of ham radio operators.

Beginning of Amateur Radio

Clay Laster, W5ZPV
The Beginner's Handbook of Amateur Radio—3rd Ed.

August 1924 QST

The introduction of commercial wireless telegraph equipment after the turn of the century aroused the imagination and interest of people around the world. Some experimenters were content to build simple crystal detector receivers and monitor the raspy code signals transmitted from fixed or rotary spark transmitters at marine or government communications stations. Other experimenters, particularly the restless and aggressive youngsters, assembled spark-gap transmitters as well as crystal detector receivers and began sending "dots and dashes" between homes separated by a few miles. The range of these early "amateur" stations was increased by continual improvements in equipment and higher power output. Amateur wireless organizations, beginning with the Junior Wireless Club of New York in January 1909, were formed across the country. Amateur radio had come of age!

During the infancy period of radio, no government rules or regulations were in effect to govern the use of wireless operations. Wavelengths, or operating frequencies, in the vicinity of 300 to 1000 meters were selected by the users on the basis of available equipment. The inevitable conflict between commercial and amateur users abruptly surfaced when interference from amateur transmissions threatened the reliability of commercial radio communications. The U.S. Navy, which was quickly developing radio communications facilities, handled some of the administrative problems and began issuing "certificates of skill in radio communications" in lieu of licenses. By late 1910, the Navy had issued some 500 of these certificates, many to amateur operators.

By this time, the number of individuals interested in or participating in amateur radio had grown to an estimated 10,000 or more. Amateur transmitters with power outputs of several kilowatts could be heard from up to 400 miles (about 650 kilometers) away. However, most amateurs could not afford such luxury and had to be content with ranges of about five miles (eight kilometers) with occasional contacts from up to 100 miles (160 kilometers). Many wireless equipment stores had appeared by this time, selling crystal detectors, spark-gaps, induction coils, and tuners—the basic ingredients of wireless stations.

Getting into ham radio

Amateur radio includes every class and age group—from school children under 10 years of age through doctors, lawyers, kings, housewives, astronauts, nuns, lighthouse keepers, ambassadors, and entertainment personalities up to the age of 90 or more. Incidentally, the average amateur is 24 years old when he qualifies for his first license, although one-third of the newcomers are under the age of 16.

Only the most devoted enthusiasts can become amateur radio operators. Every licensed amateur in the shortwave bands had to pass a Morse code and theory examination in order to qualify for their license. In the United States, the Federal Communications Commission (FCC) in Gettysburg, Pennsylvania, sets up and administers amateur licensing. The test consists of a Morse code receiving section and a written test on FCC regulations and radio theory.

Amateur radio license classes

On shortwave, an amateur can apply for any of five different amateur licenses: novice, technician, general, advanced, and extra. The classes of licenses each require different written and Morse code qualifications (ranging from relatively easy to very difficult). Each class also carries different operating privileges (ranging from very little to the whole schebang).

- *Codeless technician license* Beginning in 1991, amateurs in the United States were allowed to receive a license without passing a Morse code test. Codeless technicians have the same privileges as other amateurs for the radio bands above 6 meters (54 MHz), but they are not permitted below 54 MHz. Thus, no codeless techs are permitted in the shortwave amateur bands. This new license has been the largest boost in new licensee amateur registration in decades.
- *Novice license* A code test of five words per minute (WPM) and a simple technical examination are required. No license fee is involved; the license is available to any U.S. citizen. It is issued for five years, is renewable, and authorizes Morse code operation on segments of the 80-, 40-, 15-, and 10-meter amateur bands. The maximum PEP output power permitted is 200 watts. The novice license is issued by a mail examination conducted by a licensed amateur with a general-class (or higher) license.
- *Technician license* A code test of five words per minute and a standard written amateur technical examination are required. The license is good for

five years and is renewable. It grants all amateur privileges on at least half the frequencies in each amateur band.
- *General license*   A code test of 13 words per minute, including numerals and punctuation marks, and the standard written test are required. This license, issued for five years, is renewable, and grants all amateur privileges on at least half of the frequencies in each amateur band. The general is very popular because it is the easiest-to-obtain license that allows voice communications on the amateur shortwave bands.
- *Advanced license*   A code test of 13 words per minute and the standard written test, plus an additional test on amateur voice operation, are required. Applicants already holding a general-class license need only pass the additional written test to qualify for an advanced-class license. The advanced-class license grants all amateur privileges on all amateur frequencies, except in 25-kHz segments of the 80-, 40-, 20-, and 15-meter CW bands and in 25-kHz and 20-kHz segments of the 75- and 15-meter phone bands, respectively.
- *Extra license*   In this case, "extra" refers to a class of license, not a duplicate. A code test of 20 words per minute and a comprehensive written examination are required. This license grants all amateur privileges on all amateur frequencies.

If you are interested in becoming a radio amateur, contact your local ham club for information; on the Internet, try http://www.arrl.org/divisions. They are always interested in receiving new members to their ranks, and they will be very helpful in the steps along the way toward your license. Many ham clubs will even conduct both code and theory classes for prospective hams and for those who want to upgrade their licenses.

Contacting your local amateur radio club is very easy via the Internet. Just do a few searches for "amateur radio" with the name of your closest large town or city, and chances are good that a page or two of information will turn up. If you are not on the Internet, contact the ARRL at 225 Main St., Newington, CT 06111. They should be able to provide the contact information. It sure beats chasing people around that have their callsigns printed on their cars' license plates!

Ham radio and the SWL

So far, this chapter has only covered amateur radio as a hobby unto itself. However, it *is* possible to enjoy monitoring ham transmissions without actually getting licensed. For example, some of the shortwave broadcast listeners that I know enjoy tuning in to the transmissions from the AM-mode hams that congregate around 3875 to 3885 kHz. Many of these guys have their transmitter audio running at professional broadcast standards. Because more than a few have been involved in the broadcasting field, the topic of conversation is often broadcasting.

Another hot amateur frequency for shortwave listeners in Eastern North America is 7240 kHz LSB. Every Sunday morning at 10 am (EDT or EST), this is the home of the shortwave listener's net. Here, amateur radio operators who are also listeners exchange their shortwave broadcast loggings and tips. To help out listeners who aren't ham operators, the gateway stations will announce their telephone numbers so that you can call and leave your loggings.

Other amateur communications that aren't necessarily related to shortwave listening or broadcasting are also fun to simply listen to. Sometimes when I get bored with listening to CDs or the major broadcasters, I will tune across the high-frequency amateur bands (20 meters and above) and listen for hams from around the world (Fig. 11-1). Because of the shortwave propagation at these frequencies, often it is possible to hear amateurs from around the world with good signals, yet not hear the person that they are talking to (who is often within several hundred miles of your location).

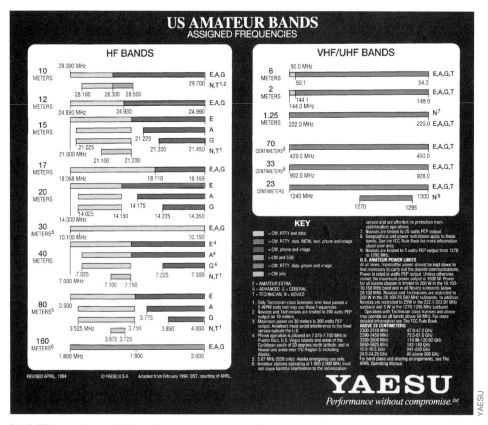

11-1 The amateur radio bands.

If you want to tune in amateur radio transmissions, you need a decent receiver with a BFO or SSB capabilities. Nearly all shortwave broadcast stations (with very few exceptions) transmit in the AM mode because the audio for music and sound effects is much better, because less-expensive receivers can tune it in, and because tens or hundreds of kilowatts output are no problem. Amateurs are limited to using hundreds of watts (200 watts PEP output maximum for novice- and technical-class licenses, and 1500 watts PEP output for the other classes), so every watt of power and bit of intelligibility is essential (Fig. 11-2).

11-2 G4VZO doesn't actually transmit from a ship. The operator just happens to be a fan of Radio Caroline, which is depicted here.

Most amateur operators maximize the strength and intelligibility of their transmissions by using either SSB (single sideband) or CW (carrier wave, Morse code) modes. SSB is covered lightly in Chapter 2, but here's a recap. An AM radio signal consists of three basic components: a carrier (50% of the transmitter power), an upper audio sideband (25%), and a lower audio sideband (25%). The carrier is just a big, wide signal that "carries" the audio signal. The carrier contains no information on its own. The two sidebands each consist entirely of the audio signal.

You can notice the components of an AM signal by tuning across one: Often the audio will be the strongest slightly to either side of the center of the frequency. If you tune right on the frequency, the signal will be strongest, but the audio will be slightly weaker. When you tune to the center of the signal, you are tuning the bulk of the carrier. By moving slightly to either side of the frequency, you are receiving more of each audio carrier.

In the 1950s (Fig. 11-3), amateur radio operators began to filter out the carrier, leaving the upper and lower sidebands. These signals were called *double sideband (DSB)*. They allowed hams to save 50% of the transmitter power. A few years later, most hams began switching to SSB because it saved an extra 50% over DSB. The benefits of using SSB were that the transmitter could emit four times as much audio power as AM alone, the signals were much narrower (so many more transmissions could simultaneously occur in the same amount of radio space), and the intelligibility with SSB was higher (the narrower signal is less prone to fading). Combine the differences between power and intelligibility between AM and SSB, and the result is sixfold: a SSB transmitter with an output power of 100 watts (Fig. 11-4) will have the same "talk power" as an AM transmitter with an output power of 600 watts!

11-3 Lance Johnson, K1MET, in his vintage "dream shack." The two main stations shown use 1950s-era equipment made by Collins and E.F. Johnson.

The problem with SSB is that (as mentioned earlier) you must use a SSB receiver or a receiver with a BFO to receive it. With an AM receiver, people sound somewhat ducklike and music sounds like a blur of seagulls screeching over a garbage dump. For this reason, SSB is sometimes called "single slopbucket" by those who put fidelity first.

11-4 The Kenwood TS-450S transceiver: a superb 100-watt radio.

Tuning in

Although hams often have 70-feet high dipoles and large towers mounted with very large beam antennas, you don't need a great antenna to receive their signals. Whatever you use to receive shortwave broadcast stations (unless you can only receive two or three) should be fine for receiving amateur transmissions. At times, I have simply used a portable receiver with a whip antenna. You probably won't receive any rare ham DX this way, but it's fine for tuning around and getting a feel for what's going on.

One important factor for tuning in amateur stations is to listen to the right band at the right time. The best guideline is similar to casual shortwave listening: listen to the amateur bands above 10 MHz (such as 20 and 15 meters) during the daylight hours and listen to those below 10 MHz (such as 80 and 40 meters) at night and the twilight hours. Of course, signals are audible on these bands at other times, but much fewer are there at much weaker strength. As mentioned earlier, on the frequencies above 10 MHz, only one side of the conversation is typically audible. So, if you want to hear long-distance transmissions, listen to these bands, but if you want to hear complete conversations, listen to the frequencies below 10 MHz.

As you tune across the ham bands for the first time, it might sound to you like mass confusion and meaningless garble. However, don't give up! Concentrate only on the strongest stations, those that you can hear clearly and distinctly. At times, even this can be difficult in some areas when you consider that the United States alone has hundreds of thousands of licensed amateur radio operators. Sometimes so many stations are on the air that, tuning across the band, one transmission will seem to bleed right into the next.

Some ham operators view shortwave listeners as "amateur radio wannabes." They see shortwave listening as simply being that step where you get familiar with shortwave radio until that glorious day when you may finally become a ham. These people are either ignorant of all the fascinating shortwave (non-amateur) stations to listen to or they assume that everyone likes to talk as much as they do. The shortwave-listening and amateur radio hobbies overlap, but are still separate.

Callsigns

To figure out who's who on the amateur radio bands, the governing bodies have established systems by which each ham is licensed with a different callsign. The International Telecommunications Union (ITU) has assigned each country in the world one or more blocks of letters to be assigned by the country's licensing authority for every station that uses radio communications. For example, all radio callsigns in the United States begin with the letters W (including WA to WZ), K (including KA to KZ), N (including NA to NZ), and A (including AA to AZ). Next comes a number followed by one, two, or three letters. A few examples of callsigns in the United States include: W1AW, W3FGQ (Fig. 11-5), KE6DN, N6CSW, N2IFY, WA0NUH, etc.

11-5 A QSL from W3FGQ.

The United States is divided into 10 callsign areas (on the Internet, see http://www.arrl.org/divisions). If a station you hear has the callsign W6AAA, you know immediately that the station is in the United States (W prefix) and in California (6th call area). Some, but not all, countries issue their callsigns in such a geographically structured manner, but you can still determine what country you have heard by the prefix of the callsign.

When checking the callsign locations, remember that a particular country might allow its hams to use different prefixes than those that are normally used. For example, an operator from Florida (4th call area) might be visiting Iowa (10th call area). Even though the callsign is W4, the station is not in that area. In other, special cases, Nigeria allowed its operators to use the prefix 916 during the month of its sixth birthday, and Australian hams used the prefix AX during its bicentennial year. So, you can see that prefixes can and do change. If you hear a prefix that is strange to you, consult the list in Appendix A to see which country it is.

Codes

Now that you can identify the country by its prefix, consider some of the other things that you might hear in a ham conversation that you would not typically hear out on the street. Through the years (as mentioned in Chapter 1), hams have developed shortcuts and abbreviations to help them communicate more rapidly—especially when sending and receiving Morse code. Many of these terms have carried over to voice communications. This coding, as a whole, is called the *Q code*. A few of the more common Q signals follow, but a more comprehensive list appears in Appendix A.

- *QRM*   In CW: "Is my transmission being interfered with?" or "Your transmission is being interfered with." In voice, QRM is a noun or a verb: "There's alotta QRM on this frequency" or "That signal QRMed mine."
- *QSL*   In CW: "Can you acknowledge receipt?" Or "I am acknowledging receipt." In voice, QSL is a noun or a verb: "Will you QSL my report?" or "Please send me a QSL." Although QSL means "I acknowledge receipt of a message," it also represents a physical object (QSL card) that proves that you heard a particular radio station.
- *QSY*   In CW: "Shall I move to another frequency?" or "Change to another frequency." In voice, QSY is a verb: "I will QSY to 40 meters."

The Q code is not used exclusively by hams. Commercial press stations and the military also uses them, but there might be some variations in the two letters that follow the Q. Another system that is somewhat similar to the amateur Q code is the CB/public service 10 code.

Some other codes that are used in amateur radio relate to the quality of the signal that is being received. The standard shortwave-listening signal codes are covered in greater detail in Chapter 15. The system that is typically used in amateur radio is *RST*: *readability*, *signal strength*, and *tone*. Under R, the numbers run from 1 (uncopyable) to 5 (perfectly copyable); under S, the numbers run from 1 (faint, signals barely perceptible) to 9 (extremely strong signals); and under T, the numbers run from 1 (very rough) to 9 (perfect tone, nice audio with no perceivable distortion). The complete RST code is also presented in Appendix A. A perfect signal report would be RST599.

Hams and CBers often modify the RST code to just RS. The ratings for R and S are the same as those used for the RST code. For a perfect signal, the other amateur would say something like "Beautiful signal down here, Bryn. I've got you 5 by 9."

Another measure of receiving signals is by the S meter on a receiver. The name *S meter* is used because it's a meter that measures the direct signal strength from the receiver. Typically, a reading of S9 is about midscale, and the top half of the meter is calibrated in decibels. A signal that would move the needle over to the 40 on the high end of the scale (often referred to as "40 over 9" or simply "40 over") would put it at 40 dB over S9—an extremely strong signal! As mentioned in Chapter 2, the S-meter scales are calibrated differently on most radios, so the S reading is often only helpful to you, and not another listener or station operator.

Other interesting aspects of amateur radio

The amateur radio hobby began on the mediumwave and shortwave frequencies. Shortwave voice and CW operation have been the mainstays of amateur radio for most of this century. However, for the past few decades, some other accepted means of communications have developed within amateur bands, including repeater operation, television transmissions, and packet radio. Also, radio "foxhunts" and balloon chases have become popular for the amateur radio direction finder.

VHF repeater operation

The 1970s and 1980s saw the growth of what might well be the most popular mode of operation today: FM with repeaters. For reliable local communications, nothing can match them. In fact, the repeater operation is very similar in coverage and operation to cellular telephones. Although these communications are not within the shortwave bands and require amateur radio equipment or a scanner, they are included here because they are such a popular aspect of amateur radio.

Before getting into the theory and operations (very lightly) of repeaters, remember that radio signals on the VHF bands cover an area in an essentially line-of-sight manner. This prevents interference from distant stations under normal conditions, but it limits the range of these signals—especially those that are located in valleys. For years, amateur operation on the 2-meter VHF band was limited because hams needed to install tall beam antennas in order to "get out."

To circumvent the range problems, yet accentuate the strong points of VHF operation, amateurs radio clubs began to establish repeater networks. Repeaters, as the name implies, are remote transceivers. Here's how it works. A ham calls into the repeater on its calling frequency. The receiving portion of the repeater picks up the signal from the amateur and keys the transmitter. The transmitting section of the repeater retransmits the signal for miles around.

The advantages of repeaters are that amateur radio clubs typically put their funds together to purchase the high-powered repeater, the mountain-top land where the antenna and repeater will be located, and the tall antenna and mast. Because of the cooperation of many people, everyone benefits by having an excellent antenna and powerful transmitter for very low cost. Also, the tall antennas of the repeater allow the amateurs to use portable, low-power transceivers for communications. In fact, the most common transceivers for use with repeaters are small, handheld models that are very similar in size and design to walkie-talkies (Fig. 11-6).

So, with a relatively inexpensive transceiver and an amateur radio license, you could have something that would be somewhat similar to free cellular telephone service. However, of course, no commercial calls are allowed on the amateur radio bands, and many hams in the area are listening to your contacts.

Packet radio

Packet radio emerged in amateur radio circles in the late 1970s and early 1980s. This communications medium is very popular with new and younger hams because it mixes amateur radio with computers. Because of the different protocols, software, and hardware for computers, packet radio has many different possibilities and complications. This facet of amateur radio alone has generated a handful of books over

11-6
An Alinco DJ-160 VHF FM handheld transceiver.

the past decade. This section very superficially looks at packet radio. For more information on the topic, see *Packet Radio* by Jonathan Mayo, *The ARRL Operating Manual*, *Your Gateway to Packet Radio—2nd Ed.* by Stan Horzepa, *Your Packet Companion* by Steve Ford, and http://www.tapr.org/tapr/html/pkthrome.html.

In order to access the Internet or a BBS, you need a computer, communications software, a modem, a working telephone line, and another computer to link up with. When you make a connection, you type commands into the computer to activate the modem, which makes the telephone call.

Packet radio is essentially a wireless version of the computer online communications. Typically, a packet station operates via a packet communications program on a computer that is connected to a terminal node controller (TNC). The TNC is connected to an amateur radio transceiver, which is in turn connected to an amateur radio antenna. The TNC is essentially a computer modem that is constructed specifically for operating with a radio. Instead of a telephone line, an amateur radio transceiver and an antenna are used.

When communicating via packet radio, you first call the station that you want to contact on the proper frequency (the one that would be monitored by the station that you are attempting to reach). If the other computer is "listening," your computer will link up, just like connecting to the Internet or to a BBS. At this point, nearly anything can be accessed and nearly any data can be passed. For example, you can even send e-mail or access a BBS via packet radio.

Some packet radio stations and networks are elaborate—even operating over repeaters so that somewhat local operators can freely access the BBS or even the Internet. Thus, it is very inexpensive to access a vast wealth of information, and you don't even have to worry about the unreliability of the telephone system (especially important if you live in a country or region that lacks local Internet access or that has unreliable telephone systems).

The modern-day drawback of packet radio is speed. Current modem speeds are now 28.8K. Because this technology has improved, much of the information and software that you can download is more complicated, maxing out those K. However, because of atmospherics and man-made interference, the typical speed for shortwave packet communications is a mere 300 baud, although the VHF/UHF TNC speed jumps up to 1200 baud.

Television

Amateur television has existed for decades in two general formats: *fast-scan TV* (*FSTV*, also known as *ATV*) and *slow-scan TV* (*SSTV*). FSTV operations typically only occur in the UHF frequencies, so they are not covered in this book. However, the technical operations and format are very similar to North American commercial TV, so use that as a guideline. As is the case with amateur radio and TV, no broadcasting is ever allowed; expect video of an amateur in his shack or maybe some clips from a vacation to the beach, not reruns of "Hawaii 5-0."

Slow-scan TV, however, is used throughout the shortwave amateur radio bands. The bandwidth required for a real-time video signal is very wide; a typical FSTV signal would wipe out a large chunk of shortwave spectrum. So, the amateurs were forced to compensate and develop a format that would have a 3-kHz bandwidth (the same as a standard SSB voice signal). The result is SSTV, which is TV sent at much lower resolution (typically about ¼ of the resolution of standard TV).

Another technical difference between FSTV and SSTV is the frame-refresh rate. A typical TV screen in North America consists of 525 lines of tiny dots. The TV scrolls through the lines and illuminates each line, one-by-one. When it is finished, the screen blanks out and starts over. This process occurs 30 times per second. The human eyes and brain can't keep up with the changes, and everything is essentially blurred together. The result is that we see a picture that appears to be fluid and continuous. The screen of an SSTV signal is only refreshed once every eight seconds! Needless to say, the content is pictures, not video.

Because of the boom in video and computer technology, SSTV and FSTV are on the rise. Personally, I would rather download still images from the Worldwide Web than decode SSTV, but I suppose that part of the fun is the challenge. FSTV still has many technological advantages over attempting to play real-time video over the Internet, however. Speaking of the Internet, one SSTV-dedicated Web page is http://www.ultranet.com/~SSTV.

Hamfests

Hamfests grab the most public attention of any amateur radio events because they are open to the general public. Hamfests are part convention, sale, and flea market, depending, in part, on the particular hamfest (Fig. 11-7). Some have hourly forums

42nd Annual BREEZESHOOTERS HAMFEST & COMPUTERSHOW

LARGEST HAMFEST IN THE TRI-STATE AREA

GENERAL HAMFEST CHAIRMAN LARRY POLESNAK WB3KJG (412)367-7933

BUTLER FARM SHOW GROUNDS
SUNDAY • JUNE 2, 1996 • 8AM-4PM

****

TABLE SELECTION IS ON A FIRST COME FIRST SERVED BASIS, SO APPLY TODAY!!!

CONTACT: GEORGE ARTNAK N3FXW
3350 APPEL ROAD
BETHEL PARK, PA 15102

(412) 854-5593

CALL TO RESERVE INDOOR
TABLES AT $15.00 PER TABLE
CUT OFF DATE MAY 20th

$2.00 PER PERSON
ADMISSION DONATION

- INCLUDES SMALL PRIZE DRAWINGS DURING THE HAMFEST
- CHILDREN UNDER 12 YEARS OF AGE ADMISSION IS FREE
- TAILGATE / FLEAMARKET SPACES AVAILABLE AT $5.00 PER SPACE
- FARM SHOW GROUNDS AVAILABLE TO VENDORS/TAILGATERS AFTER 5 PM SAT. JUNE 1, 1996

AMATEUR PRIZES BY ICOM AMERICA

COME SEE PRODUCT DEMONSTRATION BY REPRESENTATIVES OF ICOM AMERICA

MAIN PRIZE DRAWING AT 4:00PM - WINNERS NEED NOT BE PRESENT TO CLAIM MAIN PRIZES

1st IC-775DSP NEW HF XCVR
All Mode, Built-in Tuner & PS, 200W

2nd IC-738 HF XCVR W / PS
All Mode Built-in Tuner, 99 Memory Chan.

3rd IC-706 HF/50/144MHz XCVR
All Band, 100W HF, 10W 2M, Remov. Front

4th IC-707 HF MOBILE XCVR
100W, Gen. Rcv Coverage, Dual VFOs

5th IC-2350H 2M/70CM MOBILE
50W VHF, 35W UHF, 110 Memories

6th IC-Z1A DUAL BAND HT
2M-70CM, Detachable Front Panel

7th IC-W31A DUAL BAND HT
2M-70CM, Alphanumeric Memory Names

8th IC-2GXAT 2M HT
Durable Constr, 40 Non-Volatile Memories

9th IC-2000H 2M FM MOBILE
50W, 50 Channel Memory

10th UNIDEN 2510 MOBILE
10 Meter, Multi Mode Mobile

11 HALF HOUR PRIZES OF ICOM IC-2GXAT 2M HAND HELDS
STARTING AT 10:00AM AND DRAWN ON THE HALF HOUR

PENTIUM 100 COMPUTER RAFFLE

INTEL PENTIUM 100 PCI with 8MB Ram, 256K Cache, WINDOWS 95, 1GB EIDE Hard Drive, 104 Keyboard 14"SVGA Color Monitor, 1MB PCI Video adapter, 3.5" Floppy drive, 2 Serial Ports, 1 Parallel Port, Mouse 4X-CD ROM, Sound Card, Fax-Modem and a Ink Jet Printer.

11-7 A flyer from the 42nd annual Breezeshooters Hamfest & Computershow in Butler, Pennsylvania.

with well-known speakers, others concentrate on having well-known amateur radio companies display their new products, and still others are just a big radio flea market. The largest hamfests typically concentrate on the first two, and the smallest on the latter two.

I love hamfests because I am fascinated with junk, history, and low prices. I do much of my shopping for light-duty electronics, books, and components at hamfests because the prices are often the best that you can find and because you never know what you'll discover (Fig. 11-8). Also, it's fun to see the manufacturer's new equipment first-hand—a great way to get up to date. Finally, the information from the forums is excellent.

11-8 A hamfest table with books, software, a complete Heathkit SB-104 amateur radio station, a linear amplifier, an antenna tuner, and more!

If you want to go to a local hamfest and you live in the United States or Canada, look in the monthly magazines *QST*, *Nuts'n'Volts*, and *CQ*. *Worldradio* and *73 Amateur Radio* also contain hamfest listings, but they aren't nearly as comprehensive. For long-range hamfest information, see http://www.arrl.org/hamfests.html.

A general guideline for hamfests is that the two-day hamfests are the largest events. Also, those that have large magazine ads are typically large, too. Bigger isn't always better, but the more vendors and flea market sellers that are there, the more you can see and select from.

One strange aspect of hamfests is that those that are "out in the sticks" aren't always the smallest. For example, the largest hamfest in the world, with a regular attendance of more than 30,000, is located in Dayton, Ohio. Also, although Philadelphia is the fourth largest city in the U.S., it has no large hamfests.

"Foxhunts"

A foxhunt in amateur radio circles has nothing to do with chasing and killing foxes. Instead, the sport involves direction-finding radio signals to locate a hidden transmitter.

Many amateur radio clubs sponsor foxhunts at hamfests and other club functions. Those involved with the contest place a very small, low-power VHF beacon transmitter in an inconspicuous location (usually placed within a few miles of the starting location). Those who are involved with the foxhunt are readied at a starting position with their radio direction-finding (also known as *RDF* or *DF*) equipment. To DF the beacon transmitter, the hobbyists use sensitive, portable receivers with highly directional loop antennas. The first person to locate the transmitter wins.

Foxhunts can be fun radio events—a physical activity that doesn't involve the proverbial ham shack. Another positive aspect of the contests are that they are a primer for amateurs to track down radio signals in real-life situations. Occasionally, amateur radio equipment is stolen and used by the thieves, or amateur radio transmitters are used maliciously (to jam repeaters, etc.). Foxhunting provides amateurs with experience to go out and find the sources of the interference (in the latter case) or locate the equipment (in the former example).

For more information on foxhunts, see *Transmitter Hunting: Direction Finding Simplified* by Joseph Moell and Thomas Curlee. I think this book is now out of print, but it is still available from a few amateur radio stores and from some libraries. On the Internet, try http://spitfire.ausys.se:8003/hsn/rpo.htp.

Sending QSL reports

Consider for a moment that hundreds of thousands of amateur radio operators are licensed worldwide. Then, consider the hobby of writing for and receiving QSL cards from around the world (Fig. 11-9). Mindboggling, huh? Not only would the total number of hams be impossible to QSL, but the postage costs would be tremendous! If you sent 200,000 reports to amateur operators in the United States (assuming that your location is in the U.S.), 30,000 to hams in Canada, and 200,000 to amateurs overseas, the total postage bill alone would tally $197,800. An equal amount of return postage would double the cost to $395,600. Now that's a lot of money to spend on a hobby!

Fortunately, amateur radio operators are resourceful, energetic, and thrifty. To allow amateurs to actively pursue QSL collections without blowing their life savings on postage, QSL bureaus were organized and established.

To better understand how a QSL bureau works, take the example of a special-event station. Special-event stations often operate from a special or historic location for a single day with a special QSL card. As a result, many other amateur stations pursue these operations for the special QSLs. After the special-event station has its day on the air, the operators go through the logbook and fill out hundreds of QSL cards for those that they were in QSO with.

When the cards are all filled out, instead of sending each card to each separate ham station, the whole batch of cards are sent to an outgoing QSL bureau. Here, the package is opened and volunteers sort out the cards by callsign prefix. After sorting out packages like this from other stations, the piles of sorted QSLs will be sent to the proper incoming QSL bureaus. For example, all of the QSLs for stations with the VE3 prefix (Ontario, Canada) will be sent to the VE3 incoming QSL bureau. VE3 ama-

11-9 An impressive QSL card from the amateur radio station at the United Nations Headquarters.

teurs send SASEs (self-addressed stamped envelopes) to their incoming QSL bureau. The volunteers sort through the QSL cards and file them in the envelopes to be sent to their final destinations.

If you are a shortwave listener and you want to QSL some amateur stations, you can still benefit from this service if you live in the United States (this type of service varies from country to country). Currently, Mike Witkowsi is the ARRL incoming QSL bureau head for shortwave listeners. For more information on this service, send a SASE to: 4206 Nebel St., Stevens Point, WI 54481.

Amateur radio online

Aside from the more specialized amateur radio online Web pages, a number of more general pages are also posted.

The ARRL has been the dominant amateur radio organization in the world for nearly a century. It just makes sense that their Web page would be one of the first to check out: http://www.arrl.org.

If you are a Canadian amateur radio operator, take a look at the Radio Amateurs of Canada Web page at: http://www.rac.ca.

One of the biggest annoyances is looking up callsigns, searching for amateur radio addresses, etc. *The Radio Amateur Callbook* is a massive listing of amateur radio callsigns and addresses. It's a great resource, certainly one that you must have if you're an active amateur. However, the North American edition and the International editions are the size of two New York City phone books. Worse yet, you have to buy new copies every year to stay up to date. Obviously, the following amateur radio callsign database Web pages are popular:

```
http://www.valr.edu/~hamradio/index.html
http://www.mit.edu:8001/callsign
http://www.qrz.com/callbook.html
http://www.mcc.ac.uk/cgi-bin/callbook
```

An interesting variation of the callsign listing is an FTP site that features e-mail addresses of hams on the Internet: ftp://ftp.cs.buffalo.edu/pub/ham-radio/hams_on_usenet.

If you need recent information and news about the amateur radio hobby, see the Amateur Radio Newsline and Ham Radio Online Magazine:

```
http://www.acs.ncsu.edu/hamradio/news.html
http://206.13.40.11
```

CHAPTER 12

FM and TV DXing

Aside from electricity, two of the most taken-for-granted technologies are FM radio and television. Typically, you just expect them to work, and they always perform the same way—there's no mystery or unpredictability. Mediumwave radio varies noticeably between day and night, and shortwave radio seems completely bizarre and random. However, beneath the plain-vanilla veneer of FM and TV monitoring are some freak propagational conditions that can make this type of DXing fun.

You might not want to spend hours hunting around the FM band, changing TV channels, or writing reception reports—just for a handful of verification letters on station letterhead and some stickers. That's okay, too. Even if you don't ever plan to DX on these broadcast bands, some of the information included in this chapter will enable you to install a system that will improve your listening and viewing.

Note: In this chapter, TV and FM signals are often collectively called *VHF signals*. Although the FM broadcast band and TV channels 2 to 14 are in the VHF band, TV channels 14 to 69 are in the *UHF band*. If something applies directly to the UHF frequencies, it is called out as such. Otherwise, expect that *VHF* includes both the VHF and the UHF TV channels.

Propagation

It's the p-word again. This time, typical propagation at the VHF frequencies prevents these signals from traveling vast distances. Radio propagation is covered in Chapter 5, but here's a little recap.

Radio waves travel through the air until they are either absorbed or refracted by the layers of ionized particles in the atmosphere (known collectively as the *ionosphere*). Signals at the lowest (LF) frequencies are completely absorbed by the lowest layers of the ionosphere; only the surface wave of the radio signal is heard. At the top of the LF and throughout the mediumwave (MW) band, the surface wave is still important, but some signals are refracted by one of the lower levels of the ionosphere at night, when the layers are least ionized. Moving up to the shortwave frequencies below 10 MHz, the surface wave is very small—often less than 100 miles, but the radio signals slice through the lower layers of the ionosphere and are refracted by the upper layers, resulting in signals that can be heard hundreds or thousands of miles away at night. Above 10 MHz, the radio signals will often "skip" thousands of miles during the daylight hours, but pass right through the ionospheric layers at night. In addition, essentially no surface wave exists at these frequencies.

FM and TV (VHF) frequencies are above 10 MHz (88 to 108 MHz for FM, and roughly 46 to 850 MHz for TV). You can deduce that (1) if radio waves have the ability to cut through more layers of the ionosphere as the frequency increases and (2) shortwave signals above 10 MHz are typically only refracted by the ionosphere when it is strongly ionized, then the VHF frequencies will pass right into space any time. That's a correct deduction, Watson. The VHF frequencies have no reliable propagational characteristics. Reception is assumed to be *line-of-sight*. That means that, if you can see the transmitting antenna with binoculars or a telescope, you can probably see or hear the station. If you live in a hollow or have a mountain between you and the transmitting antenna, you won't receive the station.

I grew up a few miles from the highest point in Pennsylvania—no obstructions east until Europe and no obstructions west until the Rocky Mountains of Colorado. Our TV reception was fantastic—we could regularly receive (with anywhere from weak to strong signals) stations from Steubenville, Ohio (100 miles away); Wheeling, West Virginia (80 miles away); Pittsburgh, Pennsylvania (70 miles away); State College, Pennsylvania (100 miles away); Hagerstown, Maryland (70 miles away); and Washington, DC (125 miles away). Not bad: five states/districts with just typical reception. However, many people in some of the lower-lying towns could only receive two or three of the closest stations because the mountains were in the way.

So, with virtually no obstructions, why couldn't I receive TV stations from Denver and Dublin? Probably the biggest reason is the curvature of the Earth. You can only see so far on even the clearest day because the Earth curves away from you. Even if those TV signals are "shot" right along the Earth's surface, they will eventually slice out through the ionosphere at a low angle. I would assume that if the Earth was flat, the signals could travel much further (considering that FM radio and TV signals travel into space), but this point is more than a bit irrelevant (Fig. 12-1).

Excellent VHF propagation

Thus far, this section on VHF propagation has covered the only ways that these radio waves typically behave. However, some really cool, if strange, things can happen to these signals that carries them hundreds or thousands of miles away. These modes of propagation include tropospheric propagation, sporadic E skip, F2 skip, auroral propagation, and meteor scatter.

Tropospheric propagation

Tropospheric (often just called *tropo*) *propagation* is one of the most basic ways to receive VHF DX. With tropo propagation, the laws of shortwave propagation are pitched into the cosmic file 13. It has nothing to do with the ionosphere or even ionized particles. It's all weather, so put on your meteorological thinking cap and have a seat by the chalkboard.

FM and TV DXing 267

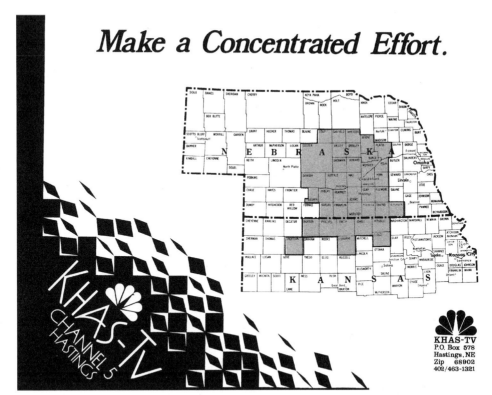

12-1 A coverage map of KHAS-TV, Channel 5, in Hastings, Nebraska.

As you know, the weather on the Earth consists of constantly moving warm fronts and cold fronts that bring us rain, snow, clear skies, etc. Very often, a warm front will push out a cold front (or vice versa) or collide to cause a nasty storm. The 1994 blizzard in northeastern North America that many people have called "the storm of the century" was caused when a cold front from the Northwest collided with a very wet, warm front from the Southeast. Many feet of snow fell, and millions of people spent a few days digging out.

In addition to causing vicious storms, the warm- and cold-front combos also do weird things to VHF radio signals. When a warm air mass is above a cold air mass, the colliding edge of the masses can refract signals back to the Earth; when the warm and cold air masses are reversed, the colliding edge will reflect signals back! The end effect for the DXer is that the weather pattern brings in a portable ionospheric layer so that you can receive distant stations. The signals from these tropo effects can easily travel hundreds of miles.

The ultimate in tropo reception occurs when sandwich of cold and warm air masses appears in the atmosphere. For example, if there is a high warm air mass, a lower cold air mass, then a lowest warm air mass, the radio signal could travel up into the middle of the air mass, refract down through the cold air mass, bounce back up from the colliding edge of the lower warm air mass, etc. The effect is called *ducting*, and signals will typically get trapped in a duct until they finally return to the Earth—hundreds or thousands of miles away.

Because tropo reception depends on bands with different temperatures of air, the best reception will occur at the edges of storms and when there are rapid temperature fluctuations. As a result, spring and autumn are the best seasons for tropo reception, but if the weather conditions are optimal, good tropo can occur during the winter and summer months, too.

Tropo is best heard in the UHF frequencies (such as TV channels 14 through 69), although it also occurs throughout the VHF spectrum. However, the "skip" methods of propagation occur no higher than the VHF frequencies. So, if you are watching the static on Channel 25 in Minnesota and suddenly WHAG-TV from Hagerstown, Maryland fades in for about five minutes, you can bet that the reception was via tropospheric propagation.

Skip

The other standard method by which you can receive VHF signals is *skip*. Skip is the same type of ionospheric refraction and reflection that occurs throughout the short-wave frequencies. The two basic types of VHF skip are *sporadic E* and *F2*. Both of these are named after the layers of the ionosphere by which they are refracted.

F2 skip is incredible and rare to receive, but the process by which it happens isn't terribly exciting. At times, when the 11-year solar cycle is at its peak, the *maximum usable frequency* (the frequency by which signals will be refracted by the ionosphere) will sometimes reach as high as 60 or 70 MHz. However, the current solar cycle won't peak until 2006 or 2007, just in time for the 7th edition of *The Complete Shortwave Listener's Handbook*! Until that time draws nearer, don't worry about F2 VHF skip.

The other type of skip, sporadic E, is very common, but it is unpredictable and only occurs for minutes at a time. Nobody really seems to know what causes sporadic E skip, but patches of the E layer of the ionosphere suddenly become strongly polarized—enough so to refract VHF signals, which usually pass right through the E layer. These polarized patches only refract VHF radio signals for a few minutes at a time.

The most common times for receiving stations via sporadic E skip is in the middle of your summer, regardless of what hemisphere you are located in. Like tropo propagation, stations via sporadic E will sometimes suddenly appear on your radio or TV with amazing clarity, then quickly fade away. However, sporadic E differs in the sense that it rarely affects frequencies higher than the FM band (88 MHz).

Other propagational modes

The other modes of propagation are more freakish events than anything that you can depend on. They are listed here more because they are interesting than for any practical value:

- *Auroral skip* Auroras (also known as the *northern lights* or the *southern lights*) are fantastic visual events that are caused by massive solar flares. Across the shortwave bands, auroras will often destroy reception—polarizing the ionosphere and absorbing most any signal that comes their way. In the VHF ranges, the aurora-polarized ionospheric layers will often refract signals back to Earth, making DX possible.

- *Meteor scatter* When meteors fall to Earth, they can ionize the layers of the ionosphere and cause VHF signals to be refracted back to Earth. With meteor scatter, it's time to take off that meteorological cap and put on the astronomical (*not* astrological) touque. If you want to DX via meteor scatter propagation, you should check astronomical news for meteor forecasts.
- *Man-made object skip* I've heard reports of people receiving VHF radio signals that were skipped back to Earth via such objects as flying airplanes. Although these claims seem somewhat plausible, how could you be sure that the signal that you just received was reflected by an airplane? After all, it could have been retransmitted by a UFO. . . .

Equipment

You can receive good DX with nearly any radio or TV because the signals from sporadic E skip and tropo propagation are often strong. The most important conditions are that you have good information and that you monitor the frequencies regularly. Even though mediocre equipment won't exclude you from catching some good DX, a decent radio or TV will improve your chances of receiving distant stations with marginal signals.

FM

You can find FM radios in everything from boom boxes to portable stereos to car stereos to computers. Despite being so widespread, most of these radios are little more than toys—an inexpensive IC chip and a few discrete components that pull in a few FM signals. To find good FM performance, you need to follow the guidelines for good mediumwave performance:

- Many of the more-expensive, recently manufactured car stereo head units have good, sensitive receivers. The problem is that these units don't have separate record out or headphone jacks.
- The GE Superadio models are the choice of most broadcast band (nonshortwave) listeners. This radio looks quite simple, but it is very sensitive and selective. It is also quite important that the receiver has connections for external antennas and jacks for recording the programs and listening via headphones. One downside to the Superadio models is that they are mono only. They are also in a tall, wide cabinet: not at all compatible with a component stereo system. Another is that these radios aren't digital, so you only have a close approximation of where you are tuning.
- A nice alternative to buying a new radio specifically for FM broadcast DXing is to use a receiver from a component stereo system that you already have. These receivers are well built and typically contain a stereo amplifier. Because they have been developed for high-performance applications, they contain such features as stereo line outputs, headphone jacks, external antenna jacks, a ground jack, station memories and presets, etc. The radio portion of these receiver/amplifiers will vary from model to model, so be sure to ask other FM DXers for advice and do comparison testing at stereo shops.

- The final option for picking out a better-than-average FM broadcast receiver is to purchase a communications receiver with the FM broadcast option. These receivers also typically vary in quality for FM band reception, but most are superb, with digital display, excellent sensitivity and selectivity, station memory and presets, audio output jacks, external antenna jacks, etc.

TV

Choosing a good television for DXing is much different than choosing a good radio receiver. TVs are idiot boxes: even though the parameters for receiving TV signals are very complicated, they are intended to be turned on, not played with. As a result, you won't find DSP controls to help you reduce the "snow" interference in a picture, noise blankers, synchronous detectors, and notch filters. You certainly won't find a tuning knob with a digital frequency display. TVs are sold by the size of their picture tubes, the sharp picture, good color, or by the design of the cabinet.

So, how do you find a good TV for DXing? For the most part, it's difficult, if not impossible, to accurately rate all of the TVs on the market for quality of reception. About the best that you can do on your own is to take a look at the features on the TVs at a local appliance store. Unfortunately, if the store is part of a large chain, it will almost certainly have the TVs connected to cable TV. To rate the sensitivity of a receiver, explore smaller appliance and furniture stores.

Instead of testing every TV yourself, follow a few general guidelines:
- Color reception is not important.
- The size of the picture tube is basically irrelevant.
- If you have a family, you don't want to annoy everyone by constantly flicking through the channels.

The basic conclusion here is: Get a relatively small (13"), black-and-white TV. These TVs are intended for apartment dwellers with poor antennas, so they seem to be more sensitive than many of the larger models. Another great feature is that they're cheap; I often see these models selling for $10 to $25 at yard sales and hamfests.

Antennas

Of course, a good antenna will help you hear plenty of FM and TV DX, but unlike longwave, mediumwave, or shortwave, you can still catch great DX without a good antenna. As described in the propagation section, when a sporadic E or tropo ducting event occurs, the signal that you receive will often be as strong (or stronger) than a local station. My best TV DX catch (more than 1000 miles away) came on a portable black-and-white TV that had the telescopic antenna broken off (Fig. 12-2)! I wonder what else I could have heard that day with an excellent antenna?

DXers, however, will typically use the best (often the most expensive) large Yagi or log-periodic antenna they can find for this type of DXing. Both of these antenna types consist of a basic dipole antenna with a number of other director and reflector elements that are used to "focus" the signal. The result is very high gain: the antenna can receive very well in one direction only. The better the antenna, the narrower this reception area is (but the signals are stronger).

P.O. Box 578
Hastings, NE 68902
402-463-1321

An NBC Affiliate

John T. Benson
General Manager

KHAS-TV 5
HASTINGS

16 June 1989

Andrew Yoder

3007R 4th Ave.

Beaver Falls, PA 15010

Dear Mr. Yoder,

This is to verify your reception of KHAS-TV Channel 5 in Hastings, Nebraska on the 30th of May 1989 at 7:59 - 8:05 PM CDT.

Thank you for your report, we have an ERP of 100 KW with an antenna height of 767 feet.

Long-haul conditions such was occurring on May 30th should be quite common during the next few years due to the 11-year solar cycle being at it's peak. Low-Band VHF stations such as ours will be coming into your area by means of E-Layer skip propagation.

Good luck with your TV and SWL DXing!

Sincerely,

Roger Book, Assistant Chief Engineer

12-2 A QSL letter from KHAS.

Likewise, you should buy the best antenna that you can afford (if FM DXing is your hobby, get a good FM-only antenna) and install it as high as you can on a rotor. The rotor will turn the antenna so that you aren't limited to receiving in just one narrow direction.

Warning: Never, ever install your antenna near power lines! If you are planning to install a mast, do *not* place it or your ladder—even temporarily—in a position where it could fall over and contact the power lines. These lines can sizzle you crispy faster than a fish stick in a Fry Daddy. Every year, a number of people are electrocuted while installing TV, shortwave, and amateur radio antennas. Be careful—even if it means settling for a lesser antenna.

DXing

You must track a few different sources of information from those for shortwave and mediumwave DXing to effectively discover when is the best time for DXing. However, as with those types of monitoring, it is always good to check around sunrise and sunset for possible gray-line DX. Some of the other times to tune around include:

- Very broadly, in the summer months because the chances of sporadic E skip increases. This is great for mediumwave and shortwave DXers because the summertime static crashes typically eliminate most good DX.
- If the forecasters accurately predict the collision of a warm front and a cold front, you might receive some good DX via tropo.
- I don't know how common astronomical reports are, but one was featured on our local public educational TV station for years. If you can determine when meteors will be falling, you might catch some interesting signals.

If you are serious about DXing FM and TV stations, the best way to know how and when to listen is to use a scanner in the low-band area (30 to 50 MHz). Sometimes this region is quiet, but at other times, the band will be loaded with public service and commercial communications from hundreds of miles away. When the latter occurs, you know that some good conditions are causing the stations to come in—and if the stations can be heard on the lower frequencies, there's a good chance that the higher frequencies will also be active with distant broadcasts (Fig. 12-3).

12-3 Here's some rare FM DX in North America and Europe: A bumpersticker from Olimpica Stereo in Bogota, Colombia.

FM

The most helpful FM DXing tips relate directly to your information. The first thing that you need to do is learn the radio band in your general area. Write down the stations, frequencies, slogans, locations, and formats of the stations that you can regularly receive in your area. If you have a very directional antenna and can receive a number of stations on one frequency, it would also be helpful to note what direction the antenna is pointing. Once you have these basic guidelines drawn up, you can more readily identify which stations are not typically received in your area.

A must-have guide for FM DXers is the annual edition of *The M Street Directory*, which features radio station callsigns, formats, locations, lists sorted by callsign, lists of radio station fax numbers, etc. This book is available from most of the large shortwave mail-order dealers.

In some parts of the world, the FM frequencies vary. For example, radio stations in the United States are licensed to operate only on every odd tenth of a megahertz (for example, 92.1, 92.3, 92.5, etc.). However, in Mexico, stations are only assigned to operate on even tenths of a megahertz (for example, 92.2, 92.4, 92.6, etc.). Depending on where you live, you can generally determine the location by the frequency. As you can see from the stickers in Fig. 12-4, some countries (in this case, Germany) even allow stations on both odd and even tenths of a megahertz. If you want to use the frequency to help you determine the location, you need to know the FM band and the frequency allocations for nearby countries.

Some countries use different specifications for FM broadcast bands. For example, Russia and Eastern Europe use two FM broadcast bands: the *VHF low band* and *high band*. The low band runs from 66 to 73 MHz, and the high band covers from 100 to 108 MHz. You can try for these stations if you have either a good scanner or a VHF receiver. You can forget listening here if you live in North or South America, Oceania, and most of Africa. If you live in Western Europe and you can hear a station or two broadcasting from 66 to 73 MHz, you can bet that you have some pretty good DX.

TV

TV DXing is much more complicated than almost any other form of DXing because you can receive both audio and video signals. Very often with DX, one of these signals will be received better than the other.

A few tips for TV DXing include:
- Have a Polaroid camera handy. You can click off a quick photo of an ID screen, which is typically used between advertisements.
- If you are receiving a network station, tune to another closer station in the same network and see how the advertisements differ.
- Typically, more than half of the ads will be aired nationally. These are useless. When you see a local ad on the local station, that's your cue to get back to the DX station and listen/watch for telephone numbers, town locations, and station IDs.
- Record the stations that you DX onto video tape. Then, you can watch the ads and IDs and listen for key locations without the worry of missing something while you're writing down all of the information.
- If you have a good VHF/UHF receiver (not a scanner), try also connecting this to your TV antenna and listen to see if you can pull a better audio signal from the radio.

Depending on where you live, you might be able to receive stations from different countries. Unfortunately, many of the readers of this book will only be able to catch stations from within the United States. However, many countries use different standards for their broadcasts, which cannot be received on all TVs. For example, nearly all countries in North and South America use the NTSC system of video transmission, but most other countries in the world use the PAL system. If you live in Ireland and a TV signal from Boston, Massachusetts is coming in on your standard Irish TV, the station won't be decodeable. The opposite is true, too.

12-4 Some stickers from FM stations in Germany. Notice that the frequencies end in both even and odd digits.

If you live in a coastal region or near other countries that use different standards for their TV systems, you can still DX these broadcasts if you have a good scanner or a VHF/UHF receiver. Of course, you won't receive the video portion of the broadcast, but the audio portion will still be copyable.

QSLs

For more information on QSLs and the verification process, see Chapter 15, which exclusively covers that aspect of the hobby. It is often more difficult to receive QSLs from FM broadcast and TV stations than from mediumwave and shortwave (especially) stations because these stations' signals do not typically travel very far and the personnel are not accustomed to receiving QSL reports. Thus, very few of these stations have their own QSL cards printed; most only have the station's letterhead (Fig. 12-5).

SCOTLAND'S BRIGHTEST
RADIO STATION

Q96 FM LTD
26 Lady Lane
Paisley PA1 2LG

Tel: 0141 887 9630
Fax: 0141 887 0963

With Compliments

12-5 Letterhead from Q-96, 96.3 MHz in Paisley, Scotland.

Even though FM and TV stations very rarely receive reception reports, your chances of receiving a QSL might be greatly increased because many of the engineers are amateur radio operators or even TV DXers themselves. Even so, it is a good idea to explain that your hobby is receiving distant radio and/or TV stations and that the verification is an important aspect of the hobby. Of course, be courteous and polite when asking for the QSL or verification letter (Fig. 12-6). Although many stations will respond regardless, you will have a better chance of receiving a reply if you include a self-addressed, stamped envelope (SASE).

FM and TV resources

Aside from *The M Street Directory*, several other resources are helpful for the VHF broadcast DXer. Some of the best sources of information concerning FM and TV

#5 Television Drive, P.O. Box 480 Tel: 304-623-5555
Bridgeport, West Virginia 26330-0480 FAX: 304-842-7501

July 31, 1989

Mr. Andrew Yoder
3007R 4th Avenue
Beaver Falls, PA 15010

Dear Mr. Yoder:

I am in receipt of your letter with a post mark dated June 28th regarding your ability to receive our signal on Tuesday, June 27, 1989 in the A.M. time period between 8:23 AM ET to 8:36 AM ET.

I apologize for the tardiness of our response to your letter. With vacations of our personnel on a constant basis, it has made many of us doing double duty and leaving very little time for other matter at hand.

We appreciate hearing from you in regard to our signal reception. WDTV transmitter site is located just north of Weston, WV high atop Fisher Mountain with studio location off Interstate I-79 East of Clarksburg, WV.

Thank you and warmest regards.

Sincerely,

Nick E. Pellegrin
Program Director

NEP/cb

12-6 A QSL letter from WDTV, channel 5, in Bridgeport, West Virginia.

broadcast stations is on the Internet. Unfortunately, most of the stations currently on the Internet are from the United States and Canada, although a number of broadcasters from Europe and Australia are also appearing. Very limited numbers of stations from Asia, South America, and Africa currently have Internet Web sites. Because of the sheer numbers of broadcast sites on the Internet, they can't all be listed here, but some of the sites that list other broadcast links are included here.

FM radio Internet sites

The Radio/TV Dial Pages list hundreds of U.S. and Canadian radio and TV broadcast links at:

http://metronet.lib.mi.us/mikel/radio

The BRS Radio Directory bills itself as "the most comprehensive directory of radio stations on the Web." The listings of stations and formats for U.S. and Canadian stations are some of the most complete that I've seen, but the international listings are a bit limited. In fact, RTL Luxemburg was listed as broadcasting from the city of Luxembourg, Germany! Be sure to check it out at:

http://www.brsradio.com/stations

One of the best DXer sites is World Wide Radio. Here, you can run searches by callsign, location, frequency, etc. This is excellent if you heard a particular format on a frequency, but only received a partial callsign. With the search function, you can probably figure out what you heard. This site contains only radio broadcasting in the U.S.:

http://wwradio.com/home1.htm

A couple of good college radio listings are provided at Cyberdog Databases and KZSU, Stanford (worldwide listings):

http://www.radzone.org/tmd_srch.html
http://kzsu.stanford.edu/other-radio.html

TV Internet sites

One of the best sites for locating Web pages, e-mail addresses, postal addresses, etc. for TV broadcast stations around the world is UTV. This excellent page contains, not only the typical list information, but also Nielson ratings for U.S. markets around the country. Unfortunately, this site is also a bit weak with the small nations. Check out the U.S. site at:

http://www.utv.net/tv/us.htm

The UTV international site is at:

http://www.utv.net/worldtv/worldtv.htm

A nice listing of U.S. public/educational/community TV stations is at:

http://www.actwin.com/itvs/ptv1.html

If fast download time is a must, see the quick'n'simple TV station listing at:

http://www.bcpl.lib.md.us/~williams/stations

Last, but certainly not least, Euro TV is an excellent page (http://www.eurotv.com) dedicated to . . . well . . . European TV. The emphasis here is more fan-based, like a *TV Guide* of the Internet. You can search keywords for programs to watch at any given time in Europe, receive scheduling information, and even talk TV trivia!

13
CHAPTER

Radio-related collections

Baseball cards, cereal boxes, even photos of mushrooms and funguses (no athlete's foot photos!)—you name it, and I'll collect it! It's not that I collect things because I'm bored and need a hobby, not because I can make some money on my collections, and not because someone on a TV faux news show told me that it's cool to collect it. It's just that there are so many neat and exciting things in this world, and our lives and memories are intangible. It's also excellent to have a scrapbook or photo album full of mementos of a hobby that you love—a chronicle of fun.

Perhaps it's no surprise that I take lots of photos at home and have a shelf full of photo albums. Likewise, I made a big scrapbook/photo album from my workplace. Other co-workers thought it was great and wanted a copy for when they left. A few months later, the entire company was closed, leaving me with the only chronicle. As fast as life changes, if you want to collect a tangible record of these events, you've got to do it right now . . . as it is happening.

Value

One really great aspect of collecting radio artifacts is that most have almost no monetary value. Those words would send waves of terror down the spines of antique dealers, pot'o'gold seekers, and profiteers. "Worth nothing?! Then, why collect it?!" Because it's fun, interesting, and it means something to you personally.

The awesome aspect of collecting virtually worthless items is that you can enjoy the hobby and trade for items that you want without having to worry about investors running the prices of the items beyond your limits.

This isn't to say that everything in radioland has no monetary value—some items (such as old radios) do, but their value is much lower than what they originally sold for.

QSLs

QSLs are the perennial favorite shortwave radio collectible. Although QSLs are cherished in the shortwave and amateur radio communities, only a tiny percentage of the population knows what they are. A QSL is a tangible verification (usually a letter or card) that you have heard a radio station or heard/saw a TV station. Typically, such a card will feature the station's logo or a nice local photo postcard and will state the time that you heard the broadcast (in UTC, for shortwave stations; in local or UTC time for any regional TV stations), the frequency on which you heard the broadcast, and the date of the broadcast. Without this information, you have a nice postcard; with the information, you have a nice postcard and an achievement!

Surprisingly, the whole QSLing hobby goes back to the early days of local MW broadcasting. In the 1920s and 1930s, thousands of listeners sat by their crystal and TRF ("coffin" shaped) radios with headphones on, scouring the dials for stations that were a far distance away (the radio term *DX* means distance). During this time, many local stations freely announced their QSL information and how to get listener cards and DXing was a hot hobby. In fact, the radio fad even got to the point that mainstream organizations had listening contests, and movie and radio stars were known to be active DXers. It was something like the 1970s, when the CB movie *Convoy* was out—and CB was everywhere from Avon bottles to paperback novels to collectible stickers in bread bags.

The appeal of QSL collecting is this mixture of achievement and cool station postcards. The result is a handful of rabid QSL collectors. Like a crazed pack of accountants, these radio addicts methodically search the shortwave bands, looking for new prey. They send out hundreds and thousands of reception reports—ever seeking to close the gap between the number of stations QSLed and the number of active stations (and countries). Although I admit that I am poking a bit at the compulsive nature of some QSL fiends, it's nothing compared to many baseball card collectors that I've seen!

Of course, the best reason for collecting QSLs is (as mentioned earlier) because it's fun. I've got a few hundred QSLs in albums that I take along to shortwave-listening outings, conventions, etc., and many people seem to get a kick out of paging through those books. Likewise, I have a great time looking through other hobbyists' albums ("Hey, remember this station . . . ?").

In order to enjoy your QSLs, you need to be able to see them. One idea for saving your QSLs is to place them in albums (as mentioned). The easiest way to see your QSLs is to cover your walls with them. The least time-consuming method is to dump them all in a box, which makes them difficult to see!

I have tried all of these methods at different times and for various reasons. When I started shortwave listening, I used tape, thumbtacks, and rubberish adhesive to stick the QSLs, stickers, and pennants all over my walls and around the doorframe of my room. Big mistake! The tape got old and yellowed or tore my QSLs, the thumbtacks put little holes in them, and the rubberish adhesive stained the paper.

A few years later, I moved my most cherished QSLs into a "magnetic" photo album for safe keeping. Another big mistake! The adhesive on the page permanently attached itself to some of the QSLs, and either the adhesive or the plastic cover sheet yellowed them. So, I tore out some of my favorite, yellowed QSL cards and letters and started over.

Now, I have my best QSLs in a few different large three-ring binders. To hold the QSLs in place, I have mostly used plastic binder pages that contain a black construction paper center. Only friction is used to hold the cards in place, so they scoot around a bit. I suppose that sometime, I will need to hold them in place with photo corners or make tiny cuts in the sheet to fit each QSL into. Gone are the days of sticking QSLs all over my walls, but I think I might begin *carefully* thumbtacking up my nicer pennants.

So what is the best way to store your QSL collection? It's a personal matter, so you will need to choose a method that satisfies you. However, considering the time and expense required to amass even a small stack of QSLs, it is worth the effort to store them in a manner that is the least destructive as possible. With some time, you can come up with a really cool collection.

No matter how cool you think your QSL collection is, yours will never match that of the CPRV. The CPRV is the Committee to Preserve Radio Verifications. The Committee was created in 1986. The reserve now contains nearly 30,000 QSL cards and letters, primarily from shortwave and mediumwave broadcast stations. The CPRV consists of QSLs that have been donated by the families of deceased shortwave listeners or those listeners who have decided that their collections are better off where they can be preserved well beyond their lifetimes.

All of the CPRV QSLs have been sorted and are electronically indexed. They are currently in storage at the facilities of WCSN—the shortwave outlet of the Christian Science Monitor in Boston, Massachusetts. For a number of years, the historic QSL cards were featured in special "CPRV pages" in many of the North American radio club bulletins. The CPRV pages were discontinued in early 1996, but with any luck, the concept will soon be revived.

If you want to be sure that your QSL collection is preserved for history, you don't have to ship off your cards now. Rather, contact the CPRV at: 38 Eastern Ave., Lexington, MA 02173. They will be happy to register your collection and provide you with stickers to place on the boxes or albums where you keep your QSLs. These stickers state that, in the event that you can no longer enjoy your QSLs, you would like them to be donated to the CPRV. It's fantastic that the early days of radio will be preserved for the years to come.

Pennants

I think that radio pennants are some of the greatest radio collectibles out there—like high-class versions of stickers. A radio pennant is typically a relatively small pennant (anywhere from about 3" to 15" long) in either a vertical or horizontal format that is made of some type of light material (usually satin, nylon, or light cotton) with the station's logo. Radio pennants are often very beautiful and bright, with custom station silkscreening.

Pennants are sent out by many stations as promotional material. You rarely need to request them—if you sent a nice, accurate reception report and the station has some pennants on hand, you might receive one. Sounds easy.

Pennants are sent out by some of the major shortwave broadcasters who understand the desires of DXers who collect pennants. Some radio stations probably even have a few DXers on staff to push to have pennants printed for listeners. However, you'll be lucky if you ever see a U.S. mediumwave or FM station handing out pennants to its listeners. Pennants are primarily used by the Latin American stations. This makes receiving pennants quite a bit more difficult.

Latin American pennants are much more difficult to receive now than they were in the 1960s and 1970s. Many countries in Latin America have experienced serious economic problems, and this makes the cost of producing and mailing large, multicolor pennants prohibitive. Also, many of the larger commercial radio stations from the cities of Latin America have dropped their shortwave service in favor of the more-profitable FM and mediumwave broadcasts. Today, a very large number of shortwave stations are on the air from Latin America, but many more of these are rural, low-budget operations than was the case several decades ago.

As mentioned, pennant collecting is more difficult than it was a few years ago, but it is still entirely possible to amass a bunch in the late 1990s. If the economic trends continue in Latin America and shortwave broadcasting in general, you might be glad that you searched for the pennants while you did.

If you are on the Internet, two radio pennant collectors are on line with lists of pennants for trade. Fred Kohlbrenner's pennant page is just fantastic, featuring hundreds of beautiful pennants (Figs. 13-1 and 13-2) from over the course of several decades. The URLs are:

```
http://ourworld.compuserve.com/homepages/fkohl  (Fred Kohlbrenner)
http://www.srg-ssr.ch/prv/index.html  (Giavanni D'Amico)
```

Tapes

Some radio hobbyists collect radio programs on cassette. In general, these collectors are fans of old-time U.S. commercial radio—programs such as "The Shadow," "The Lone Ranger," "Amos & Andy," and the many soap operas of the day. Fortunately, these programs were archived. Literally hundreds of different programs are available from radio program dealers.

For more information on purchasing these programs, contact:

BRS Productions
P.O Box 2645
Livonia, MI 48151

Wayne Caldwell
P.O. Box 831183
Richardson, TX 75083

Satellite Broadcasting, Inc.
4580 E. Mack Ave.
Frederick, MD 21701

A fantastic Internet Web page with links to many other old-time radio tape dealers and traders is the Original Old-Time Radio (OTR) WWW Pages. Also, this site contains plenty of history and information about these programs. This is a must-see for any fan of old-time radio:

```
http://www.old-time.com/vendors.html
```

Radio-related collections 283

13-1 Just a few of Fred Kohlbrenner's beautiful pennants; these are from Latin America.

13-2 More beautiful pennants from Fred Kohlbrenner's collection.

I know of almost no one who collects shortwave radio programs. I have traded with some radio hobbyists for recordings of "Axis Sally" and "Tokyo Rose," but that is about the extent of international shortwave broadcasts on tape. However, tape collecting among pirate radio hobbyists is booming. Some listeners (including myself) have snippets or full-length programs from hundreds of stations on cassette. Although some of the recordings are off-air, many are copies of the studio program that was originally broadcast. If you are interested in tape trading (unless you have some really rare, preferably pre-1980 pirates), please don't contact me, try the audio page that is affiliated with the Free Radio Network Internet page:

http://w3.one.net/~folk/tapes.html

Stickers

Collecting radio station stickers is popular among radio listeners, in part because the stickers are often nice looking and because so many radio stations have stickers. Most stickers wind up on the bumper of someone's car or notebook, so sticker life and numbers aren't nearly as permanent as pennants. In a sense, I suppose that makes them even more collectible. Although many hobbyists collect radio stickers casually, no one really seems to catalog their collections, like the collectors of radio programs, QSLs, receivers, and magazines.

T-shirts

Radio t-shirts are very collectible because they are expensive to manufacture; few are distributed. Most shirts from mediumwave and FM stations are given away in contests, but most from shortwave stations are sold as fundraisers.

Like stickers, t-shirts usually don't last long. The only good way to display the shirts is to wear them. After a few years, they will be worn out. So, a radio t-shirt from the 1970s is a real rarity. I don't know of anyone who regularly collects radio station t-shirts, but it's an interesting idea.

Photos

Photographs of shortwave radio stations are much more difficult to come across because they aren't traditional promotional material, as they are for mediumwave or FM radio stations. Maybe the shortwave broadcasters are uglier, but it probably has something to do with the general lack of advertising and commercial programming on shortwave.

In general, the only way to get photographs of shortwave stations is to either go visit them or send a bunch of photos of yourself and postcards of your area and hope that someone will respond with a station photo or two. Chances are that they won't.

Some of the nicest and most interesting radio station photos that I've seen are the result of Don Moore's many adventures while visiting shortwave broadcasters in Latin America. Fascinating! If you want to see a number of photos from his collection, either wait for one of his articles to appear in *Monitoring Times* or check his Internet Web page at:

http://www.mcrest.edu/moore/latinam.html

Radios

Of the different categories here, collecting old vacuum tube radios is by far the most common. It can also be expensive: Communications receivers typically remain in the $100 to $400 range, but rare and novelty mediumwave radios from the 1930s and 1940s can easily exceed $1000.

The prices for old radios are driven up by the profitable antiques and collectibles trade. Just a few years ago, old tube radios were just considered to be junk. Now they are filling stalls at antique malls around the U.S. This sudden interest in collecting radios seems to be spurred on by the recent popularity of antiques/collecting and also in art deco designs and pop culture.

At least a dozen books have been written on the subject, including some large-format, color coffee-table books. One of the nicest books for collecting standard mediumwave radios is *The Collector's Guide to Antique Radios—2nd Ed.* by Marty and Sue Bunis. If you are interested in buying or selling old-time mediumwave radios or memorabilia, the ultimate resource is *Antique Radio Classified*, which is a very large monthly newsletter that consists almost entirely of radio classified ads! For more information, write to:

ARC
P.O. Box 802
Carlisle, MA 01741

or see their Internet Web site at:

`http://www.antiqueradio.com`

Communications receivers

Because collecting antique radios is such a large and multifaceted subject, this section just focuses on communications and amateur shortwave receivers.

The performance and purchasing aspects of some of the best shortwave receivers—such as the Collins R-390A and the Hammarlund SP-600 and HQ-180—were covered in Chapter 2. Today's shortwave radio manufacturers are just a blip on the radio timeline. The only communications receiver manufacturer that has survived the transition from tube radios to present is R.L. Drake, which began at the end of the tube era in the early 1960s. For decades, if you wanted a high-quality shortwave receiver, chances are that it was manufactured by Collins, Hallicrafters, Hammarlund, National, or Drake. In Europe, some of the best receivers were manufactured by Rohde & Schwartz (Germany) and Racal (England), but they were rarely seen in North America.

Hallicrafters

The all-time most popular shortwave receivers had to be those manufactured by Hallicrafters of Chicago, Illinois. Hallicrafters seemed to find the perfect blend of everything: quality components, performance, style, and price (Fig. 13-3). Their radios are among the most collectible because they filled everyone's needs so well. Hallicrafters radios spanned from very inexpensive beginner's rigs (such as the S-20R and the S-38 series) to some of the very best (such as the ill-fated DD-1 and SX-88).

Even though the beginner's radios were very inexpensive for their day (and were certainly not top performers), they were all built solidly and they looked nice—

13-3 Few current manufacturers of any products could boast of this sort of toughness! The ad for the Hallicrafters S-20R Sky Champion is from 1942.

like a radio you could be proud of. Hallicrafters made the most of the nice packaging; they often were better looking as the price went up. For example, the SX-99, SX-96 (Fig. 13-4, upper left), and SX-100 were the Hallicrafters low-, medium-, and high-end radios in the late 1950s. All had somewhat similar cabinet designs, but the SX-99 (the least expensive of the lot) *looked* like a beginner's radio. I think that the SX-96 (the mid-range radio) is the prettiest of the three, but the SX-100 is still sharp *and* looks like a hard-working DX machine.

One of the most popular collector radios is the Hallicrafters SX-28/SX-28A. These were among the top shortwave receivers used by the U.S. in World War II, yet the company still worked extensively on their look. For such a performance-oriented radio, the company could have saved money by making its appearance cheap (painted front-panel lettering, plain cabinet, etc.). Instead, the top receiver was set into an art deco cabinet with chrome side strips, styled knobs, and a thick steel-plate front panel that had the lettering and fake leather grain stamped into it. Beautiful, and possibly the best DX receivers built up until that time (Fig. 13-5).

Hammarlund

A not-quite-opposite approach was taken by the other H-company, Hammarlund of New York, New York, and later Mars Hill, North Carolina. The radios were solid, high performance blocks of steel with a totally spartan appearance. Instead of working the concepts of slick marketing and beautiful receivers, Hammarlund cared only about per-

13-4 Some classic receivers (counterclockwise from upper left): Hallicrafters SX-96, Hammarlund HQ-180C, Motorola R-390A, Heath RX-1 Mohawk, Collins R-392, and the matching Hammarlund speaker for the HQ-180C.

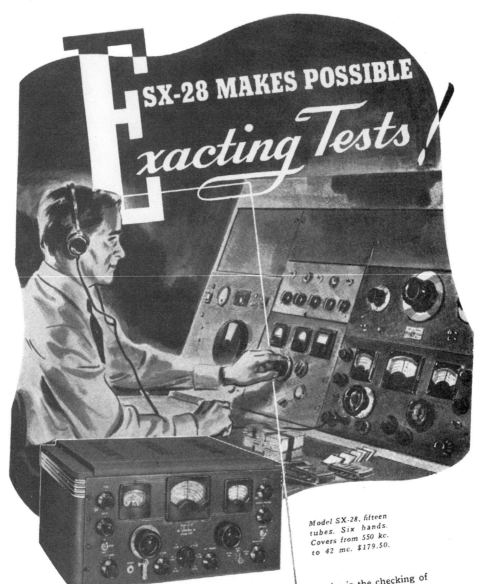

13-5 A classic in every way: A 1942 ad for the Hallicrafters SX-28.

formance. According to *Communications Receivers: The Vacuum Tube Era—3rd Ed.* by Raymond S. Moore, the Hammarlund Manufacturing Company was created in 1910, making it one of the oldest and longest-running radio manufacturing companies.

Right from the beginning, Hammarlund took a spartan approach. Nearly every radio ever produced by the company has the same general stylings, with only a few minor changes. In fact, during the 1960s, they released several different receivers with the same cabinet, knobs, and general knob placement! From more than a few feet away, the reaction might be "Nice Hammarlund, is that an HQ-145, HQ-160, HQ-170, or HQ-180?" This was the ultimate volksradio for the DXer. Forget status, it was a great radio, and *you* knew it!

Hammarlund receivers ranged in price from fairly expensive to very high priced. The low-end and mid-range models were affordable for most shortwave listeners and amateurs. The high-end models, such as the SP-600, were well beyond the reach of most people until they were eventually sold as military surplus rigs. In tribute to its excellent design, the SP-600 was on the market for more than 20 years, making it the longest-selling communications receiver (and probably of any type of radio). The SP-600 and HQ-180 (Fig. 13-4, middle left) were among the best tube receivers ever manufactured. They are covered more in Chapter 2.

Visiting Radio's Past

If you are interested in radio collectibles and the history of radio, you will certainly want to visit one or more of the radio museums. Most of these museums are small, private collections dotted around the country. The curators are often radio hobbyists, maybe amateur radio operators, and are usually very knowledgeable about radio. The driving force behind the museums is often simply radio, not money, so you can expect a good time.

When you go to visit a radio museum, be sure to take along a camera (ask on arrival if photos are allowed), and be prepared to ask a lot of questions. Also, be sure to either buy a bunch of postcards or grab a bunch of brochures to send to your friends or along with your reception reports. They make a great conversation piece and help personalize a reception report.

The following list includes information for a number of radio-related museums:

ARRL Headquarters
225 Main St.
Newington, CT 06111
203-594-0200 (tel)
http://www.arrl.org
Schedule: Open Monday through Friday, 9:00 A.M. to 5:00 P.M.

Antique Wireless Association Museum
59 Main St., Village Green
Rt. 5 and 20
Bloomfield, NY 14469

716-657-6260 (tel)
http://www.vivanet.com/freenet/a/awa
n2rsm@frontiernet.net
Schedule: Open at various schedules through the year.

Atwater-Kent Museum
15 S. 7th St.
Philadelphia, PA 19106
215-922-3031 (tel)
Schedule: Open Tuesday through Saturday, 10:00 A.M. to 4:00 P.M.

Bellingham Antique Radio Museum
1315 Railroad Ave.
Bellingham, WA 98225
206-734-4168 (tel)
http://www.antique-radio.org/radio.html
jwinter@pacificrim.net
Schedule: Open Wednesday through Saturday, 11:00 A.M. through 4:00 P.M. or by appointment.

CBC (Canadian Broadcasting Corp.) Museum
P.O. Box 400, Stn. A
Toronto, ON M5W 1E6
Canada
416-205-3700 (tel)
http://www.cbc.ca/aboutcbc/tbc/museum/museum.html
Schedule: Open Monday through Friday, 10:00 P.M. to 4:00 P.M.

Centre for the History of Defence Electronics (CHiDE)
Bournemouth University
12 Christchurch Rd.
Bournemouth BH1 3NA
United Kingdom
01202-503879 (tel)
http://chide.bournemouth.ac.uk/default.html
chide@bournemouth.ac.uk
Schedule: By appointment.

Library of American Broadcasting
University of Maryland
Hornbake Library
College Park, MD 20742
301-405-9160 (tel)
http://www.itd.umd.edu/ums/umcp/bpl/bplintro.html
bp50@umail.umd.edu
Schedule: Open Monday through Friday, 9:00 A.M. to 5:00 P.M.

Manitoba Amateur Radio Museum
Box 10

Austin, Manitoba R0H 0C0
Canada
204-728-2463 (tel)
http://www.mbnet.mb.ba/~donahue/austin.html
ve4xn@ve4bbs.#hwd.mb.can.na
Schedule: Open June through October 1, Monday through Friday, 9:00 A.M. to 5:00 P.M.

Museum of Broadcast Communications
Chicago Cultural Center
Michigan Ave. at Washington St.
Chicago, IL 60602
312-629-6000 (tel)
http://webmart.org/mbc
Schedule: Open Monday through Saturday, 10:00 A.M. to 4:30 P.M.; Sunday, 12:00 P.M. to 5:00 P.M.

Museum of Radio and Technology
1640 Florence Ave.
Huntington, WV 25701
304-525-8890 (tel)
http://www.library.ohiou/museumr&t/museum.htm
Schedule: Open Friday and Saturday, 10:00 A.M. to 4:00 P.M.; Sunday, 1:00 to 4:00 P.M.

Museum Radio-Wereld
Achterstraat 9
7981 AS Diever
The Netherlands
31-521-592386 (tel)
http://www.xxlink.nl/tourism/museums/493.htm
Schedule: By appointment.

Museum of Television and Radio (Beverly Hills)
465 N. Beverly Dr.
Beverly Hills, CA 90210
310-786-1000 (tel)
http://www.mtr.org
Schedule: Open Wednesday through Sunday, 12:00 P.M. to 5:00 P.M.; Thursday, 12:00 P.M. to 9:00 P.M.

Museum of Television and Radio (New York)
25 W. 52nd St.
New York, NY 10019
212-621-6600 (tel)
http://www.mtr.org
Schedule: Open Tuesday through Sunday, 12:00 P.M. to 6:00 P.M.; Thursday, 12:00 P.M. to 8:00 P.M.; Friday, 12:00 P.M. to 9:00 P.M.

The U.S. Army Communications-Electronics Museum
Bldg. 275, Kaplan Hall
Ft. Monmouth
Red Bank, NJ 07703
908-532-9000 (tel)
Schedule: Open Monday through Friday, 12:00 P.M. to 4:00 P.M.

The Western Heritage Museum of Omaha
801 S. 10th St.
Omaha, NE 68108
402-444-5071 (tel)
Schedule: Open Tuesday through Saturday, 10:00 A.M. to 5:00 P.M.; Sunday, 1:00 P.M. to 5:00 P.M.

National Radio

The National Radio Company of Malden, Massachusetts had a hybrid background. They were one of the earliest radio manufacturers, and their approach was very similar to Hammarlund in the 1930s and early 1940s. After their main engineer left to form his own company, the company's receiver line and marketing became very similar to that of Hallicrafters.

The early National lines were the AGS and HRO series. Both were very expensive, plain black boxes. Unlike Hammarlund's early Comet Pro and Super Pro lines, National's were updated frequently. Several dozen different HRO receivers were released over the 30 years that the series was produced, most varying only slightly from one or more other models.

National offered tough competition to the SX-28A in World War II with the NC-120 and the RAO series (Fig. 13-6). It was even discovered that German Field Marshall Edwin Rommel used captured National Radio equipment instead of his own country's rigs. In the late 1940s and 1950s, National had some very pretty, top-notch shortwave receivers, plus a large number of mediocre low-end models. Some of the good radios included the NC-173/NC-183 series, and the NC-200 series. The high-end radios were the HRO-50 and HRO-60; these rigs defied one of the basic rules of tube radio quality: more tubes = better radio. The HRO-50 contained 16 tubes, but performed better than the 18-tube "improved" HRO-60.

Collins

The last major manufacturer of shortwave receivers was Collins, which is sometimes called the Rolls-Royce of communications equipment. Collins emerged from Cedar Rapids, Iowa, in the early 1940s with one receiver on the market. The approach taken was similar to that of Hammarlund, except with no limits on the cost. They made the best radios, and if you needed to sell your car, house, dog, and kids to buy one, so be it. In 40 years of manufacturing, Collins only sold about a dozen different receivers—some available for more than a decade.

In actuality, many of the Collins receivers were comparable to the best that other companies had to offer. Many listeners would rate the Hammarlund SP-600 and HQ-180,

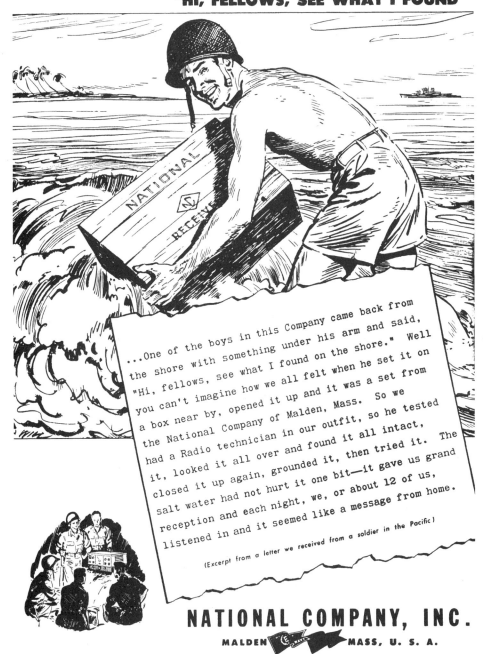

13-6 More tales of the tough: A 1944 National radio ad.

and the National HRO-50 and HRO-500 in the same league as most of the Collins rigs (some would say better). The difference is that Collins *never* released a bad radio—all were superb. So, when you paid over $1000 for a 51J-4 in 1956 (gasp!), you knew that it was a *Collins*, and you paid a bit more for the name. Today, Collins equipment is very expensive for the market; you can get much more radio for the money by looking for something from one of the other companies. The ultimate tube radio is the Collins-designed R-390A (Fig. 13-4, bottom left), the ultimate refinement in heavy-duty vacuum-tube shortwave listening. Fortunately, at least 60,000 of these receivers were manufactured, so the demand/number of receivers ratio is excellent for the shortwave listener.

The end

In the 1960s, shortwave receivers became smaller. The successful amateur radio companies began designing transceivers, rather than transmitter/receiver combinations. Also, foreign competition began to undercut the U.S. market with equipment that had lower-quality parts and construction and lesser performance. Also, National and Hammarlund were bought and sold a number of times in the 1960s. This surely contributed to their ultimate decline. By the early 1970s, Hammarlund and Hallicrafters were selling their remaining receivers, National was manufacturing radio parts, and Collins was living off its reputation.

What astounds me is not that the companies eventually failed or changed product lines, but that no one seemed to care. Even though four amateur radio magazines were operating through the 1970s, I don't recall ever seeing a single article that lamented the passing away of Hammarlund, Hallicrafters, or National or that discussed the history of these companies or their radios. It's almost tragic that these events occurred without comment; their advertisements simply disappeared.

Replacing these companies were Drake, Yaesu, Heath, Icom, Trio (later known as Kenwood), and Swan. Of these, only Drake made a serious commitment to the shortwave listener in the 1970s with the R-4 series (hybrid tube and transistor technology) and the R-7. Yaesu, Icom, and Kenwood developed a reputation for shortwave receivers in the 1980s.

If you have any interest in vacuum tube receivers and amateur radio transmitters or transceivers, an excellent resource is *Electric Radio*, a large (often 56 pages) monthly radio bulletin that is dedicated to vintage radio gear and the history of radio. It often contains interviews with old radio designers, articles on radio restoration, and in-depth reviews of classic radio equipment. Equally important is the 15 pages or so of classified ads. *Electric Radio* is available from 14643 County Rd. G, Cortez, CO 81321.

Aside from *Communications Receivers* (mentioned earlier), two other excellent antique radio books are *Radios by Hallicrafters* by Chuck Dachis and *The Zenith Transoceanic* by John H. Bryant and Harold N. Cones. Both are fantastic coffee-table books, filled with plenty of color glossy photos. Both go into serious detail. Even if you can't afford everything that I've plugged as a "must-have" in this book, you really owe it to yourself to visit a Borders or Barnes & Noble store and just look at these!

Radio magazines and bulletins

I just love mass media. Books, magazines, audio (records, cassettes, CDs, etc.), video, radio, posters—I think it's all great. So, what could be better than mass media about mass media? Cheap or free mass media about mass media!

Because radio is such a popular hobby and because its rise parallels the expansion of magazines, you can find a vast amount of print information on the topic. After working in electronics publishing for six years and attending plenty of hamfests (packratting all the while), I have accumulated hundreds, maybe thousands, of magazines. Most of my finds have been fairly recent and common, but sometimes I'll find something that I've never seen before for a good price.

What is a good price? It depends on what you want and how much you need it. Most post-1950 radio magazines sell for anywhere from 30¢ to $1 per issue (Fig. 13-7). Those from 1930 to 1949 will typically go for 50¢ to $10 per issue (Fig. 13-8). The really early magazines from 1910 to 1929 often sell for $5 to $30 per issue. Of course, these prices are based on condition, how common the magazine is, if any famous radio stars are featured within, etc.

The following is a brief list of some of the more common U.S. magazines that have been available over the years. Newsletters, even some of the largest recent ones (such as NASWA's *The Journal* and *Electric Radio*), rarely turn up at hamfests.

- *73*   Amateur radio. This magazine has been published since 1961, and it emphasizes amateur radio construction projects.
- *CQ*   Amateur radio. Since 1945, *CQ* has been a very popular amateur radio magazine.
- *Electronics Illustrated*   Electronics projects. This 1950s and 1960s magazine was very similar to *Popular Electronics*.
- *Electronics Now*   Electronics projects. This is the 1990s name for *Radio-Electronics*.
- *Elementary Electronics*   Electronics projects and radio monitoring. *Elementary Electronics* was a 1960s and 1970s magazine that occasionally featured some interesting radio articles. *EE* also did spin-off annuals, such as *Communications World*, *Electronics Hobbyist*, and *Budget Electronics*.
- *Ham Radio*   Amateur radio. This 1970s and 1980s amateur magazine was absorbed by *CQ*.
- *Hands-On Electronics*   Electronics projects and radio monitoring. *Hands-On Electronics* was a 1970s and 1980s magazine that occasionally featured some interesting radio articles.
- *Modern Electronics*   Electronics projects. A 1980s magazine that rarely covered radio or communications.
- *Monitoring Times*   All-band monitoring. Dating back to 1981, *Monitoring Times* features a large amount of radio information and historic articles. Not yet collectible, so get the older issues while you can!
- *Popular Communications*   All-band monitoring. Since its inception in 1982, *PopCom* has featured a large amount of radio information and historic articles. Not yet collectible.
- *Popular Electronics*   Electronics projects. This sister publication to *Electronics Now* features more radio-related articles. Hank Bennett, the author of the first *Complete Shortwave Listener's Handbook*, was a long-time columnist for *Popular Electronics*. *PE* has absorbed a number of other magazines, including *Electronics World* and *Hands-On Electronics*.
- *QST*   Amateur radio. Very collectible. *QST* is by far the oldest surviving magazine in the entire electronics/communications field, dating back to 1915!
- *RADEX*   Shortwave radio and DXing. Very collectible. *RADEX*, from the

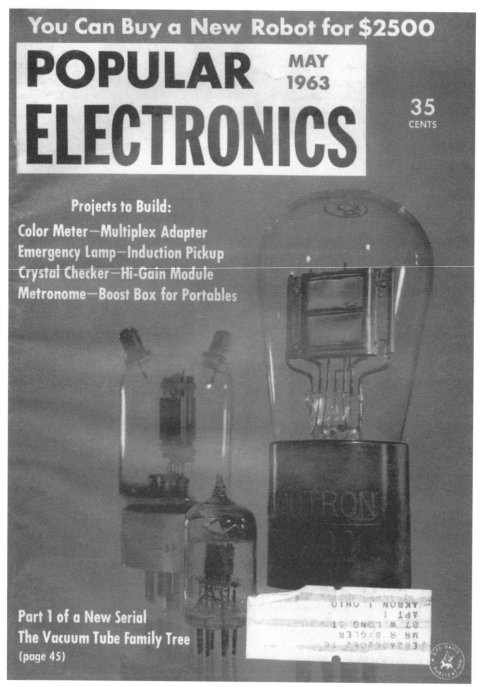

13-7 *Popular Electronics* was one of the only magazines from 1950–1980 that included shortwave radio information. Columnist Hank Bennett and the SWL registry program greatly improved the popularity of shortwave radio in North America.

13-8 The source of information about mediumwave and shortwave radio in the 1920s and 1930s, *Radio News* is a must-have for collections of radio nostalgia.

1920s and 1930s, featured tons of mediumwave and shortwave tips and hardcore DX articles—all illustrated with promotional photos of radio stars!
- *Radio!*   All-band. A short-lived quarterly beginner's monitoring magazine produced by Radio Shack in the 1990s.
- *Radio Craft*   Radio projects and monitoring. This 1920s and 1930s magazine later changed its name to *Radio-Electronics*.
- *Radio-Electronics*   Radio projects and monitoring. One of the oldest and all-time best-selling radio and electronics magazines. Changed its name to *Electronics Now*.
- *Radio News*   Radio/fan magazine. Lots of information about radio stars and listening to the radio. Very collectible. Became *Radio TV News* in the 1950s and later *TV News*.
- *Radio Experimenter*   Radio projects and monitoring. This hobbyist magazine became *Radio-TV Experimenter* in 1951, and became *Science & Electronics* in 1971. *RTVE* often featured hobby articles from the well-known C.M. Stansbury II and Tom Kneitel in the 1960s.
- *Radio World*   Professional broadcasting. A radio-industry newspaper that also features some mediumwave- and shortwave-listening articles.
- *S9*   CB radio. This CB magazine became *S9 Hobby Radio* in 1980 and featured many offbeat shortwave-related articles.
- *Shortwave Guide*   Shortwave listening. An early 1980s monthly newspaper that merged with *Monitoring Times*.
- *Short Wave Craft*   Shortwave radio projects and monitoring. This 1930s magazine from the initial shortwave radio craze mostly featured shortwave radio construction projects, but it also contained a few interesting radio broadcasting articles.
- *Short Wave Radio*   Shortwave radio. This 1930s magazine from the initial shortwave radio craze featured some interesting beginner's radio articles.
- *Worldradio*   Amateur radio. A current newsprint magazine.

If you start collecting old magazines, such as these, be sure that you have lots of storage space in locations that are neither too hot nor too damp and protected from insects, if possible. The daily heating and cooling of an attic will damage paper, but it is preferable to a damp basement, where mildew can ruin pages and covers. You need to be especially careful with some of the magazines that are printed on poor-quality paper, such as *Radio TV Experimenter* and *Electronics Illustrated*. This stuff is like a coarse, pulpy version of newsprint, and only 30+ years later, it is already self-yellowing and beginning to become brittle. By far the best of the bunch is the type of acid-free paper used by *QST* while it was in its small format (pre-1975). It takes years of mildew and water to destroy the pages, and I have never seen them turn yellow. The only drawback is that a stack of old *QST*s seems as heavy as a stack of bricks!

A few very collectible annual books are the *World Radio TV Handbook*, *Passport to World Band Radio*, and the *ARRL Radio Amateur's Handbook*. All of these are common, but the pre-1980 *WRTHs* and pre-1950 *ARRL Handbooks* are very difficult to find.

14
CHAPTER

Computers and radio

If you stay in touch with the radio community—attend hamfests, read radio magazines, talk with other listeners, etc.—you will often hear the opinion that "computers are killing radio." Many of the oldtimers voicing this thought will note that amateur radio was thriving in the 1950s and 1960s, but now the kids would rather play video games than experiment with the radio.

There is a kernel of truth to these statements. Fifty years ago, radio was the only game in town. If you wanted to be involved in electronics or communications, you became interested in radio. The fact that World War II was in full swing and tens of thousands of Army, Navy, Air Force, and Marine recruits from around the world were trained to operate two-way radio equipment added to this interest and the domination of the hobby. Upon entering civilian life in the late 1940s, these people automatically took to amateur radio as their primary hobby. In the 1950s, if you were young and bright, you were probably either interested in amateur radio or model railroading.

Competition for the youth of the world began in the late 1950s—both television and rock music became widely accepted. Then, more kids began to sit around and watch TV or listen to records than listen to the radio. The CB radio craze of the late 1970s attracted more shortwave listeners and amateur radio operators, but that fad only lasted a few years. With the 1980s began the so-called MTV generation and the personal computer revolution. In the 1990s, the great social upheaval of the Internet occurred, leaving many people saying "Will there be a place for radio in the 21st century?"

As the focus of youth education and culture changed to computers and rock music, radio as a hobby has been left behind. In the 1980s and 1990s, I have rarely seen amateur radio or the shortwave-listening hobby represented as a regular service to the general public. Very occasionally, an amateur radio club will have a display in a mall or will be featured in a newspaper article (often something to the effect of "Local man networks with thousands worldwide"). Unfortunately, the shortwave broadcasters don't advertise; even though the Voice of America is a massive radio network, it is virtually unknown in North America.

So, with little attention and sudden competition from the Internet and DBS satellite TV, is there any room for shortwave? I certainly believe that shortwave broadcasting and two-way operations will continue into the 21st century. Why?

1. Shortwave radios can be inexpensive, but Windows-compatible computers aren't.
2. Internet connections are expensive—typically a minimum of $15 per month, yet shortwave radios operate for free.
3. Internet connections are complicated and are very difficult to protect. Shortwave radio transmitting antennas are easy to install, and the radio signal can't be tampered with as it travels through the sky.
4. The less that the shortwave spectrum is used by the general public, the more it will be used by those who want their communications to be kept secret.

Because of these factors, I believe that shortwave radio will continue to be used regularly—even if some of the large shortwave stations and small shortwave networks in developing nations diminish. Shortwave will continue to be used by powerful nations who want to influence people around the world, by people in locations that are difficult to access (such as the mountains of Peru), by organizations who want their communications to be relatively secure (utility stations), and by those who want to reach niche audiences (such as pirate stations and those who buy airtime on commercial shortwave stations).

Clearly, shortwave radio can't compete as a regular broadcasting service in developed nations, where people have access to hundreds of mediumwave and FM radio stations, TV and cable channels, newspapers, and Internet broadcasters and Web pages. However, for the niche market in the Western nations, shortwave radio is tough to beat. With shortwave, very little power can cover many listeners over thousands of miles. Even if the density of listeners per square mile is mighty thin, a large overall audience can still be attained. If the word gets out to a particular niche market that shortwave is the place to be, those people will buy radios and listen.

A perfect example of this niche-market situation has occurred with U.S. stations WWCR and WRNO. Both of these broadcasters air a wide variety of right-wing programs that are paid for by different groups. Before long, most everyone in the populist, patriot, Nazi, and other conservative, small-government groups knew that their programs were on shortwave. They bought shortwave radios, and the popularity of these programs spawned more paid programs. Because of the often-extreme viewpoints of these programs, they are very controversial within the shortwave-listening scene. Thoughts of the shortwave bands being taken over by political extremists are not overly pleasing, but not too long ago, Radio Moscow seemed to be broadcasting every 10 kHz with plenty of communist propaganda. An awful lot of right-wing programs will need to buy airtime before they can equal the broadcasting output of Radio Moscow in its heyday!

One less-controversial example of a niche market is ZXLA, Print Disabled Radio in New Zealand. Instead of operating with music, news, sports, etc., like most stations, ZXLA is a service for the blind. They transmit such programming as audio books, aired daily. ZXLA is a low-power station intended for a national audience only; I'm sure that many visually impaired people around New Zealand began listening to shortwave, just for ZXLA's programs, which would be too costly to air over all of New Zealand on mediumwave or FM. Shortwave radio is the ultimate medium to reach the blind, and perhaps other Print Disabled Radio stations will be instituted around the world in the next decade.

Of course, some segments of the shortwave radio community will die off as we pass into the 21st century. Some people will predict the death or the commercial failures of shortwave. However, for the right nations, companies, groups, or individuals, shortwave radio will remain a gold mine of enormous opportunities. Millions of people around the world will continue to listen as shortwave adjusts to this new world, including me.

Over the years, I've heard quite a bit of "us against them" sentiment related to computers and radio. The radio-heads look grumpy when the computer geeks talk about how computers and computerized audio will eliminate radio and all of its associated static and fading. However, after a virus or disk crash takes the computer down for a few hours or days, the computer geeks are a bit more solemn. The big question is: Will computers eliminate radio or will both find their own niche and peacefully coexist?

Or is this the appropriate question after all? I think that the real questions should be: How can inexpensive computer technology be used to improve radio communications? How can radio be used to improve computer communications? Instead of thinking us vs. them, maybe we should be thinking about the new world of communications that is being created and what elements should be chosen to create some truly excellent technologies.

Receiver-control programs

For a few years, some of the high-end receivers have had computer interface boards available as accessories. Some of the most popular models are the Drake R-8, Kenwood R-5000, and Icom R-71A. After you purchase a board and plug it in, you can connect the computer to your shortwave radio. Like everything else with computers, simply making the connection will yield nothing; you need to purchase the proper software to make the system work.

Enter the private, third-party computer software designers who create different types of receiver-control software. Now, not only do you have a radio and a computer, but with the magic of software, you can alter the radio's ergonomics and some of its features. Amazing!

Because many of the radios use different specifications and so many of the computer software designers are small businesses, many companies only specialize in software for one or two receivers. Be sure to check out the software so that you know what the functions are and if it is compatible with your receiver. The best ways to do this are to check into the Internet Web page of the companies and see what each piece of software will do. Some, such as the FineWare page, will allow you to download shareware or beta versions of the software so that you can get a real feel for what the software can do.

What does it do?

Most receiver-control programs have many of the same basic features. The differences typically lie in how well the program is organized, how fast you can accomplish a task, and how easy the software is to learn.

The following information covers WiNRADiO software (more on this in the next section), but it is what you can expect from many of the different receiver-control programs.

After you load up the software, the computer screen will display what looks like the front panel of a communications receiver (Fig. 14-1). It has a large frequency display, a tuning knob, signal strength display, reception mode buttons (AM, SSB, narrow or wide FM), and control memories and scanning.

14-1 The front panel of the WiNRADiO.

WiNRADiO comes with three basic scanning modes:
- Direct scan (use one of the two "immediate scan" buttons)
- Range scan (set up one or several scanning ranges)
- Memory scan (scan within one or more memory groups with a priority channel)

The radio can be manually tuned in several different ways:
- By typing in a frequency directly from the keyboard
- By turning the knob using mouse or keyboard (the rotation can be accelerated by simultaneous pressing of certain keyboard keys)
- By tuning in selected step sizes using stepping keys (clicking by mouse or from keyboard)
- By using keyboard-assigned "hot keys"

The "WiNRADiO WorldStation Database Manager" is also available. This program is fully integrated with standard WiNRADiO software and makes it possible to tune WiNRADiO directly from the database (by clicking on a database item) or to locate the details for an unknown transmitter to which you are currently listening. The database also has extensive importing facilities and comes with a list of over 300,000 worldwide radio stations.

For more information

Some of the best receiver-control software companies are:
- *TRS Consultants*   Produces receiver-control programs for a wide variety of receivers—all with different features (Table 14-1). TRS Consultants is widely respected for the accuracy of its shortwave database program. In fact, many of the other companies' control programs are made to be compatible with it! The TRS Web page is excellent, with plenty of other radio-related information (even some recent audio clips of neat DX catches!) and links. Be sure to preview it at http://www.trsc.com.

Table 14-1. The features of the receiver-control programs from TRS Consultants

	HF-1000	HF-150	JST-245 NRD-535	NRD-525	R-5000	R8A
Database Management of Receiver Memories	X	X	X	X	X	X
VHF Coverage				X	X	
Bandscan	X	X	X	X	X	X
Graphical Bandscan with S-meter Readings			X			
Spectrum Analysis			X		X	
Frequency Range Scan	X	X	X	X	X	X
Event Management	X		X		X	X
Control Panel Display	X	X	X		X	X*

*Limited functionality

- *Computer Aided Technologies*   Produces Scancat Gold and Scancat 6.0 to control "most radios by AOR, Drake, Kenwood, Icom, Yaesu, and JRC," plus PRO-2005, PRO 2006, 2035/OS456, Lowe HF-150, and Watkins-Johnson HF-1000. In addition to some of the standard receiver-control tasks, Scancat can search between any two frequencies, search by any increment, create disk files, and perform spectrum analysis. Also sells Copycat and Copycat-Pro to control the Universal M-7000 and M-8000 communications decoders. Check their Web page at http://www.scancat.com.
- *FineWare*   Produces Smart R8 DOS and Windows receiver-control software (Fig. 14-2) for the Drake R-8 and R-8A, which can tune it in various increments (down to 0.01 kHz). It contains an editable station broadcast database, has an onscreen UTC clock, etc. Also produces other interesting shortwave-related software. With one program, SWBC Interval Signals, you can play 70 different shortwave interval signals. It's a neat program and can help you to learn and identify the different signals that you hear on the shortwave bands. Another program, Smart Audio Control (Fig. 14-3), allows you to control the tone and volume of the receiver via computer. Better yet, Smart Audio Control also contains a fully functional audio oscilloscope and spectrum analyzer. Find the Web page, which contains downloadable

shareware and beta releases of the software, at http://www.crosslink.net/~mfine.

- *Spectrum Systems*   Produces FirstRate Windows and Macintosh receiver-control programs for the Drake R-8 and JRC NRD-535. This software includes the standard receiver-control functions, a frequency database, logging capabilities, a world grayline map, and a graphical MUF (maximum usable frequency)/LUF (lowest usable frequency) calculator. Like the FineWare site, you can download shareware versions of this software from their Web page at http://www.infi.net/~dharvey/firstrate.htm.
- *KC4ZGL Ham Software*   Produces Scan Manager 1.1 Pro, a Windows-based program with an editable database, onscreen UTC clock, mouse- or keyboard-controlled scanning, printable database reports, etc. Write to 1548 Cedar Bluff Trail, Marietta, GA 30062 or check their Web page at http://www.atl.mindspring.com/~tony/kc4zgl.html.

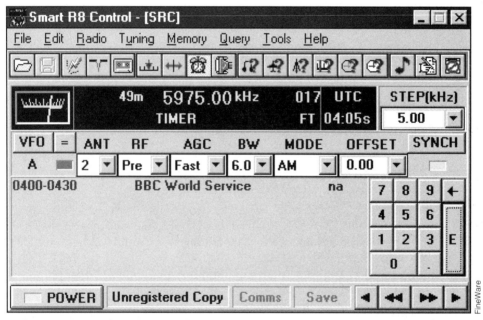

14-2   The display of the Smart R-8 receiver-control program from FineWare.

Computerized receivers

In order to fully utilize the computer-control programs, you need an excellent receiver, a computer interface board for the receiver, and a computer. Admittedly, many people already have personal computers, but is there a way to reduce some of the steps in the process, thereby lowering the cost and the "stuff" needed for a top-notch receiving set-up?

A step in this direction occurred at Dayton Hamfest in 1996. An Australian firm, Rosetta Laboratories, launched a product called WiNRADiO, which claimed to be

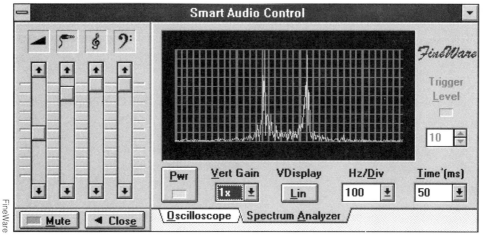

14-3 The display for the Smart Audio Control from FineWare.

"the World's first wide-band communications receiver card on a PC card." Although I had seen a computerized HF transceiver in the past (I haven't seen any ads for it in a few years, so the company apparently went bust), I'm not aware of any other wide-band (in this case, meaning from 0.5 to 1300 MHz) PC card receiver.

Placing a sensitive receiver inside a computer seems to contradict common sense. What are these wacky Aussies going to do next, tell me to bury my antenna underground? As mentioned in Chapter 6, the noise interference generated by the PC is enough to wreck your radio listening. Yet, Rosetta Labs opted to eliminate this interference by putting an entire shielded radio inside a personal computer (Fig. 14-4). Provided the shielding is good enough, the effect should be much the same.

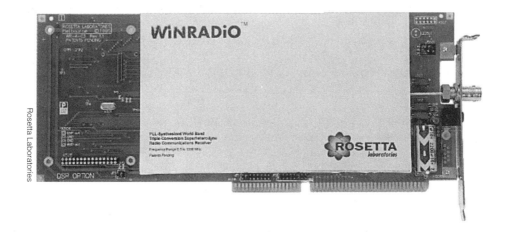

14-4 The actual WiNRADiO hardware.

Several advantages of having a communications receiver inside a PC are:
- The user interface (in WiNRADiO's case, running under Windows) can be very user-friendly and can contain many more features and functions than could be fitted on a stand-alone fixed-panel receiver.
- Some very advanced features, such as spectrum analysis, expensive on a stand-alone receiver, can be made cheaply with a PC-based radio.
- The PC mass storage facilities (hard disk) mean practically unlimited memory capacity for frequency storage. A direct integration with a database is also possible.
- A single radio card can have many different software "personalities," developed to suit your particular preference or application.
- A close integration of a radio and PC can mean an easier development of enhanced functions (for example, real-time signal enhancement and decoding facilities).

I won't get into many of the technical details, but here is a rough sketch of how the receiver fits in with your computer system. WiNRADiO is available on a standard full-size PC-bus (ISA) card, which requires a single slot on your PC. A BNC connector is used to connect an antenna, and a standard audio jack is used to connect speakers or headphones (so, a sound card is unnecessary).

I received word from Rosetta Labs that they are planning to release a specialized shortwave receiver in the upcoming months. If you're on the Internet, be sure to visit their Web page for new releases:

http://www.winradio.net.au

Shortwave receivers of the future?

Because radio add-ons to computer systems could utilize existing microprocessor and DSP hardware, these receivers could well be the super receivers of the future. Or perhaps an enterprising receiver manufacturer will purchase thousands of obsolete boards from 80386, 80486, and low-end Pentium computer systems and start building radios around them.

I think that the receivers of the future (insert space music here) will be either a radio that is built around these surplus computer parts or one that is a receiver board to be plugged into a computer. I haven't yet tried the WiNRADiO to see how it handles computer noise, but perhaps the super computer receiver will be a plug-in box for a laptop computer (which produces much less RFI than a desktop model).

The basic concepts of radio design have remained the same for nearly 60 years. The only major change in that time was in the 1960s, when vacuum tube designs were switched over to solid-state. The 21st century will revolutionize radio systems with all-digital technology. Shortwave will be an interesting mixture of radio and computer equipment—ever more capable than that of today. Pull out your compusceiver, and get ready for the future!

15
CHAPTER

Reporting and verifications

Shortwave listening can be accomplished with any radio receiver capable of tuning the shortwave frequency spectrum—all or any part of a band of frequencies between 1600 kHz and 30 MHz. And, as mentioned in the first chapter, shortwave listening in the general sense is not confined only to those frequencies. It can also include the mediumwave frequencies of 540 to 1600 kHz, where local stations can be found, as well as on long wave from 540 kHz on down to 50 kHz or lower, even though the frequencies from 1600 kHz to 50 kHz are not considered to be shortwave. Even a novice SWL can tune in the larger overseas broadcasting stations— the BBC in London, Radio Nederland in Hilversum, Holland, Radio Australia in Melbourne, Deutsche Welle in Cologne, Germany, RAE in Buenos Aires, Radio Cairo, the Voice of America, and others. Can't you imagine the excitement of tuning and monitoring some of the low-power stations in Asia, Africa, South America, or some other distant land?

Shortwave listening is a hobby second to none for adventure and excitement. You can enjoy the challenge of tuning and listening to shortwave broadcasting stations around the world, from Europe, Asia, Africa, to the vast Pacific, South America, the North Atlantic, and points beyond.

You are probably wondering how to tune in a distant station when atmospheric static, interference from other broadcasting stations, and other peculiarities are often present. Remember, with patience and a determination to tune them, it can be done. Don't be discouraged if the station you are tuning is not heard the first time. Try again.

This chapter was prepared, in part, by Mr. John Beaver, Sr. of Pueblo, Colorado, who was a long time member and officer of the Newark News Radio Club.

Report preparation

One of the purposes of shortwave listening is keeping the stations informed on how good a job they are doing. A report of reception to a station monitored can tell, in a few words, or in a more detailed discussion of the transmission, just how well the signals are received, and the overall quality of reception. Some shortwave broadcasting stations have paid monitors in select countries throughout the world, but most stations depend on reception reports sent them by listeners (Fig. 15-1). This, then, will tell you how you can take part in this hobby in an active manner and, at the same time, join in with thousands of other letter writers who unselfishly give of their time to continually let the overseas broadcast stations know how they are being heard. Further, these reception reports often guide the station programmers in the selection of future programs that appeal to most people.

A sincere, honest, and detailed report is of value to various departments of the station receiving a report of reception. Keep this in mind when preparing a report on what was heard and just how well you received it through your radio set. A good report is welcomed. On the other hand, a poorly prepared and carelessly written report has little or no value to the station and the personnel who check and analyze reception reports against their station log.

A case in point: Several times throughout the years, Hank Bennett, as editor of one of the shortwave columns for the Newark News Radio Club, received letters of complaint from high-ranking officers of shortwave stations. These letters often stated that many listeners not only do not include even the sketchiest information (such as date and time), but they *demand* that the station verify their report. Reports of this type are of use only to the post office because they sold the stamps that were placed on the envelope. He replied to the stations and suggested that any such reports be immediately placed in the nearest trash can. Another complaint from the stations indicates that some people stoop so low as to actually copy certain items of program material—verbatim—that appear from time to time in the bulletins of various radio clubs. Some of these listeners were not aware, apparently, that a number of the shortwave stations also subscribe to these radio club bulletins!

So be honest when preparing a reception report. Write clearly and legibly. Remember, a little time spent writing a good report receives merit. Use a computer or typewriter if possible so it will be easier and neater than a hand-written report. When your report is hand written, block letter printing is preferred for legibility.

SWLs who send reception reports to stations usually are in one of four categories: those who listen for pleasure and entertainment only; those who listen to gain a better understanding of other countries or to learn a foreign language; those who listen for the purpose of reporting and collecting verifications and QSL cards; and those who listen and monitor for the purpose of reporting technical data of value to the engineering department of the station heard.

Verifications, or QSLs as they are more popularly known, are cards or letters sent to the listener by the station, after the listener has supplied satisfactory proof of reception. Some QSL cards are very colorful, others are rather plain; they are all verifications and serve the purpose intended. Many QSLs have become rare collectors items down through the years, and some are no longer available; thus, they are irreplaceable to the owner.

Verifications are no more than a few words on a card or in letter form. They might be quite elaborate and attractive. Some are accompanied by the schedule of the station. In a few cases, a souvenir from the country of the verifying station might be sent—a pennant or an item historically connected with the country. Some sta-

RADIO EASTERN HIGHLANDS
"KARAI BILONG KUMUL"

National Broadcasting Commission

Phone 72 1733
Telex 72 603

P.O. Box 311
Goroka. EHP.
Papua New Guinea.

Ref:

12th March 1993

Dear Mr Yoder

I thank you so much for your reception report dated 27th September 1992. My sincere apologies for the delay in me replying to it.

I am very glad to inform you that the report was found to be correct. The frequency you tuned into was our shortwave frequency of 3395 kilohertz (3.395 MHZ). The programme details you provided were checked against our programme log and were also found to be correct.

Radio Eastern Highlands is transmitting on an output power of 10 kilowatts. The aerial or antenna design is co-phased, half-wave dipole. The transmitter (SW) is a Japanese made NEC together with its associated equipment. Apart from the shortwave transmitter the station operates a medium wave (MF) and an FM transmitter. The MF has an output power of 2.5 kilowatts while the FM has 100 watts. The programme source for these two transmitters come from our headquarters in the nation's capital, Port Moresby. The shortwave programmes are produced and broadcasted here in the station, that is why most are village oriented. The programmes here are meant to educate, entertain and inform our provincial listeners in and around the Eastern Highlands Province. Our station, Radio Eastern Highlands is one of 19 provincial radio stations in PNG. We have a staff strength of 15 of which 6 are females. From the 15, 5 are casual and the remainder are permanent officers.

15-1 A superb QSL from Radio Eastern Highlands in Papua New Guinea.

Apart from the broadcasters we have a journalist and two technicians including myself on the station. Having a journo and at least a technician on a station is a must for all provincial stations in the country.

Now a bit on our country Papua New Guinea. PNG as you may have known is one of the island countries in the South West Pacific. It is a tropical country sharing the international border with Australia, Indonesia and Solomon Islands. We do not have seasons like summer, autumn, spring and winter like you do, but have what we call WET and DRY. Our wet season begins around November and goes on until April. From May to October we experience a pretty dry sunny season.

PNG is a democratic country just like the United States. Our system of government is Westminister by which the people elect their representatives to the National Parliament. The representatives elect their leader who becomes the Country's Prime Minister.

Coffee, Copra and Cocoa are our main export commodities. At the moment we are experiencing a mineral boom. It is estimated that PNG is going to be one of the richest countries in the world having many mineral deposits all around the country.

PNG is also very proud of its natural environment. Maybe it is one of the few countries in the world that has all its natural environment still in place. We have some of the rare animal species found in no other parts of the world.

Any way, if you would like to know more about PNG you could always get in touch with any of our sister stations in the country who also operates on the frequencies along the 90 metre shortwave band.

Thank you so much once again.

Yours sincerely

IGNAS YANAM
Technical Service.

15-1 A superb QSL from Radio Eastern Highlands in Papua New Guinea. Continued.

tions have been known to send the listener a small gift in return for a good reception report. Recently, I received a beautiful pennant and an extremely nice letter from a station not known to send pennants at all.

Report data

If the shortwave listener sending a report to a station qualifies only under the first two categories above, the report seldom contains any technical information that might be helpful to the station. The listener most likely will comment on a particular program, ask for information about the country, and possibly inquire as to language courses that might be offered by the station. If you are really interested in getting a verification, note carefully and pay attention to details to include in your report of reception.

When reporting for verifications, it is necessary to report reception over a period of time, at least one-half hour, when possible. This listening time period can, of course, be adjusted as conditions and transmission time of the station dictate. Longer reporting periods are preferred, and accordingly, are much more useful to the station, particularly if the report contains pertinent and technical information. It's not unusual to monitor and report on reception over a period of one or even two hours during a station's transmission time. However, many station schedules might run for a shorter period of time; therefore you must judge for yourself how long you can monitor the transmission of any particular station.

In addition to knowing exactly which station you have tuned in, you should determine as accurately as possible the frequency used by the transmitting station. Listen carefully to station announcements. These are a valuable aid. However, when no frequency announcement is given, and the language is not familiar, there are other methods of determining the frequency. A listing of stations by frequency is a valuable guide in this department. The station in question could be operating on possibly two or three frequencies in a particular waveband during a particular period of time. It is very important to report on the right frequency when tuning a particular wavelength of the broadcasting band.

When a station has several transmitters operating at the same time on parallel frequencies, the report should indicate the specific transmitter being heard. Readability quality should also be noted for each frequency being reported. This is not difficult when the station is transmitting the same program on two or more frequencies through the shortwave broadcast band, and you are able to hear signals on parallel frequencies during the same time period.

Be sure to indicate the date and time period during that transmission was heard and the time zone being used in your report. Simply stating, for example, "I heard you at 12:00 hours," is not enough. If local time, or another time zone is used, specify which one so that the person analyzing your report can compute the time to match that of the station log.

You would do well to use a clock with a sweep second hand. Clocks with built-in conversion scales can also be obtained, and are very useful when tuning and reporting on your listening adventures.

A good report of reception should show a listing of music selections heard (the title or a brief description of it), a short summary of news items, talks, or other programming heard, and any other information that will verify your tuning of a particular station. If the programming is in other than English, and you are not familiar with the language, you should listen carefully for anything that will assist them in reporting your reception and list this with the corresponding time that it was heard.

Program details are important and should be accurate and to the point. Reception reports received by the station are checked against the station log for the time period indicated. Insufficient information can rule the report of little or no value; it could be discarded by station personnel. This is a great loss to both the station and to the listener sending in a report for an intended verification.

Remember, a report of reception is of little value unless it contains needed information. Merely listing times and program items does not necessarily make a good reception report, although there are a few stations who will verify on those points alone.

Reception details, in addition to programming, should include a report on the signal strength and the readability quality of a particular transmission being reported for verification. Remember that the station is interested in knowing how well they are being received in your area. They want to know how their signals perform during the time period covered in the reception report.

They would also like to know how their signals compare with those of other known stations in the same frequency range. Did their signal tend to fade at intervals or was it strong and heard at an easy listening level? If there were intervals of fading, indicate how much and if it was a slow fade or of a rapid flutter quality.

A readability report should include just how the signal was received from the listening level standpoint. Was the signal completely readable, or were there times (list your comments on this) when the copy level was difficult to understand? Were the effects of static or other interference from stations on or near the station in question observed? If so, identify and list the station(s) if possible.

Complete and accurate information does make for a good report and will be appreciated by the personnel of the station that review the report from the listener. It will assist the station in future program planning, to adjust schedules and frequencies if reception is unsatisfactory over a period of time in the listener's area, and to enable the station's engineering department staff to have a more full realization of what their signal is doing and where improvements can be made. You, the listener, will benefit with better programming and better overall reception. The engineering department is vitally concerned with supplying the best signal and reception possible for your entertainment and enjoyment.

Include comments on the programming heard, what you liked and didn't like, but remember to be constructive in your criticism. Don't unjustly criticize the program without some thought of offering suggestions of preference, rather than a particular program transmitted. Shortwave broadcasting stations serve thousands, if not millions, of listeners, and it is impossible to suit the ears of everyone who tunes in their programs. Constructive criticism and suggestions are welcomed by the stations and enable their program planners and staff personnel to plan for future programs that will be most enjoyed by the majority of the listening audience.

Be sincere in the suggestions that you make and don't demand that a particular program be scheduled. If you prefer a certain type of music, talks, features, or other programs, mention those; then, let the staff of the station be the judge as to what a majority of their listeners would like to hear during their transmission time periods.

In preparing the actual report, list all information in letter form (Fig. 15-2). Do not send reports on post cards. They simply cannot accommodate enough of the necessary information to be of value to the station. If you have your own SWL card, send it along with your report. But do not use it alone for reception reporting purposes (Fig. 15-3.)

Brief mention of the equipment you used for monitoring should be included in your report. List the make and model number of your receiver with the number of tubes or transistors, and the type and length of antenna that you are using. You should also include a brief comment on your weather and temperature, but do not go into lengthy detail.

P.O. Box 109
Blue Ridge Summit, PA 17214
USA

Hello!

I am pleased to report reception of the Sierra Leone Broadcasting Service on October 22, 1996 from 2258-2328 UTC on 3316 kHz. If this report is correct, could you please verify it with one of your verifications/QSL cards? Now onto the report...

//
The details of the report are on the cassette.
\\

I listened to your broadcast on a Kenwood R-5000 communications receiver connected to a 125-foot longwire antenna. The antenna's orientation is basically East/West. The signal was very weak. Last winter, I heard your station with a much better signal, so hopefully I will hear you better in the next two or three months. My apologies for not sending a report earlier, when your signals were much easier to listen to.

The town where I live is named Mont Alto and it is about a 15-minute drive from where I receive my mail in Blue Ridge Summit. Mont Alto is in the south central portion of the state of Pennsylvania. Pennsylvania is in the northeastern portion of the United States. In the summer, the temperatures are normally in the 80s Farenheit, and in the winter, they are typically in the 20s or 30s Farenheit (although last year, the temperature dropped to -32 degrees Farenheit one very cold night!). The area around Mont Alto is agricultural. The area is filled with farms that raise cows, corn, soy beans, etc. Washington, DC is about 80 to 100 miles south of us.

A bit about myself: I am 29 years old and work as book editor and freelance writer. My wife is a substitute teacher at the local school district. We have a 4-year old son, Corbin, and a 4-month-old daughter, Bryn. Some of my hobbies include: reading, hiking, listening to shortwave, writing, playing a variety of sports, listening to a variety of music, camping/hiking, traveling, etc.

Thank you very much for your time. I hope you find this report interesting. I look forward to hearing from you!

73s,

Andrew Yoder

15-2 A sample reception report.

15-3 A sample reception report form from Radio Japan.

Codes

Reporting codes are very useful to you as well as to the station receiving the report. You will find this method of reporting to be simple yet comprehensive. Every report sent to a station for QSL purposes should include a strength and readability evaluation of some sort; just what type you happen to prefer is up to you, but we strongly suggest that you use one of the universally recognized codes. Through the years, there have been a number of reporting codes in use, but of all of them perhaps the best known and most widely used are the *QSA-R code*, the *555 code*, and the *SINPO code*.

The QSA-R code is one of the earlier codes used mainly in ham radio. QSA is one of the internationally known Q signals that means, "The strength of your signals is . . . (1 to 5)." R stands for "readability" and this scale runs from 1 to 9. The R is actually an abbreviation of the Q signal QRK, which means, "The readability of your signals is ... "In both cases, the higher the number used, the better the signal is judged to be; thus, a report of QSA5 R9 is utter delight. In recent years, through general usage, the R has become better known as S and the perfect report here would be simply Q5 S9.

The RST and SINPO codes have proven to be the most widely used as well as the most popular with SWLs. A chart for each code appears in Tables 15-1 and 15-2. Let's break down these two codes so that you can see and better understand how they work. The RST code is a system that is believed to have begun with the British Broadcasting Corporation. This code is simple and comprehensive. The novice listener, in particular, can well find this reporting code useful as a guide when tuning and monitoring any station for reporting purposes. The SINPO code is by far the most widely used. Many shortwave broadcast outlets ask their listeners and monitors to report to them in this code. We urge you to acquaint yourself with this code, because it is the one that you will most likely be using frequently.

Table 15-1. SINPO code

Signal strength (QSA)	Interference (QRM)	Noise-atmospherics (QRN)	Propagation disturbance (QSB)	Overall merit (QRK)
5 Excellent	5 None	5 None	5 None	5 Excellent
4 Good	4 Slight	4 Slight	4 Slight	4 Good
3 Fair	3 Moderate	3 Moderate	3 Moderate	3 Fair
2 Poor	2 Severe	2 Severe	2 Severe	2 Poor
1 Barely audible	1 Extreme	1 Extreme	1 Extreme	1 Unusable

Table 15-2. 555 code

Signal strength (QSA)	Interference (QRM)	Overall merit (QRK)
5 Excellent	5 None	5 Excellent
4 Very good	4 Slight	4 Very good
3 Good	3 Moderate	3 Good
2 Fair	2 Severe	2 Fair
1 Poor	1 Extreme	1 Poor
0 Inaudible	0 Total	0 Unusable

For example, a report to a station whose signal is loud and clear, free of interference and static, and no fade level, would be written SINPO 55555. Do *not* run the letters and numbers together in this fashion: S5 I5 N5 P5 O5. This will only tend to confuse the person who reviews your report. Another example of reception with a good signal, slight interference, slight static level, and a moderate fade level, would be written SINPO 44434, or a similar merit depending on the degree of interference, static, and fading of the signal. An additional example of a signal that was received with only a fair level, moderate interference, moderate static level, and a marked degree of fading would be written SINPO 33322. In conclusion, should the listener be monitoring for a station on a regular basis, and an occasion would come about when the signal of the station is not audible, due to interference or other adverse phenomena, the reporting code possibly would be written: SINPO 02320.

No matter which reporting code you prefer, always be honest when evaluating the signal received. The more detailed your report, the more value it will have for the station receiving it, and the more credit the listener will reflect on the SWL hobby overall. Careful preparation of the report will merit you the desired verification, or QSL.

Remember at all times to be courteous in requesting a QSL. Some listeners have demanded that their report be verified, then wondered why the station, on occasion, discarded their reception report. It is much more effective to suggest that the station verify your report using the following example as a guide:

"Should this report of reception on the dates, times, and frequency check accurately with your station log, and if it is found to be of value to your engineering staff, verification would most sincerely be appreciated."

This method is particularly important when writing to a station that is known for not usually verifying reception reports. They will offer reply with a QSL if a report is well prepared and contains needed information—and if you are courteous in your request.

Now that I have gone into great detail about using the SINPO codes, remember that they aren't useful for most DX stations. Most regionals don't receive piles of mail from DXers; they are just providing a regular broadcasting service. They, like the personnel at your local FM station, will wonder what those numbers are that you put on your letter. When you are writing to any international broadcaster, use the SINPO code; however, when writing to any tiny regional stations, be sure to describe the strength, fading, etc. instead.

Postage

It is always proper to include return postage when sending reports overseas. Remember, the station is not obligated to verify your report. Many stations indicate that return postage is not required, but the majority appreciate receiving it. Some stations in the shortwave spectrum are government-owned or operated, in which case return postage is not needed. When in doubt, it is always best to include it with your reception report.

Return postage for foreign countries can be sent in the form of international reply coupon (IRCs) and these are available from your post office. A few countries that do not belong to the International Postal Union will not honor IRCs and the postal clerk will have this list in his postal manual.

Do not use IRCs when sending mail to any station in a smaller country that is not a government station. Even though a country might officially accept IRCs, many small post offices have never seen them and will not cash them in. I've known this from personal experience; I've gone to about a dozen post offices in the United States, and only one knew what IRCs were! So, if you send the IRC to a station in the

Peruvian Andes, where it is just a useless little piece of paper, you can bet that you won't hear a reply!

If you would prefer to send return postage in the form of mint (new and unused) stamps of the country in which the station is located, this is okay too. But where can you obtain mint stamps of foreign countries? One source is stamp dealers, but be sure they are currently in use in the foreign nation you're sending the report to. You might also want to try the various "DX Stamp Services" offering stamps to ham radio DXers.

By using stamps from foreign countries, you are taking a bit of a risk. The stations will typically respond to you more reliably than by any other method; however, if a coup takes out the government or if rising inflation suddenly devalues the currency, your stamps could be worthless.

My favorite method of enclosing return "postage" is to send a U.S. dollar bill. The dollar is enough to cover postage, has value in any country, is cheaper than sending mint stamps or IRCs, and the station might have a little money left over to cover the printing cost of your QSL, pennant, etc. By sending cash, you risk the chance of having the letter stolen for the cash, so be sure to conceal it if you decide to try this method.

Above all, do not send U.S. stamps except as mementos when sending reception reports to any other country. They aren't valid in any country other than our own and you'll be wasting your money.

Time

It is important to indicate reception time in a standard manner when reporting on reception of any broadcasting station. While you might prefer to use local time (if so, this should be noted in your report), a time conversion chart is a most useful object to have in your listing post.

Because shortwave stations are scattered throughout the world, it has become common practice to report reception in terms of Universal Time Coordinated (UTC).

To equate our 24-hour day with the geographic picture of the surface of the earth, keep in mind that an increment of one hour occurs with each 15-degrees change in longitude. UTC is simply the time at the point of zero longitude, which happens to pass through Sussex, England.

The 24-hour clock system is generally understood and accepted around the world. In this system, the hours from 1:00 AM to 11:00 AM are expressed as 0100 to 1100 hours. Noon is referred to as 1200 hours. From 1:00 PM to 11:00 PM, the times are expressed at 1300 to 2300 hours. Midnight is referred to by some as 2400 hours, but is most popularly known as 0000 hours; this latter form is preferred when reporting to stations. A few other examples:

8:40 AM is 0840 hours
11:01 AM is 1101 hours
12:10 PM is 1210 hours
2:36 PM is 1436 hours
6:45 PM is 1845 hours
11:59 PM is 2359 hours
11:59 PM plus 2 minutes is 001 hours

A time conversion chart has been included for many points around the world in the appendix. With it you will be able to convert to and from UTC from the standard American time zones and from foreign points.

Like using the SINPO codes, using UTC time is also a waste when writing to smaller stations. To be safe, I usually write the time in the local time of the station and follow it with the UTC time in parenthesis. By doing so, I'm hoping that the people who don't understand UTC will ignore it (and just use the local time), and those who do understand UTC will use it (and not wonder 9 PM where?). Regardless, don't use just UTC for the tiny distant stations (UTC is preferred by pirate stations, however).

Tape-recorded reports

In recent years, the practice of sending tape recorded reception reports has gained rapidly in popularity, and it has been proven to be very effective in obtaining QSLs. The station receiving a taped report is able to judge far more accurately how well its signals are being received. This method of reporting is proving to be far superior to more conventional written reports of reception, if the listener records the transmission exactly as it is being received through his radio set. It is important to remember not to change any equipment settings on your radio when recording on tape, regardless of how the signal is received, unless the volume of the signal rises to a point where overload to your recording equipment might occur.

The station is interested in knowing how their signal strength and readability quality varies over a period of time and how their signal is affected both by other stations and electrical or ionospheric disturbances. Care in recording, and carefully tuning your station before switching your recorder on, will merit a good report on tape and be of vital use to the engineering department of the station in question. Don't hurry, be patient, and record exactly what is being received through your radio set. This is very important. Don't rush. A carefully recorded report cannot be done haphazardly.

Before starting to record on tape, tune the station in as accurately as possible. Remember to adjust the audio gain (volume control) of the receiver so that it is sufficient for recording level. Keep in mind that a taped report should be long enough to be of value to the station. Taped reports of less than 10 minutes are of little use. A taped report of 20 to 30 minutes is suggested, to provide the station with the information that they desire.

Do not permit outside noise to be recorded on the magnetic tape. Connect the recorder directly to the receiver headphone jack or from the speaker terminals by an input cable or patch cord. Never try to record programming directly through the microphone of the recorder. The chances are too great of picking up outside noise, which will cause the tape to be useless for verification purposes.

If you want to speak on the tape, do so only after completion of the actual taped report, preferably toward the end of the second track.

When sending a taped report, be sure to enclose a letter that includes all of the standard information: date, time, frequency, your name and address, and that you would like to receive a QSL or verification. Also be sure to include the other important information: details about yourself, family, and interests, what you liked about the program, etc. To be sure that my tape isn't completely lost in a station, I write all of the time, date, frequency, and station name on a label on the cassette. I also tape down one of my mailing labels on the cassette, so they know who sent it to them.

I really like sending taped reception reports to stations that don't broadcast in English. Rather than me trying to decypher their language on the air, then goof up with the Urdu-language dictionary and tell them that I heard the announcer say "I love pigs," I send a tape. It's much easier for me, and potentially less embarassing. The price can be a bit steep for cassettes, so I get used tapes for 10¢ to 30¢ apiece from yard sales or Salvation Army stores. Sometimes friends give me nonworking or "eaten" tapes that I fix.

I think that taped reports are the best way to go, if you can afford it. However, if the station is really poor and you have to sift the details, don't send a tape. Sometimes, you can catch a detail out of three minutes of noise. If you send a written report, the operator will see a detail. If you send the tape, they will hear the three minutes of noise; if you're lucky, they will hear the detail. Overall, this is just an annoyance for someone who, if you're lucky, will verify your report. Also, not everyone can sift information through the static. I play tapes back for my wife, saying "Listen to this, isn't it great?" And all she hears is static. All you can do is go with a judgement call—a tape might or might not work better than a written report in weak-signal conditions.

Foreign-language reports

English is, in large measure, the international language of the world today. An English report to virtually all major stations or government-owned stations can thus lead to a reply. However, Latin America, Asia, and Africa are full of smaller stations that don't have anyone who speaks English at hand. In such cases, a reception report will have a better chance if written in a language spoken locally. Throughout most of Latin America, Spanish is the universal language; Brazil, on the other hand, uses Portuguese, and Guiana uses French. African stations would be most likely to have French-speaking staff, although Portuguese is favored in Guinea-Bissau, Mozambique, and Sao Tome E Principe.

If, like most gringos, you can't write shortwave reports in Spanish, Portuguese, or French, there is no need to despair. Tables 15-3 to 15-11 were prepared by William Avery of University Park, Maryland, who in addition to being an SWL, also is professor emeritus of classical languages and literature at the University of Maryland. By using these tables, you can prepare a presentable reception report in any of these three languages. Just be sure to note the accent marks on the various words and phrases; if your typewriter or computer printer lacks the appropriate marks, write them in with a pen or pencil.

Many stations in Latin America send out pennants along with their QSL cards or letters. Because most SWLs welcome such pennants, a request for a pennant has been included in the report form.

One important point to remember is that these reception report forms are only basic guides for sending letters to stations. Don't use them as-is because it contains no personal information. To put the form letters here in better perspective, imagine how much you hate receiving impersonal "junk mail" (if you don't hate it, you're one of the few). Then, imagine if you were trying to get your work finished at some understaffed station and you kept receiving requests from different people, but the letters all read exactly the same! If I was in that position and I didn't care about DXing, my reaction would probably be to pitch the letter and pocket the extra postage or cash.

Table 15-3. Spanish language reception report form

Estimado Sr. :

Tuve el placer inmenso de sintonizar su emisora (1) en (2) kHz, onda (3), el día (4) de (5) de (6) entre las (7) y las (8), hora de (9) . Los detalles de mi recepción, los cuales espero que Vd. encuentre de interés, son indicados abajo:

Hora   Características del programa
                        (10)

Las señales de su emisora fueron (11) con (12) interferencia. Mi receptor es un (13) .

Si Vd. encuentra que este informe de recepción es correcto, ruego a Vd. se sirva enviarme una breve carta o tarjeta postal para comprobar mi recepción. Mi pasatiempo es escuchar radiodifusoras lejanas y recibir la correspondiente verificación. En consecuencia, quedaría sumamente honrado si pudiese contar con la suya. Por favor: me gustaría mucho que Vd. me enviara un banderín o gallardete de su gran emisora. Muy agradecido por la buena atención que se dignara Vd. prestar a la presente, quedo de Vd.

Sin otro particular y con los sentimientos de mi más alta consideración, me subscribo como su amigo afmo. y s. s.

(your name)

Table 15-4. Using the Spanish language report form

1. Enter the station's name or slogan, not just call letters.
2. Enter the frequency on which you heard the station.
3. If you heard the station on shortwave, enter corta. If you heard it on the broadcast band, enter large.
4. Enter the date of your reception.
5. Enter the month of your reception (see glossary).
6. Enter the year of your reception.
7. Enter the time you first began listening to the station. (Use station's local time only)
8. Enter the time you finished listening to the station.
9. Enter the country in which the station you heard is located.
10. Using glossary, list the items heard along with the time they were heard. List time under hora and descriptions under caracteristicas del programa.
11. This describes the strength of the station's signals. Use excelentes for strong, regulares for fair, and insuficientes for weak.
12. This describes the interference. Use ninguna for none, poca for little, moderada for moderate, and densa for heavy.
13. This is the type of receiver you are using. Use receptor de communicaciones for communications receiver, receptor domestico for any general home or table receiver, and receptor portatil for a portable set.

Table 15-5. Spanish phrase glossary

Commercial:	Anuncio commercial
Announcement as:	Anuncio como
Station identification as:	Identificación de su Emisora coma
The announcer mentioned:	El locutor mencionó
Instrumental music:	Música ligera
Folk music:	Música folklórica
Classical music:	Música clasica
Religious music:	Música religiosa
Choral music:	Música coral
American popular songs:	Canciones populares se los EE. UU.
Newscast:	Boletín de noticias
Sports broadcast:	Programa deportive
Song by a man/a woman:	Canción cantada por un hombre/una mujer
Chimes:	Toque de campanas
Gong:	Gong
Tone:	Tono
Applause:	Aplausos
Sound effects:	Efecto de sonido
Echo effects:	Efecto del eco
Sign on:	Comienzo de las transmisiones
Sign off:	Fin de las transmisiones
National anthem:	Himno Nacional
January:	enero
February:	febrero
March:	marzo
April:	abril
May:	mayo
June:	junio
July:	julio
August:	agosto
September:	septiembre
October:	octubre
November:	noviembre
December:	diciembre

Table 15-6. Portuguese language reception report form

Estimado Senhor:

    Tivo o imenso prazer de sintonizar Vossa emissora (1) em (2) kHz, onda (3), no (4) de (5) de (6) entre (7) e (8) horas, hora de (9) . Os detalhes da minha recepção, os quais espero que o Senhor ache ter algum interêsse, estão indicados abaixo: Hora Caracteristicas do programa

    Os sinais de Vossa emissora forarn (11) com (12) interferencia. O meu receptor é um (13) .

    Se o Senhor achar que o presente informe é correto, rogo tenha a bondade de envier-me uma breve carta ou cartão postal para comprovar a minha recepção. ( ) meu passatempo é escutar radiodifusoras longinquas e receber a verificação respectiva. Por isso, eu ficaria sumamente honrado se pudesse contar com a Vossa. Por favor: eu estaria muito contente de receber de Vossa emissora uma bandeirola ou galhardete. Muito agradecido por a boa atenção que o Senhor digne-se prestar á presente, fico do Senhor atento servidor e obrigado.

_____
(your name)

Table 15-7. Using the Portuguese language report form

1. Enter the station's name or slogan, not just call letters.
2. Enter the frequency on which you heard the station.
3. If you heard the station on shortwave, enter curta. If you heard it on the broadcast band, enter larga.
4. Enter the date of your reception.
5. Enter the month of your reception (see glossary).
6. Enter the year of your reception.
7. Enter the time you first began listening to the station, using the station's local time.
8. Enter the time you finished listening to the station, using local time also.
9. Enter the country in which the station is located.
10. Using the glossary, list the items heard along with the time they were heard. List time under hora and descriptions under caracteristicas do programa.
11. This describes the strength of the station's signals. Use excelentes for strong, regulares for fair, and insuficientes for weak.
12. This describes the interference. Use nenhuma for none, pouca for little, moderada for moderate, and densa for heavy.
13. This is the type of receiver you are using. Use receptor de comunicacões for communications receiver, receptor doméstico for any general home or table receiver, and receptor portátil for a portable set.

Table 15-8. Portuguese phrase glossary

Commercial:	Anuncio comercial
Announcement as:	Anuncio como
Station identification as:	Identificação de Vossa emissora como
The announcer mentioned:	O locutor mencionou
Instrumental music:	Música ligeira
Folk music:	Música folklórica
Classical music:	Música classica
Religious music:	Música religiosa
Choral music:	Música coral
American popular songs:	Canções populares dos Estados Unidos
Newscast:	Noticiario
Sports broadcast:	Programa desportivo
Song by a man/woman:	Canção cantada por um homen/mulher
Chimes:	Toque de campainhas
Gong:	Gongo
Tone:	Tom
Applause:	Aplauso
Sound effects:	Efeitos sonoros
Echo effects:	Efeitos do eco
Sign on:	Começo das transmissões
Sign off:	Fim das transmissões
National anthem:	Ino nacional
January:	janeiro
February:	fevereiro
March:	março
April:	abril
May:	maio
June:	junho
July:	julho
August:	agosto
September:	setembro
October:	outubro
November:	novembro
December:	dezembro

Table 15-9. French language reception report form

Cher Monsieur:

    J'ai eu le grand plaisir de me mettre à l'écoute de votre poste émetteur (1) sur (2) kHz, onde (3) le (4) (5) (6), entre (7) heures et (8) heures, heure de (9) . Les détails de ma réception, lesquels j'espére que vous trouverez de quelque intérêt, sont les suivants:

    Huere Caractéristiques du programme

    Les signaux de votre poste émetteur ont été (11) avec (12) parasites. Mon récepteur est un (13).

    Si vous trouverez que ce rapport est exact, je vous prie de me faire le plaisir de m'envoyer une brève lettre ou carte postale pour confirmer ma réception. Mon passe-temps est d'écouter des postes de radiodiffusion lointains et d'avoir la vérification respective. Par conséquent, je serais fort honoré si je pouvais compter sur la vôtre. S'il vous plaît: je serais très content si vous aviez la bonté de m'envoyer une banderole ou flamme de votre important poste émetteur. Avec mes remerciements les plus sincères pour la bonne attention que vous daigniez faire à la présente, je vous prie d'agréer l'expression de mes sentiments les plus distingués.

    (your name)

Table 15-10. Using the French language report form

1. Enter the station's name or slogan, not just call letters.
2. Enter the frequency on which you heard the station.
3. If you heard the station on shortwave, enter courte. If you heard it on the broadcast band, enter longue.
4. Enter the date of your reception.
5. Enter the month of your reception (see glossary).
6. Enter the year of your reception.
7. Enter the time you first began listening to the station, using the station's local time.
8. Enter the time you finished listening to the station, using local time also.
9. Enter the country in which the station is located.
10. Using the glossary, list the items heard along with the time they were heard. List time under heure and descriptions under caractéristiques du programme.
11. This describes the strength of the station's signals. Use excellents for strong, réguliers for fair, and insuffisants for weak.
12. This describes the interference. Use pas de for none, peu de for little, de modérés for moderate, and de lourds for heavy.
13. This is the type of receiver you are using. Use récepteur de communications for communications receiver, récepteur domestique for any general home or table receiver, and récepteur portable for a portable set.

Table 15-11. French phrase glossary

Commercial:	Annonce publicitaire
Announcement as:	Annonce comme
Station Identification as:	Identification de votre poste émetteur comme
The announcer mentioned:	Le présentateur a mentionné
Instrumental music:	Musique instrumentale
Folk music:	Musica folklorique
Classical music:	Musique classique
Religious music:	Musique religieuse
Choral music:	Musique chorale
American popular songs:	Chansons populaires des États-Unis.
Newscast:	Journal parlé
Sports broadcast:	Programme de sport
Song by a man/woman:	Chanson chantée par un homme/femme
Chimes:	Jeu de sonnettes
Gong:	Gong
Tone:	Ton
Applause:	Applaudissements
Sound effects:	Bruitage
Echo effects:	Effets de l'écho
Sign on:	Commencement des émissions
Sign off:	Fin des émissions
National anthem:	Hyme national
January:	Janvier
February:	Février
March:	Mars
April:	Avril
May:	Mai
June:	Juin
July:	Juillet
August:	Août
September:	Septembre
October:	Octobre
November:	Novembre
December:	Décembre

So, try to hack together some interesting sentences with a dictionary. You might even be able to find someone, perhaps from a local high school or college, who speaks the language of the station that you are writing to. They might be willing to help you write a few sentences, if you are especially nice about how you ask.

Some languages are just hopeless for me. I can hack together a few sentences of Spanish and French that seem to be recognizable, but I can't even imagine trying to pull off some good, clean Indonesian sentences. Now it's time for some backup help. If you want to improve your foreign-language reports, get a copy of the Language Lab by Gerry Dexter (available from Tiare, P.O. Box 493, Lake Geneva, WI 53147).

This book covers everything you need to know about writing letters and reception reports to radio stations in most languages that you would need to write a report in.

Enclosures

When you write a reception report to a station, be sure to send something that you would like to receive if you were working there. Over the years, I have seen plenty of very bland reports: date, time, frequency, some details, and "please send a QSL." Some of these same people complain that some stations don't have nice QSLs, didn't send them a pennant, etc. In any situation, you can only expect to receive a bit less than what effort you put into it. If you want someone to be nice to you, you've got to be nice first.

So, if you would like to receive a three-page letter, send one. If you would like to read information about the area that someone lives in, send along a travel brochure for your area. If you like music, mention it in your letter and send a taped report, with some hand-picked music on the other side. One of the best articles to send along with a reception report is a recent photo of you or of your family. If you are writing your report in a different language and it still seems a bit impersonal, nothing personalizes a letter better than photographs.

Aside from photos, cancelled commemorative stamps are always good to send. They're light, nice looking, and from your area. Also, they're a popular collectible. Another favorite are radio stickers from around the world; bumperstickers can be expensive to mail, so try to find smaller stickers.

While this summary of reporting and verification procedures is directed primarily to the shortwave listener, the information is applicable when reporting to other stations—those in the standard broadcast band, utility stations, and possibly FM and TV DX stations. This summary is intended as a guide to assist you in obtaining QSLs from stations in all parts of the world.

In conclusion, you will get out of your hobby only the effort you put into it. When preparing a reception report, it is important to remember, again, that you owe it to yourself to be honest and sincere in your efforts, whether reporting on tape or in writing. The suggestions set forth in this chapter will benefit not only you but the station as well. With this in mind, set your goal, and go to it. Good luck and happy listening!

A
APPENDIX

Useful shortwave tables

Worldwide time chart

Newcomers to the shortwave listening hobby will find what appears to be several different forms of time used, depending on the station to which they are listening. Stateside broadcast band stations stick almost entirely to the well known AM-PM system. Shortwave stations in virtually every country of the world base their time upon Universal Time Coordinated, which is five hours ahead of the American eastern standard time. To add to the apparent confusion, most of the shortwave stations also use the 24-hour system instead of straight AM and PM. The 24-hour system is universal and is also used by most DXers in keeping their logs and in sending reception reports. Further, the 24-hour system is generally recorded in UTC, but it can be used in any time zone of the world. In the 24-hour system, 12 o'clock is known as 1200. One hour later, the time is 1300. The midevening hour of 9:00 PM is 2100. Any minutes past the hour, of course, are expressed as just that: 23 minutes past 9:00 PM is simply 9:23 PM or 2123. Midnight is known as 0000, although some sources prefer 2400; the former is far more widely used.

Following is a time conversion chart for many of the larger cities around the globe. Please note that for those persons who have not yet mastered the 24-hour system, eastern standard time is given in the 24-hour system in the second column and in the AM-PM method in the last column. A more comprehensive table is to be found in the *Radio Amateur Callbook Magazine* and in the *World Radio-Television Handbook*.

Manila Peking	Tokyo	Melbourne	Hawaii	Rio de Janeiro	Azores	Canary islands	EST a.m.-p.m.
0800	0900	1000	1400	2100	2200	2300	7:00 PM
0900	1000	1100	1500	2200	2300	0000	8:00 PM
1000	1100	1200	1600	2300	0000	0100	9:00 PM
1100	1200	1300	1700	0000	0100	0200	10:00 PM

328 *Appendix A*

Manila Peking	Tokyo	Melbourne	Hawaii	Rio de Janeiro	Azores	Canary islands	EST a.m.-p.m.
1200	1300	1400	1800	0100	0200	0300	11:00 PM
1300	1400	1500	1900	0200	0300	0400	Midnight
1400	1500	1600	2000	0300	0400	0500	1:00 AM
1500	1600	1700	2100	0400	0500	0600	2:00 AM
1600	1700	1800	2200	0500	0600	0700	3:00 AM
1700	1800	1900	2300	0600	0700	0800	4:00 AM
1800	1900	2000	0000	0700	0800	0900	5:00 AM
1900	2000	2100	0100	0800	0900	1000	6:00 AM
2000	2100	2200	0200	0900	1000	1100	7:00 AM
2100	2200	2300	0300	1000	1100	1200	8:00 AM
2200	2300	0000	0400	1100	1200	1300	9:00 AM
2300	0000	0100	0500	1200	1300	1400	10:00 AM
0000	0100	0200	0600	1300	1400	1500	11:00 AM
0100	0200	0300	0700	1400	1500	1600	Noon
0200	0300	0400	0800	1500	1600	1700	1:00 PM
0300	0400	0500	0900	1600	1700	1800	2:00 PM
0400	0500	0600	1000	1700	1800	1900	3:00 PM
0500	0600	0700	1100	1800	1900	2000	4:00 PM
0600	0700	0800	1200	1900	2000	2100	5:00 PM
0700	0800	0900	1300	2000	2100	2200	6:00 PM

A few countries scattered around the world do not conform exactly to the procedure of even hours and, therefore, have times that are so many hours and so many minutes ahead of, or behind, UTC. They are as follows:

Country	Deviation from UTC
Newfoundland	−3 hours, 30 minutes
Surinam	−3 hours, 30 minutes
Guyana	−3 hours, 45 minutes
Iran	+3 hours, 30 minutes
Afghanistan	+4 hours, 30 minutes
India	+5 hours, 30 minutes
Sri Lanka (Ceylon)	+5 hours, 30 minutes
Nepal	+5 hours, 40 minutes
Burma	+6 hours, 30 minutes
Cocos, Keeling Is.	+6 hours, 30 minutes
Malaysia	+7 hours, 30 minutes
Singapore	+7 hours, 30 minutes
Northern Australia	+9 hours, 30 minutes
Southern Australia	+9 hours, 30 minutes
Cook Islands	+10 hours, 30 minutes
Nauru	+11 hours, 30 minutes
Norfolk Island	+11 hours, 30 minutes

Commonly used SWL abbreviations

After editing a logsheet for awhile, I've had many people say "I'm interested in radio, but I just can't understand the loggings." Others have said "I enjoy your newsletter—especially trying to figure out all of the abbreviations." The following list is published here so that you can figure out what the radio newsletters are saying! It is neither a universal nor a complete list, but it does have quite a few of the abbreviations that you're bound to encounter while reading shortwave radio bulletins. The listing here is quite different from amateur radio abbreviations.

ABC	Australian Broadcasting Corporation	FEN	Far East Network
Abt	About	FFFR	Fight for free radio
ac	Alternating current	FM	Frequency modulation
ACE	Association of Clandestine radio Enthusiasts	Freq	Frequency
		Gd	Good
Add	Address	GOS	General Overseas Service
Af	Africa	Gt	Great
AIR	All India Radio	Hrd	Heard
AM	Amplitude modulation	Hvy	Heavy
Ancd	Announced	Hz	Hertz
Ancr	Announcer	ID	Identification
Anmt	Announcement	Instr	Instrumental
As	Asia	Intl	International
BBC	British Broadcasting Corporation	IRC	International Reply Coupons
Bdcst	Broadcast	IRCA	International Radio Club of America
BCB	Broadcast band	IS	Interval signal
CA	Central America	kHz	Kilohertz
CARACOL	Primera Cadena Radial Colombiana	kW	Kilowatts
		LA	Latin America
CBC	Canadian Broadcasting Corporation	Lang	Language
		Ltr	Letter
CIDX	Canadian International DX Club	ME	Middle East
Corp	Corporation	Ment	Mention
CRI	China Radio International	MHz	Megahertz
CW	Morse code (carrier wave)	Mod	Modulation
D	Daily	MW	Mediumwave
dc	Direct current	Mx	Music
de	From	Nx	News
DOC	Department of Communications (Canada)	NA	North America
			National anthem
DX	Distance	NASWA	North American Shortwave Association
ECNA	East Coast North America		
EG	English	NBC	Namibian Broadcasting Corporation
EDT	Eastern Daylight Time		
EST	Eastern Standard Time		National Broadcasting Corporation of Papua New Guinea
Eur	Europe		
FCC	Federal Communications Commission (U.S.)	NHK	Nippon Hoso Kyokai
		NRC	National Radio Club
FE	Far East	O/t	Of the or on the
FEBC	Far East Broadcasting Corporation	Pac	Pacific
		Pgm	Program

Pop	Popular	SW	Shortwave	
PWBR	Passport to World Radio	SWBC	Shortwave broadcast	
R.	Radio	SWL	Shortwave listener	
RAI	Radiotelevisione Italiana	TC	Timecheck	
Rcvd	Received	Tnx	Thanks	
Rcvr	Receiver	Tx	Transmitter	
Rdif	Radiodiffusora	TWR	Trans World Radio	
	Radiodiffusion	UK	United Kingdom	
REE	Radio Exterior de Espana	UTC	Universal Time Coordinated	
Reg	Regional or regular	Ute	Utility	
Rxd	Received	V	Varies (frequency)	
RCI	Radio Canada International	Verie	Verification, QSL	
RFE	Radio Free Europe	Vern	Vernacular	
RFO	Société Nationale de Radio-TV Française D'outer-Mer	VOA	Voice Of America	
		W	West	
Rpt	Report		Watts	
RRI	Radio Republik Indonesia	w/	With	
RST	Reporting code	WCNA	West Coast North America	
SA	South America	Wknd	Weekend	
SFX	Sound effects	WRTH	World Radio-TV Handbook	
Sig	Signal	Wx	Weather	
SINPO	A reporting code	Xmsn	Transmission	
SIO	A reporting code	Xmtr	Transmitter	
Sked	Schedule	Yr	Year	
S/off*	Sign off	&	And	
*S/on	Sign on	//	Parallel	
SSB	Single-sideband	30	End of transmission	
Spcl	Special	73	Best regards	
SRI	Swiss Radio International	73fr	Best regards and free radio	
Svc	Service	88	Love and kisses	
Stn	Station			

Language abbreviations

For years on end, SWLs have used rather self-explanatory abbreviations to indicate languages when reporting to various club bulletins. These included such abbreviations as Eng for English, Jap for Japanese, and Sp for Spanish. Within the past couple of years or so, certain clubs and members thereof have 'adopted' a new form whereby EE meant English, RR for Russian, and JJ for Japanese, among others. While both systems are good, they both, nevertheless, have glaring faults. The following list of language abbreviations was submitted by a member of the **Newark News Radio Club** and, in the opinion of the author, it is a good, foolproof system. This list is alphabetical by language, not by abbreviation.

AP	Afghan-Persian	BN	Bengali	EP	Esperanto
AK	Afrikaans	BG	Bulgarian	ES	Estonian
AB	Albanian	BR	Burmese	FN	Finnish
AH	Amharic	CB	Cambodian	FR	French
AR	Arabic	CN	Cantonese	GM	German
AM	Armenian	CR	Creole	GR	Greek
AY	Amoy	CZ	Czech	GN	Guarani

AS	Assamese	DN	Danish	GJ	Gujarati
AZ	Azerbaijani	DT	Dutch	HA	Hausa
BL	Belorussian	EG	English	HB	Hebrew
HD	Hindi	PS	Persian	TZ	Tadzhik
HG	Hungarian	PL	Polish	TG	Tagalog
IN	Indonesian	PT	Portuguese	TM	Tamil
IT	Italian	PJ	Punjabi	TL	Telugu
JR	Japanese	PS	Pushtu	TH	Thai
KR	Korean	QC	Quechua	TB	Tibetan
Kt	Kurdish	RM	Romanian	TK	Turkish
LH	Lithuanian	RS	Russian	UG	Uighur
LV	Latvian	SC	Serbo-Croatian	UK	Ukrainian
MC	Macedonian	SN	Sinhalese	UR	Urdu
ML	Malay	SL	Slovene	UZ	Uzbek
MN	Mandarin	SM	Somali	VT	Vietnamese
MG	Mongolian	SP	Spanish	YD	Yiddish
NP	Nepali	SH	Swahili	YR	Yoruba
NW	Norwegian	SW	Swedish	ZL	Zulu

RST method of signal reports

Readability

1—Unreadable
2—Barely readable; occasional words distinguishable
3—Readable with considerable difficulty
4—Readable with practically no difficulty
5—Perfectly readable

Signal strength (S)

1—Faint; signals barely perceptible
2—Very weak signals
3—Weak signals
4—Fair signals
5—Fairly good signals
6—Good signals
7—Moderately strong signals
8—Strong signals
9—Extremely strong signals

Tone (T)

1—Sixty-cycle ac or less, very rough and broad
2—Very rough ac, very harsh and broad
3—Rough ac tone, rectified but not filtered
4—Rough note, some trace of filtering
5—Filtered rectified ac, but strongly ripple-modulated
6—Filtered tone, definite trace of ripple modulation

7—Near pure tone, trace of ripple modulation
8—Near perfect tone, slight trace of modulation
9—Perfect tone, no trace of ripple modulation of any kind

Q SIGNALS

The following is not the entire list of Q signals, but these are the signals that are most frequently used. The entire list may be found in the *Radio Amateur Callbook* or from the American Radio Relay League. These are primarily used in the amateur radio service, but many of them are easily adapted to SWL as well. Most Q signals can be used in either question or statement form; for question form; simply add a question mark after the Q signal.

QRG Will you tell me my exact frequency in kilohertz? Your exact frequency is _____.

QRH Does my frequency vary? Your frequency varies.

QRI How is the tone of my transmissions? The tone of your transmissions is (1—good, 2—variable, 3—bad).

QRK What is the intelligibility of my signals? The intelligibility of your signals is (1—unintelligible to 5—perfectly intelligible).

QRL Are you busy? I am busy. Please do not interfere.

QRM Is my transmission being interfered with? Your transmission is being interfered with.

QRN Are you troubled by static? I am troubled by static.

QRO Shall I increase power? Increase power.

QRP Shall I decrease power? Decrease power.

QRQ Shall I send faster? Send faster, _____ words per minute.

QRS Shall I send slower? Send slower, _____ words per minute.

QRT Must I stop sending? Stop sending.

QRU Have you anything for me? I have nothing for you.

QRV Are you ready? I am ready.

QRX When will you call again (on _____ kHz)? I will call you again at hours on kHz.

QRZ Who is calling me? You are being called by _____ on _____ kHz.

QSA What is the strength of my signals? The strength of your signals is (from 1—barely audible to 5—extremely strong).

QSB Are my signals fading? Your signals are fading.

QSL Can you acknowledge receipt? I acknowledge receipt.

QSO Can you communicate with _____ direct or by relay? I can communicate with _____ direct or by relay through _____.

QSP Will you relay to _____ ? I will relay to _____ .

QSY Shall I change to another frequency? Change to another frequency or to _____ kHz.

QTC How many messages have you to send? I have _____ messages to send.

QTH What is your location? My location is _____ .

QTR What is the correct time? The correct time is _____ .

QRRR (Unofficial) This is the amateur distress signal and is to be used in cases of emergency only.

International morse code

Letter, number, punctuation	Code symbols	Phonetic sound
A	•—	ditdah
B	—•••	dahditditdit
C	—•—•	dahditdahdit
D	—••	dahditdit
E	•	dit
F	••—•	ditditdahdit
G	——•	dahdahdit
H	••••	ditditditdit
I	••	ditdit
J	•———	ditdahdahdah
K	—•—	dahditdah
L	•—••	ditdahditdit
M	——	dahdah
N	—•	dahdit
O	———	dahdahdah
P	•——•	ditdahdahdit
Q	——•—	dahdahditdah
R	•—•	ditdahdit
S	•••	ditditdit
T	—	dah
U	••—	ditditdah
V	•••—	ditditditdah
W	•——	ditdahdah
X	—••—	dahditditdah
Y	—•——	dahditdahdah
Z	——••	dahdahditdit
1	•————	ditdahdahdahdah
2	••———	ditditdahdahdah
3	•••——	ditditditdahdah
4	••••—	ditditditditdah
5	••••	ditditditditdit
6	—••••	dahditditditdit
7	——•••	dahdahditditdit
8	———••	dahdahdahditdit
9	————•	dahdahdahdahdit
0 (zero)	—————	dahdahdahdahdah
/ (fraction bar)	—••—•	dahditditdahdit
. (period)	•—•—•—	ditdahditdahditdah
? (question mark)	••——••	ditditdahdahditdit
, (comma)	——••——	dahdahditditdahdah
Error	••••••••	ditditditditditditditdit

334 *Appendix A*

The following are not in the FCC code tests for antenna licenses, but they are useful to know.

: (colon)	— — — •••	dahdahdahditditdit
; (semicolon)	— • — • — •	dahditdahditdahdit
( ) (parenthesis)	— • — — • —	dahditdahdahditdah
Double dash	— ••• —	dahditditditdah
Wait	• — •••	ditdahditditdit
End of message	• — • — •	ditdahditdahdit
Go ahead	— • —	dahditdah
End of work	••• — • —	ditditditdahditdah

Each code group shown is sent as a single symbol without pauses. The unit of code time is the "dit," equivalent to a short tap on the telegraph key or a quick flip of the tongue. The "dah" is three times as long as a dit; spacing between dits and dahs in the same character is equal to one dit. The spacing between letters in a word is equal to one dah. The spacing between words is equal to five to seven dits.

The NASWA radio country list

Scope This list is designed solely for those who listen to shortwave broadcast stations (SWBC).

Purpose A standard for NASWA Scoreboard rankings, awards and contests and additional recordkeeping uses desired by individuals.

Time This list has no cutoff dates. Users need consider only one date, September 1, 1945. Logging a station at a particular location can be counted for that "radio country" regardless of the date of the logging, as long as it is after September 1, 1945.

Listings Initial "radio country" references in this booklet are to each area's commonly used name in 1945. Next, in parentheses, are listed names which that "radio country" used subsequent to that date. The final listing refers to the current country name. Since some countries change their names over the years, this format eliminates the need to re-alphabetize the list for changes or additions. Since the "radio countries" are arranged alphabetically by continent, it should not prove too difficult to use the list, with a little practice.

Policy NASWA's Radio Country List Is unique. It considers the interests of a longtime listener who heard, and counted, radio countries which, in the political world, no longer exist. Even if war or revolution, conquest or annexation, split-up or merger change the world's map, he need not subtract it from his list of countries logged. And the interests of the newer listener likewise are considered because he can count loggings of stations in countries which, politically, are extinct. After all, reception of foreign shortwave stations has very little to do with political boundaries and independence dates.

NASWA's Radio Country List avoids fixed criteria, though it is broadly based on three factors: geography, politics and hobby tradition. In the final analysis, though, the five-member Country List Committee makes an arbitrary decision as to what will be included in the list and what shall not. Our decisions, hopefully, will be within the mainstream of DXer opinion.

NASWA's Radio (Country List includes "radio countries" which have or have had SWBC stations on the air at some time since the end of World War II. The list contains the names of well over 200 "radio-countries." It is a safe bet that no DXer has ever heard, much less verified, all of them. Most are still broadcasting, giving listening hobby newcomer and oldtimer alike a wide range of targets.

Additions to the list and name changes, when appropriate, will be made from time to time by the NASWA Country List Committee. For reasons noted above, there will be no deletions from the list due to a station ceasing to broadcast or geopolitical factors.

Persons interested in a more detailed explanation of "radio-country" lists in general and the philosophy involved in creating the NASWA Radio Country List are referred to "Counting Countries," an article by Don Jensen, chairman, NASWA Country List Committee, in *Proceedings 1989* (Fine Tuning Special Publications, Stillwater, OK).

Questions regarding this list and its use may be directed to the NASWA Country List Committee, c/o NASWA Headquarters.

NASWA station-counting rules

What is an SWBC station? The NASWA Country List Committee has prepared the following guidelines. Like the country list itself, these rules are a combination of logic, tradition, and "feel." The decisions behind these rules are admittedly arbitrary, but the committee has attempted to apply reason and fairness, and recognize generally accepted practices in the listening hobby.

(1) An SWBC station shall be a station whose fundamental frequency lies in the radio spectrum between 2,000 and 30,000 kHz, whose broadcasts are intended for reception by the general public. This includes standard frequency and time stations with voice announcements, satellite stations in space which transmit intelligible material and utility stations which periodically carry broadcast programming intended for general reception.

This definition excludes spurious and harmonic radiation of broadcast signals whose fundamental frequency lies outside the radio spectrum range between 2,000 and 30,000 kHz, even though the frequency of the received spurious and/or harmonic signal lies within that range. Reception of harmonics of medium wave broadcast stations heard within the defined shortwave frequency range shall not be considered SWBC stations. The definition also excludes satellite stations transmitting only telemetry data and/or CW Morse signals; excludes utility stations transmitting voice mirrors, point-to-point relays and other utility type transmissions; excludes aero, marine or public service stations carrying weather reports, ship-to-shore, military communications, etc.

(2) Each transmitter site of a broadcasting organization shall constitute a separate station. Any move in transmitter site shall create a new station. When a broadcast organization uses multiple transmitter sites, unless the organization is widely known to use a specific transmitter site on a specific frequency at a specific time,

verification should show transmitter site before credit is claimed. Although many broadcast organizations operate several transmitters at a given site, the virtual impossibility of identifying the various transmitters rules out counting separate transmitters at that site as separate stations.

(3) Different services of the same broadcast organization do not count as separate stations; i.e., the Canadian Broadcasting Corporation's Northern Quebec Shortwave Service; the Soviet Union's Radiostation Peace and Progress, etc. Thus, one cannot count the CBC Northern Quebec Shortwave Service and Radio Canada International as two stations when using the same transmitter site at different times. Nor can one count as two station Radiostation Peace and Progress and Radio Moscow when broadcasting from the same transmitter site.

(4) Point-to-point feeds of programs from an organization's main station to its relay stations will not count as SWBC stations.

(5) Transmissions of a single broadcast organization on various frequencies may be counted only once. The use of multiple frequencies does not create as many new stations as frequencies used.

(6) A change of station ownership through sale, independence of the country or revolution will constitute a new station when coupled with changes in call or slogan.

(7) In the case of Monaco, where station studios are in Monaco but transmitters are in France, albeit a few miles away, Monaco will be considered the station location. In the case of the Vatican, the NASWA Country List Committee notes that Italy recognizes Vatican sovereignty over transmitter sites outside Vatican City proper. Should similar situations arise in the future, the commitee will consider each on a case-by-case basis.

Practical tips for using the NASWA Radio Country List

In general, the factor of paramount importance is the location of a station's transmitter, as long as reception is after September 1, 1945. In counting of a "radio country," political considerations are of much less importance.

Here are some examples:

(1) Logging the station at Lusaka, Zambia, counts as NORTHERN RHODESIA (ZAMBIA), regardless of when, after September 1, 1945, you hear the station.

(2) Logging the station at Mogadishu, Somali Democratic Republic counts as ITALIAN SOMALILAND (SOMALI DEMOCRATIC REPUBLIC) whether received prior to 1960, when politically it was still Italian Somaliland, or after 1960, when it became, politically, the Somali Democratic Republic. The longtime listener can count this country on the basis of a logging made long ago, as can the less experienced listener who heard the station more recently.

(3) Logging a station at Enugu, Nigeria, counts as NIGERIA (BLAFRA) (NIGERIA) regardless of when, after September 1, 1945, you hear it. The veteran listener who heard Enugu many years ago, heard it a second time while independent Blafra existed (1967-1970) and a third time after the area was reincorporated into Nigeria will have heard three separate stations (see Station Counting Rule #6) but only a single country.

(4) Logging a station at Goa, now part of India but once a Portuguese possession with, prior to 1960, a shortwave station, counts as GOA (INDIA). Presently there is no SWBC station at this place, though one, reportedly, is planned by All India Radio. A veteran DXer who heard the staff on prior to 1960 may, of course, count Goa (India) as a country. Once the new station is active, those DXers who entered the hobby long after 1960, may log that transmitter and count Goa (India).

(5) Logging Voice of America stations at Greenville, N.C. and Delano, Calif., plus relay stations at Monrovia, Liberia, and Tinang, Philippines, counts as three countries, UNITED STATES OF AMERICA, LIBERIA and PHILIPPINES. Because of the separate station rule, these loggings count as four stations.

(6) Logging a Cable and Wireless Ltd. station, transmitting a repeated voice identification tape indicating its location as Bermuda, does not count as a shortwave broadcast station. It is a utility transmission, not SWBC.

(7) Logging standard time and frequency station WWVH, with voice announcements, counts as a shortwave broadcast station in HAWAII. (see Station Counting Rule #1). Standard time and frequency stations using CW Morse, such as RID, Irkutsk, USSR, do not. Other types of utility stations such as Hong Kong Radio, an aeronautical VOLMET station, do not count as SWBC, although they may have voice time announcements.

(8) Point-to-point or other utility type relay stations, such as those carrying VOA programming or that of Argentine medium wave Radio Rivadavia to other stations for rebroadcasting purposes, are not SWBC stations.

AFRICA

☐ ALGERIA

☐ ANGLO-EGYPTIAN SUDAN (SUDAN)
Anglo-Egyptian Sudan to 1/1/1956; Sudan since

☐ ANGOLA
Includes Cabinda exclave

☐ ASCENSION ISLAND

☐ AZORES

☐ BASUTOLAND (LESOTHO)
Basutoland to 10/4/1966; Lesotho since

☐ BECHUANALAND (BOTSWANA)
Bahuanaland to 9/30/1966; Botswana since

☐ BELGIAN CONGO (KATANGA)(ZAIRE)
Belgian Congo to 6/30/1960; Rep. of Katanga to 1963;
Rep. of the Congo to 1971; Zaire since

☐ BELGIAN CONGO (REPUBLIC OF THE CONGO)(ZAIRE)
Belgian Congo to 6/30/1960; Rep. of the Congo to 1971; Zaire since

☐ BRITISH SOMALILAND (SOMALI DEMOCRATIC REPUBLIC)
British Somaliland to 7/1/1960; Somali Democratic Republic since; Station location: Hargeira

☐ CANARY ISLANDS

☐ CAPEVERDE
Portuguese province to 7/5/1975; independent since

☐ COMOROS (FEDERAL ISLAMIC REPUBLIC OF THE COMOROS)
French possession to July 1975; independent since Station location: Moroni

☐ COMOROS (MAYOTTE)
Remained French territory after July 1975, when other Comoros became independent; active on SWBC between about 1959-1967; Station location: Dzaoudzi

☐ EGYPT

☐ ETHIOPIA

☐ FERNANDO POO (EQUATORIAL GUINEA)
Fernando Poo to 10/12/1968; part of Equatorial Guinea since: Station baton: Malabo

☐ FRENCH CAMEROON (CAMEROUN)
French Cameroons to 1/1/1960; Cameroun since; Station location: Yaoundi

☐ FRENCH EQUATORIAL AFRICA
(OUBANGI-CHARI)
(CENTRAL AFRICAN REP.)
French Equatorial Africa (Oubangi-Chari) to 8/13/1960; Central African Republic to 12/4/1960; Central African Empire to 9/20/79; thereafter Central African Republic again

☐ FRENCH EQUATORIAL AFRICA (CHAD)
French Equatorial Africa to 8/11/1960; Chad since

☐ FRENCH EQUATORIAL AFRICA (PEOPLES' REPUBLIC OF THE CONGO)
French Equatorial Africa to 8/15/1960; People's Republic of the Congo since; Station location: Brazzaville

☐ FRENCH EQUATORIAL AFRICA (GABON)
French Equatorial Africa to 8/17/1960, Gabon since FRENCH MOROCCO (MOROCCO)
French Morocco to 3/2/1956; Morocco since; Station location: Rabat, etc.

☐ FRENCH SOMALILAND (AFARS AND ISSAS)(DJIBOUTI)
French possession to 7/27/1977; independent since

☐ FRENCH TOGOLAND (TOGO)
French Togoland to 4/27/1960; Togo since

☐ FRENCH WEST AFRICA (DAHOMEY)(BENIN)
French West Africa to 8/1/1960; Dahomey to 1975; Benin since

☐ FRENCH WEST AFRICA (GUINEA)
French West Africa to 10/2/1958; Guinea since

☐ FRENCH WEST AFRICA (IVORY COAST)
French West Africa to 8/7/1960; Ivory Coast since

☐ FRENCH WEST AFRICA (MALI)
French West Africa to 6/20/1960; Mali since

☐ FRENCH WEST AFRICA (MAURITANIA)
French West Africa to 11/28/1960; Mauritania since

☐ FRENCH WEST AFRICA (NIGER)
French West Africa to 8/3/1960; Niger since

☐ FRENCH WEST AFRICA (SENEGAL)
French West Africa to 6/20/1960; Mali Federation to 8/20/1960; Senegal since

☐ FRENCH WEST AFRICA (UPPER VOLTA) (BURKINA FASO)
French West Africa to 8/5/1960; Upper Volta to 1984; Burkina Faso since

☐ GAMBIA

☐ GOLD COAST (GHANA)
Gold Coast to 8/6/1957; Ghana since

☐ ITALIAN SOMALILAND (SOMALI DEMOCRATIC REPUBLIC)
Italian Somaliland to 7/1/1960; Somali Democratic Republic since; Station location: Mogadishu

☐ KENYA

☐ LIBERIA

☐ LILYA

☐ MADAGASCAR

☐ MAURITIUS

☐ MOZAMBIQUE
Portuguese province to 6/25/1975; independent since

☐ NIGERIA
All of Nigeria except territory formerly included in Biafra

☐ NIGERIA (BIAFRA) (NIGERIA)
Part of Nigeria to 5/30/1967; Biafra to 1/12/1970; thereafter Nigeria again

☐ NORTHERN RHODESIA (ZAMBIA)
Northern Rhodesia to 10/24/1964; Zambia since

- [ ] NYASALAND (MALAWI)
  Nyasaland to 7/6/1954; Malawi since

- [ ] PORTUGUESE GUINEA (GUINEA BISSAU)

- [ ] REUNION

- [ ] RIO MUNI (EQUATORIAL GUINEA)
  Rio Muni to 10/12/1968; Equatorial Guinea since; Station location: Bata

- [ ] RUANDA-URUNDI (BURUNDI)
  Ruanda-Urundi to 7/1/1962; Burundi since

- [ ] RUANDA-URUNDI (RWANDA)
  Ruanda-Urundi to 7/1/1962; Rwanda since

- [ ] SAO TOME

- [ ] SEYCHELLES
  British possession to 6/29/1976; independent since

- [ ] SIERRA LEONE

- [ ] SOUTH AFRICA

- [ ] SOUTHERN RHODESIA (ZIMBABWE)
  Southern Rhodesia to 11/11/1965; Zimbabwe since

- [ ] SOUTH WEST AFRICA (NAMIBIA)

- [ ] SPANISH MOROCCO (MOROCCO)
  Spanish Morocco to 4/10/1958; Morocco since; Station location: Nador

- [ ] SPANISH SAHARA (MOROCCO)
  Spanish possession to 4/14/1976; Morocco since; Station location: Aaiun

- [ ] ST. HELENA

- [ ] SWAZILAND

- [ ] TANGANYIKA (TANZANIA)
  Tanganyika to 1956; Tanzania since

- [ ] TANGIER (MOROCCO)
  Tangier to 1956; Morocco since

- [ ] TRISTAN DA CUNHA

- [ ] TUNISIA

- [ ] UGANDA

- [ ] ZANZIBAR (TANZANIA)
  Zanzibar to 4/26/1964; Tanzia since

ANTARCTICA

☐ ANTARCTICA

ASIA

☐ ABU DHABI (UNITED ARAB EMIRATES)
One of the Trucial States to 1971; United Arab Emirates since

☐ ADEN (DEMOCRATIC REPUBLIC OF SOUTH YEMEN)(YEMENI REPUBLIC)
Aden to 11/30/1971; Democratic Rep. of South Yemen to 5/22/1990; Yemeni Republic since; Station location: Aden

☐ AFGHANISTAN

☐ ANDAMAN AND NICOBAR ISLANDS (INDIA)

☐ ARMENIAN S.S.R.
U.S.S.R. union republic

☐ ASIATIC R.S.F.S.R.
Asiatic part of U.S.S.R. union republic

☐ AZERBAIJAN S.S.R.
U.S.S.R. union republic

☐ BALI (BALI, INDONESIA)
Indonesian island

☐ BHUTAN

☐ BORNEO (KALIMANTAN, INDONESIA)
Indonesian island

☐ BRITISH NORTH BORNEO (SABAH, MALAYSIA)
Malayan part of Borneo Island; British North Borneo/Sabah to 9/16/1963; Malaysia since

☐ BRUNEI
Part of Borneo Island; British protected sultanate

☐ BURMA (MYANMAR)

☐ CELEBES (SULAWESI, INDONESIA)
Indonesian island group

☐ CEYLON (SRI LANKA)
Independent since 1948; Sri Lanka since 1972

☐ CHINA (PEOPLE'S REPUBLIC OF CHINA)
Mainland China

☐ CHINA (TAIWAN)
Formosa Island; also known as Republic of China

- ☐ CYPRUS
  Station location: Limassol

- ☐ CYPRUS (TURKISH REPUBLIC OF NORTHERN CYPRUS)
  Station location: Lefkosa

- ☐ DODECANESE ISLANDS (RHODES, GREECE)

- ☐ DUBAI (UNITED ARAB EMIRATES)

- ☐ FRENCH INDO CHINA (DEMOCRATIC PEOPLE'S REPUBLIC OF VIETNAM)
  Former French Indo China; State of Vietnam from 7/1/1949; Geneva Agreement of 7/21/1954 partitioned Vietnam; also formerly known as North Vietnam to 1975

- ☐ FRENCH INDO CHINA (REPUBLIC OF VIETNAM) (D.P.R. OF VIETNAM)
  Former French Indo China; State of Vietnam from 7/1/1949; Geneva Agreement of 7/21/1954 partitioned Vietnam; independent to 1975; also formerly known as South Vietnam to 1975

- ☐ FRENCH INDO CHINA (CAMBODIA)
  French Indo China to 11/9/1953; independent since

- ☐ FRENCH INDO CHINA (LAOS)
  French Indo China to 7/19/1949; independent since

- ☐ GEORGIAN S.S.R.
  U.S.S.R. union republic

- ☐ GOA (INDIA)
  Portuguese territory until Indian annexation 12/18/1961

- ☐ HONG KONG

- ☐ INDIA

- ☐ INDIA (EAST PAKISTAN) (BANGLADESH)
  India to 8/14/1947; Pakistan to 4/1971; Bangladesh since

- ☐ INDIA (PAKISTAN)
  India to 8/14/1947; Pakistan since

- ☐ IRAN

- ☐ IRAQ

- ☐ JAPAN

- ☐ JAVA (JAWA, INDONESIA)
  Indonesian island

- ☐ KASHMIR (AZAD KASHMIR)
  India to 1/1/1949; Pakistan control since

- ☐ KASHMIR (INDIA)

- ☐ KAZAKH S.S.R.
  U.S.S.R. union republic

☐ KIRGHIZ S.S.R.
U.S.S.R. union republic

☐ KOREA (DEMOCRATIC PEOPLE'S REPUBLIC OF KOREA)
Korea to 5/1/1948; D.P.R. Korea since; also known as North Korea

☐ KOREA (REPUBLIC OF KOREA)
Korea to 5/1/1948; Republic of Korea since; also known as South Korea

☐ KUWAIT

☐ LEBANON

☐ LESSER SUNDA ISLANDS (NUSA TENGGARA, INDONESIA)
Indonesian island groups, including all islands in main archipelago from Lombok to the western half of Timor

☐ MALAYA (MALAYSIA)
Malaya to 9/16/1963; Malaysia since

☐ MALDIVES

☐ MANCHURIA (PEOPLE'S REPUBLIC OF CHINA)
China to 9/21/1949; People's Republic of China since; includes Heilonjiang, Jilin and Liaoning provinces

☐ MOLUCCA ISLANDS (MALUKU, INDONESIA)
Indonesian island group

☐ MONGOLIA

☐ NEPAL

☐ OMAN
Sultanate of Muscat and Oman to 1970; Oman since

☐ PALESTINE (ISRAEL)
Palestine to 5/14/1948; Israel since

☐ PHILIPPINES

☐ PORTUGUESE TIMOR (TIMOR TIMUR, INDONESIA)
Portuguese province to 1976; Indonesia since

☐ QATAR

☐ RYUKYU ISLANDS (JAPAN)
Control transferred from U.S. to Japan in 1972

☐ SARAWAK (MALAYSIA)
Sarawak to 9/16/1963; Malaysia since

☐ SAUDI ARABIA

☐ SHARJAH (UNITED ARAB EMIRATES)
One of the Trucial States to 1971; United Arab Emirates since; active on SWBC in 1970-1971

- ☐ SINGAPORE (MALAYSIA) (SINGAPORE)
  Singapore to 9/16/1963; Malaysia to 8/9/1965; independent Singapore since
- ☐ SUMATRA (SUMATERA, INDONESIA)
  Indonesian island
- ☐ SYRIA
- ☐ TADZHIK S.S.R.
  U.S.S.R. union republic
- ☐ THAILAND
- ☐ TIBET (PEOPLE'S REPUBLIC OF CHINA)
  Tibet to 12/20/1953; Xizang province of the People's Republic of China since
- ☐ TRANS-JORDAN (JORDAN)
  Trans-Jordan to 4/26/1949, Jordan since
- ☐ TURKEY
- ☐ TURKMEN S.S.R.
  U.S.S.R. union republic
- ☐ UZBEK S.S.R.
  U.S.S.R. union republic
- ☐ YEMEN (YEMENI REPUBLIC)
  Yemen to 5/22/1990; Yemeni Republic since

EUROPE

- ☐ ALBANIA
- ☐ ANDORRA
- ☐ AUSTRIA
- ☐ BALEARIC ISLANDS
  Part of Spain; active on SWBC in early 1950s
- ☐ BELGIUM
- ☐ BULGARIA
- ☐ BYELORUSSIAN S.S.R.
  U.S.S.R. union republic
- ☐ CHANNEL ISLANDS
  Pirate occasionally active; Station location: Jersey
- ☐ CZECHOSLOVAKIA
- ☐ DENMARK
- ☐ ENGLAND

- ☐ ESTONIAN S.S.R.
  U.S.S.R. union republic
- ☐ EUROPEAN R.S.F.S.R.
  European part of U.S.S.R. union republic
- ☐ FINLAND
- ☐ FRANCE
- ☐ GERMANY (GERMAN DEMOCRATIC REPUBLIC AND EAST BERLIN) (GERMANY)
  Germany to 10/7/1949; German Democratic Republic to 1990 when Germany reunified
- ☐ GERMANY (FEDERAL REPUBLIC OF GERMANY) (GERMANY)
  Germany to 9/1/1949; Federal Republic of Germany to 1990 when Germany reunified
- ☐ GERMANY (WEST BERLIN) (GERMANY)
  Germany to 9/1/1949; Federal Republic of Germany to 1990 when Germany reunified
- ☐ GREECE
- ☐ HUNGARY
- ☐ ICELAND
- ☐ IRELAND (EIRE)
- ☐ ITALY
- ☐ KARELO-FINNISH S.S.R. (EUROPEAN R.S.F.S.R.)
  U.S.S.R. union republic to 1945; part of European R.S.F.S.R. since
- ☐ LATVIAN S.S.R.
  U.S.S.R. union republic
- ☐ LITHUANIAN S.S.R.
  U.S.S.R. union republic; declared independence in 1990
- ☐ LUXEMBOURG
- ☐ MALTA
- ☐ MOLDAVIAN S.S.R.
  U.S.S.R. union republic; active on SWBC until mid-1960s
- ☐ MONACO
  Stations may be counted as Monaco although transmitter locations in France
- ☐ NETHERLANDS
- ☐ NORTHERN IRELAND
  Pirates occasionally active

- ☐ NORWAY
- ☐ POLAND
- ☐ PORTUGAL
- ☐ ROMANIA
- ☐ SCOTLAND
  Pirates occasionally active
- ☐ SICILY
  Part of Italy; Station location: Caltanisetta
- ☐ SPAIN
- ☐ SWEDEN
- ☐ SWITZERLAND
- ☐ UKRAINIAN S.S.R.
  U.S.S.R. union republic
- ☐ VATICAN
- ☐ WALES
  Pirates occasionally active
- ☐ YUGOSLAVIA

NORTH AMERICA

- ☐ ALASKA
- ☐ ANTIGUA
- ☐ BAHAMAS
  Active on SWBC until 1946
- ☐ BARBADOS
  Active on SWBC in 1950s and early 1960s with occasional sports broadcasts
- ☐ BONAIRE (NETHERLANDS ANTILLES)
- ☐ BRITISH HONDURAS (BELIZE)
- ☐ CANADA
- ☐ CANAL ZONE
  VOA "Courier" shipboard relay in 1952
- ☐ COSTA RICA
- ☐ CUBA
- ☐ CURACAO (NETHERLANDS ANTILLES)
  Active on SWBC until about 1960

- ☐ DOMINICAN REPUBLIC
- ☐ EL SALVADOR
- ☐ GREENLAND
- ☐ GRENADA
  Active on SWBC until about 1983
- ☐ GUADELOUPE
  Active on SWBC until about 1957
- ☐ GUATEMALA
- ☐ HAITI
- ☐ HONDURAS
- ☐ JAMAICA
  Active on SWBC until about 1958
- ☐ MARTINIQUE
  Active on SWBC until about 1973
- ☐ MEXICO
- ☐ MONTSERRAT
  Deutsche Welle relay active 1977–1988
- ☐ NEWFOUNDLAND (CANADA)
  Crown colony of Great Britain to 1949; Canada since
- ☐ NICARAGUA
- ☐ PANAMA
  Active on SWBC until 1972
- ☐ SWAN ISLAND
  Active on SWBC in the 1960s
- ☐ TRINIDAD
  Active on SWBC until about 1958
- ☐ TURKS AND CAICOS ISLANDS
  Occasional SWBC broadcasts; Station location: Grand Turk
- ☐ UNITED STATES OF AMERICA

OCEANIA

- ☐ ADMIRALTY ISLANDS (PAPUA NEW GUINEA)
- ☐ AUSTRALIA
- ☐ BOUGAINVILLE ISLAND
  Australian-administered island in the Solomons group to 1974; part of Papua New Guinea since

- [ ] COMMONWEALTH OF THE NORTHERN MARIANAS
  Station location: Saipan
- [ ] COOK ISLANDS
- [ ] FUJI ISLANDS
  Active on SWBC until 1972
- [ ] GILBERT ISLANDS (KIRIBATI)
- [ ] GUAM
- [ ] HAWAII
  WWVH standard time/frequency station here; former VOA relay site
- [ ] MARSHALL ISLANDS
- [ ] NETHERLANDS NEW GUINEA (IRIAN JAYA, INDONESIA)
  Netherlands New Guinea to 5/1/1961; Indonesia since
- [ ] NEW BRITAIN (PAPUA NEW GUINEA)
- [ ] NEW CALEDONIA
- [ ] NEW GUINEA TERRITORY (PAPUA NEW GUINEA)
- [ ] NEW HEBRIDES (VANUATU)
- [ ] NEW IRELAND (PAPUA NEW GUINEA)
- [ ] NEW ZEALAND
- [ ] PAPUA TERRITORY (PAPUA NEW GUINEA)
- [ ] SOLOMON ISLANDS
- [ ] TAHITI
- [ ] TONGA
- [ ] WESTERN SAMOA
  Active on SWBC until mid 1950s

SOUTH AMERICA

- [ ] ARGENTINA
- [ ] BOLIVIA
- [ ] BRAZIL
- [ ] BRITISH GUIANA (GUYANA)
  British Guiana to 5/26/1966; Guyana since
- [ ] CHILE
- [ ] COLOMBIA
- [ ] ECUADOR

☐ FALKLAND ISLANDS
In many Spanish-speaking countries called Islas Malvinas
☐ FRENCH GUIANA
☐ GALAPAGOS
Pacific islands belonging to Ecuador
☐ PARAGUAY
☐ PERU
☐ SURINAM
☐ URUGUAY
☐ VENEZUELA

Totals

	Countries heard	Countries verified	Stations heards	Stations verified	Best QSL
Africa					
Antarctica					
Asia					
Europe					
North America					
Oceania					
South America					
Total					

Amateur callsign allocations of the world.

AAA-ALZ	United States of America	HIA-HIZ	Dominican Republic
AMA-AOZ	Spain	HJA-HKZ	Colombia
APA-ASZ	Pakistan	HLA-HMZ	Korea
ATA-AWZ	India	HNA-HNZ	Iraq
AXA-AXZ	Australia	HOA-HPZ	Panama
AYA-AZZ	Argentina	HQA-HRZ	Honduras
A2A-A2Z	Botswana	HSA-HSZ	Thailand
A3A-A3Z	Tonga	HTA-HTZ	Nicaragua
A5A-A5Z	Bhutan	HUA-HUZ	El Salvador

Amateur callsign allocations of the world. (Continued.)

Callsign	Country
BAA-BZZ	China
CAA-CEZ	Chile
CLA-CMZ	Cuba
CNA-CNZ	Morocco
COA-COZ	Cuba
CPA-CPZ	Bolivia
CQA-CRZ	Portuguese Overseas Provinces
CSA-CUZ	Portugal
CVA-CXZ	Uruguay
CYA-CZZ	Canada
C2A-C2Z	Nauru (Republic of)
C3A-C3Z	Andorra
DAA-DTZ	Germany
DUA-DZZ	Philippines
EAA-EHZ	Spain
EIA-EJZ	Ireland
EKA-EKZ	USSR
ELA-ELZ	Liberia
EMA-EOZ	USSR
EPA-EQZ	Iran
ERA-ERZ	USSR
ESA-ESZ	USSR (Estonia)
ETA-ETZ	Ethiopia
EUA-EWZ	USSR (Belorussia)
EXA-EZZ	USSR
FAA-FZZ	France, Overseas States, and Territories of the French Community
GAA-GZZ	United Kingdom
HAA-HAZ	Hungary
HBA-HBZ	Switzerland
HCA-HDZ	Ecuador
HEA-HEZ	Switzerland
HFA-HFZ	Poland
HGA-HGZ	Hungary
HHA-HHZ	Haiti
SSA-SSM	Egypt
SSN-STZ	Sudan
SUA-SUZ	Egypt
SVA-SZZ	Greece
TAA-TCZ	Turkey
TDA-TDZ	Guatemala
TEA-TEZ	Costa Rica
TFA-TFZ	Iceland
TGA-TGZ	Guatemala
THA-THZ	France, Overseas States, and Territories of the French Community
TIA-TIZ	Costa Rica
TJA-TJZ	Cameroun
HVA-HVZ	Vatican City
HWA-HYZ	France, Overseas States, and Territories of the French Community
HZA-HZZ	Saudi Arabia
IAA-IZZ	Italy and United Nation Mandates
JAA-JSZ	Japan
JTA-JVZ	Mongolia
JWA-JXZ	Norway
JYA-JYZ	Jordan
JZA-JZZ	Indonesia (West Iran)
KAA-KZZ	United States of America
LAA-LNZ	Norway
LOA-LWZ	Argentina
LXA-LXZ	Luxembourg
LYA-LYZ	USSR (Lithuania)
LZA-LZZ	Bulgaria
L2A-L9Z	Argentina
MAA-MZZ	United Kingdom
NAA-NZZ	United States of America
OAA-OCZ	Peru
ODA-ODZ	Lebanon
OEA-OEZ	Austria
OFA-OJZ	Finland
OKA-OMZ	Czechoslovakia
ONA-OTZ	Belgium
OUA-OZZ	Denmark
PAA-PIZ	Netherlands
PJA-PJZ	Netherlands West Indies
PKA-POZ	Indonesia
PPA-PYZ	Brazil
PZA-PZZ	Surinam
QAA-QZZ	International Service Abbreviations
RAA-RZZ	USSR
SAA-SMZ	Sweden
SNA-SRZ	Poland
VTA-VWZ	India
VXA-VYZ	Canada
VZA-VZZ	Australia
WAA-WZZ	United States of America
XAA-XIZ	Mexico
XJA-XOZ	Canada
XPA-XPZ	Denmark
XQA-XRZ	Chile
XSA-XSZ	China
XTA-XTZ	Upper Volta
XUA-XUZ	Khmer Republic
XVA-XVZ	Vietnam
XWA-XWZ	Laos
XXA-XXZ	Portuguese Overseas Provinces

TKA-TKZ	France, Overseas States, and Territories of the French Community	XYA-XZZ	Burma
		YAA-YAZ	Afghanistan
		YBA-YHZ	Indonesia
		YIA-YIZ	Iraq
TLA-TLZ	Central African Republic	YJA-YJZ	New Hebrides
TMA-TMZ	France, Overseas States, and Territories of the French Community	YKA-YKZ	Syria
		YLA-YLZ	USSR (Latvia)
		YMA-YMZ	Turkey
TNA-TNZ	Congo (Republic of)	YNA-YNZ	Nicaragua
TOA-TQZ	France, Overseas States, and Territories of the French Community	YOA-YRZ	Romania
		YSA-YSZ	El Salvador
		YTA-YUZ	Yugoslavia
		YVA-YYZ	Venezuela
TRA-TRZ	Gabon	YZA-YZZ	Yugoslavia
TSA-TSZ	Tunisia	ZAA-ZAZ	Albania
TTA-TTZ	Chad	ZBA-ZJZ	Overseas Territories for which the United Kingdom is responsible for international relations
TUA-TUZ	Ivory Coast		
TVA-TXZ	France, Overseas States, and Territories of the French Community		
		ZKA-ZMZ	New Zealand
		ZNA-ZOZ	Overseas Territories for which the United Kingdom is responsible for international relations
TYA-TYZ	Dahomey		
TZA-TZZ	Mali		
UAA-UQZ	USSR		
URA-UTZ	USSR (Ukraine)	ZPA-ZPZ	Paraguay
UUA-UZZ	USSR	ZQA-ZQZ	Overseas Territories for which the United Kingdom is responsible for international relations
VAA-VGZ	Canada		
VHA-VNZ	Australia		
VOA-VOZ	Canada		
VPA-VSZ	Overseas Territories for which the United Kingdom is responsible for international relations	ZRA-ZUZ	South Africa (Republic of)
		ZVA-ZZZ	Brazil
		2AA-2ZZ	United Kingdom
		3AA-3AZ	Monaco
		3BA-3BZ	Mauritius
3CA-3CZ	Equatorial Guinea	6KA-6NZ	Korea
3DA-3DM	Swaziland	6OA-6OZ	Somali Republic
3DN-3DZ	Fiji	6PA-6SZ	Pakistan
3EA-3FZ	Panama	6TA-6UZ	Sudan
3GA-3GZ	Chile	6VA-6WZ	Senegal
3HA-3UZ	China	6XA-6XZ	Malagasy Republic
3VA-3VZ	Tunisia	6YA-6YZ	Jamaica
3WA-3WZ	Vietnam	6ZA-6ZZ	Liberia
3XA-3XZ	Guinea	7AA-7IZ	Indonesia
3YA-3YZ	Norway	7JA-7NZ	Japan
3ZA-3ZZ	Poland	7OA-7OZ	Yemen
4AA-4CZ	Mexico	7PA-7PZ	Lesotho
4DA-4IZ	Philippines	7QA-7Q	Malawi
4JA-4LZ	USSR	7RA-7RZ	Algeria
4MA-4MZ	Venezuela	7SA-7SZ	Sweden
4NA-4OZ	Yugoslavia	7TA-7YZ	Algeria
4PA-4SZ	Ceylon	7ZA-7ZZ	Saudi Arabia
4TA-4TZ	Peru	8AA-8IZ	Indonesia
4UA-4UZ	United Nations	8JA-8NZ	Japan
4VA-4VZ	Haiti	8OA-8OZ	Botswana

Amateur callsign allocations of the world. (Continued.)

4WA-4WZ	Yemen	8PA-8PZ	Barbados
4XA-4XZ	Israel	8QA-8QX	Maldive Islands
4YA-4YZ	International Civil Aviation Organization	8RA-8RZ	Guyana
		8SA-8SZ	Sweden
4ZA-4ZZ	Israel	8TA-8YZ	India
5AA-5AZ	Libya	8ZA-8ZZ	Saudi Arabia
5BA-5BZ	Cyprus	9AA-9AZ	San Marino
5CA-5GZ	Morocco	9BA-9DZ	Iran
4HA-5IZ	Tanzania	9EA-9FZ	Ethiopia
5JA-5KZ	Colombia	9GA-9GZ	Ghana
5LA-5MZ	Liberia	9HA-9HZ	Malta
5NA-5OZ	Nigeria	9IA-9JZ	Zambia
5PA-5QZ	Denmark	9KA-9KZ	Kuwait
5RA-5SZ	Malagasy Republic	9LA-9LZ	Sierra Leone
5TA-5TZ	Mauritania	9MA-9MZ	Malaysia
5UA-5UZ	Niger	9NA-9NZ	Nepal
5VA-5VZ	Togo	9OA-9TZ	Zaire
5WA-5WZ	Samoa (Western)	9UA-9UZ	Burundi
5XA-5XZ	Uganda	9VA-9VZ	Singapore
5YA-5ZZ	Kenya	9WA-9WZ	Malaysia
6AA-6BZ	Egypt	9XA-9XZ	Rwanda
6CA-6CZ	Syria	9YA-9ZZ	Trinidad and Tobago
6DA-6JZ	Mexico		
A2	Botswana	FC (unofficial)	Corsica
AC	Bhutan	FG7	Guadeloupe
AC3	Sikkim	FH8	Comoro Islands
AC4	Tibet	FK8	New Caledonia
AP	East Pakistan (Bangladesh)	FL8	Somaliland (French)
		FM7	Martinique
AP	West Pakistan	FO8	Clipperton Island
BV	Formosa	FO8	French Oceania
BY	China	FP8	St. Pierre and Miquelon Islands
C2	Nauru		
C3	Andorra	FR7	Glorioso Islands
CE	Chile	FR7	Juan de Nova
CE9AA•AM, FB8Y, KC4, LA/G, LU-Z, OR4, UA1, VK0, VP8, ZL5, 8J	Antarctica	FR7	Reunion Island
		FR7	Tromelin
		FS7	Saint Martin
		FW8	Wallis and Futuna Islands
CE0A	Easter Island	FY7	French Guiana and Inini
CE0X	San Felix		
CE0Z	Juan Fernandez Archipelago	G	England
		GC	Guernsey and Dependencies
CM, CO	Cuba		
CN2, 8, 9	Morocco	GC	Jersey
CP	Bolivia	GD	Isle of Man
CR3	Portuguese Guinea	GI	Northern Ireland
CR4	Cape Verde Islands	GM	Scotland
CR5	Principe, Sao Tome	GW	Wales
CR6	Angola	HA, HG	Hungary
CR7	Mozambique	HB	Switzerland
CR8	Portuguese Timor	HB0	Liechtenstein

Useful shortwave tables

Prefix	Country
CR9	Macao
CT1	Portugal
CT2	Azores
CT3	Madeira Islands
CX	Uruguay
DJ, DK, DL, DM	Germany
DU	Philippine Islands
EA	Spain
EA6	Balearic Islands
EA8	Canary Islands
EA9	Rio de Oro
EA9	Spanish Morocco
EI	Republic of Ireland
EL	Liberia
EP, EQ	Iran
ET3	Ethiopia
F	France
FB8W	Crozet Islands
FB8X	Kerguelen Islands
FB8Z	Amsterdam and St. Paul Islands
JT	Mongolia
JW	Svalbard
JX	Jan Mayen
JY	Jordan
K, W	United States of America
KA	(See JA)
KB6	Baker, Howland, and American Phoenix Islands
KC4	(See CE9)
KC4	Navassa Island
KC6	Eastern Caroline Islands
KC6	Western Caroline Islands
KG4	Guantanamo Bay
KG6	Guam
KG6R, S, T	Mariana Islands
KH6	Hawaiian Islands
KH6	Kure Island
KJ6	Johnston Island
KL7	Alaska
KM6	Midway Islands
KP4	Puerto Rico
KP6	Palmyra Group, Jarvis Island
KR6, 8	Ryukyu Islands
KS4	Swan Islands
KS4B	Serrana Bank, Roncador Cay
KS6	American Samoa
KV4	American Virgin Islands
HC	Ecuador
HC8	Galapagos Islands
HH	Haiti
HI	Dominican Republic
HK	Colombia
HK0	Bajo Nuevo
HK0	Malpelo Island
HK0	San Andres and Providencia
HL, HM	Korea
HP	Panama
HR	Honduras
HS	Thailand
HV	Vatican
HZ	Saudi Arabia
I1, IT1	Italy
IS1	Sardinia
JA, JH, JR, KA	Japan
JD	Ogasawara Islands
JD	Minami Torishima
OA	Peru
OD5	Lebanon
OE	Austria
OF, OH	Finland
OH0	Aland Islands
OJ0	Market Reef
OK	Czechoslovakia
ON4, 5, 8	Belgium
OR4	(See CE9AA-AM)
OX	Greenland
OY	Faeroes
OZ	Denmark
PA, PD, PE, PI	Netherlands
PJ	Netherlands Antilles
PJ	Saint Marten
PY	Brazil
PY0	Fernando de Noronha
PY0	St. Peter and St. Paul's Rocks
PY0	Trindade and Martin Vaz Islands
PZ1	Surinam
SK, SL, SM	Sweden
SP	Poland
ST2	Sudan
SU	Egypt
SV	Crete
SV	Dodecanese Islands
SV	Greece
TA	Turkey
TF	Iceland

Amateur callsign allocations of the world. (Continued.)

Callsign	Country
KW6	Wake Island
KX6	Marshall Islands
KZ5	Canal Zone
LA	Norway
LA/G	(See CE9AA-AM)
LU	Argentina
LU-Z	(See CE9, VP8)
LX	Luxembourg
LZ	Bulgaria
M1, 9A1	San Marino
MP4B	Bahrein
MP4D, T	Trucial Oman
MP4M, VS90	Sultanate of Muscat and Oman
MP4Q	Qatar
UA, UK9, UV UW9, 0	Asiatic Russia
UA1	Franz Josef Land
UA2, UK2F	Kaliningradsk
UB5, UK5, UT5, UY5	Ukraine
UC2, UK2A/C/I/L/O/S/W	White Russia
UD6, UK6C/D/K	Azerbaijan
UF6, UF6F/O/Q/V	Georgia
UG6, IK6G	Armenia
UH8, UK8H	Turkoman
UI8, UK8I	Uzbek
UJ8, UK8J/R	Tadzhik
UL7, UK7L	Kazakh
UM8, UK8M	Kirghiz
U05, UK50	Moldavia
UP2, UK2B/P	Lithuania
UQ2, UK2G/Q	Latvia
UR2, UK2R/T	Estonia
VE	Canada
VK	Australia (including Tasmania)
VK	Lord Howe Island
VK	Willis Islands
VK9, N	Norfolk Island
VK9, X	Christmas Island
VK9, Y	Cocos Islands
VK9, AA-MZ	Papua Territory
VK9, AA-MZ	Territory of New Guinea
VK0	(See CE9)
VK0	Heard Island
VK0	Macquarie Island
VO	Newfoundland, Labrador
VP1	British Honduras
VP2A	Antigua, Barbuda
TG	Guatemala
TI	Costa Rica
TI9	Cocos Island
TJ	Cameroun
TL	Central African Republic
TN	Congo Republic
TR	Gabon
TT	Chad
TU	Ivory Coast
TY	Dahomey
TZ	Mali
UA, UK1, UK3, UK4, UK6, UV, UW 1-6, UN	European Russia
VP7	Bahama Islands
VP8	(See CE9AA-AM, LU-Z)
VP8	Falkland Islands
VP8, LU-Z	South Georgia Islands
VP8, LU-Z	South Orkney Islands
VP8, LU-Z	South Sandwich Islands
VP8, LU-Z, CE9AN-AZ	South Shetland Islands
VP9	Bermuda
VQ1	Zanzibar
VQ9	Aldabra Islands
VQ9	Chagos
VQ9	Desroches
VQ9	Farquhar
VQ9	Seychelles
VR1	British Phoenix Islands
VR1	Gilbert and Ellice Islands and Ocean Island
VR2	Fiji Islands
VR3	Fanning and Christmas Islands
VR4	Solomon Islands
VR5	Tonga (Friendly) Islands
VR6	Pitcairn Island
VS5	Brunei
VS6	Hong Kong
VS9K	Kamaran Islands
VS9M	Maldive Islands
VS9O	(See MP4M)
VU	Andaman and Nicobar Islands

VP2D	Dominica	VU	India
VP2E	Anguilla	VU	Laccadive Islands
VP2G	Grenada and Dependencies	W	(See K)
		XE, XF, 4A	Mexico
VP2K	St. Kitts, Nevis	XF4	Revilla Gigedo
VP2L	St. Lucia	XP	(See OX)
VP2M	Monsterrat	XT	Voltaic Republic
VP2S	St. Vincent and Dependencies	XU	Cambodia
		XV5	(See 3W8)
VP2V	British Virgin Islands	XW8	Laos
		XZ	Burma
VP5	Turks and Caicos Island	YA	Afghanistan
YB	Indonesia	4A	(See XE)
YI	Iraq	4S7	Sri Lanka
YJ	New Hebrides	4U1	International Tele-communications Union, Geneva
YK	Syria		
YN, YN0	Nicaragua		
YO	Romania	4W	Yemen
YS	El Salvador	4X, 4Z	Israel
YU	Yugoslavia	5A	Libya
YV	Venezuela	5B4, ZC4	Cyprus
YV0	Aves Island	5H3	Tanganyika
ZA	Albania	5N2	Nigeria
ZB2	Gibraltar	5R8	Malagasy Republic
ZC4	(See 5B4)	5T	Mauritania
ZD3	Gambia	5U7	Niger
ZD5	Swaiziland	5V	Togo
ZD7	St. Helena	5W1	Samoa
ZD8	Ascension Island	5X5	Uganda
ZD9	Tristan da Cunha and Gough Island	5Z4	Kenya
		6O1, 2, 6	Somali
		6W8	Senegal
ZE	Zanbabwe	6Y	Jamaica
ZF1	Cayman Islands	7G1	(See 3X)
ZK1	Cook Islands	7O	South Yemen
ZK1	Manihiki Islands	7P	Lesotho
ZK2	Niue	7Q7	Malawi
ZL	Auckland and Campbell Islands	7X	Algeria
		7Z	Saudi Arabia
		8P	Barbados
ZL	Chatham Islands	8Q	Maldive Islands
ZL	Kermadec Islands	8R	Guyana
ZL	New Zealand	8Z4	Saudi Arabia/Iraq Neutral Zone
ZL5	(See CE9AA-AM)		
ZM7	Tokelau Islands	8Z5	(See 9K3)
ZP	Paraguay	9A1	(See M1)
ZS1, 2, 4, 5, 6	South Africa	9G1	Ghana
ZS2	Prince Edward and Marion Islands	9H1	Malta
		9J2	Zambia
		9K2	Kuwait
ZS3	Namibia	9K3, 8Z5	Kuwait/Saudi Arabia Neutral Zone
1M	Minerva Reefs		
1S	Spratly Island	9L1	Sierra Leone
3A	Monaco	9M2	Malaya

Amateur callsign allocations of the world. (Continued.)

3B6, 7	Agalega and St. Brandon	9M6	Sabah
3B8	Mauritius	9M8	Sarawak
3B9	Rodriquez Island	9N1	Nepal
3C	Equatorial Guinea	9Q5	Congo Republic
3V8	Tunisia	9U5	Burundi
3W8, XV5	Vietnam	9V1	Singapore
3X (7G)	Republic of Guinea	9X5	Rwanda
3Y	Bouvet	9Y4	Trinidad and Tobago

B
APPENDIX

SWECHO FidoNet BBS list

While you are traveling, nothing (aside from shortwave) can help to keep you in touch with the world better than computer BBSs and networks. Some of the radio-related information that you can find on BBSs and computer networks includes: schedules, loggings, background information and opinions about various programs, receiver reviews and opinions, and much more.

With a laptop computer and a modem, you can download plenty of current information. In fact, in many places, you can find a radio-related BBS within reach of a local telephone call.

There are three different types of locations to receive radio information from the computer: from a radio network, from a radio-related BBS, or from a radio echo.

The *computer network* includes the Internet, America OnLine, CompuServe, Delphi, etc. These organizations have various radio forums, where listeners can post information, ask questions, etc. The Internet radio forums are by far the most important of the networks because nearly all of the major computer services (and many independent BBSs) include the Internet. Telephone numbers for the major services are not included here because you need to subscribe first, then the various numbers can be supplied by the company's information lines.

Radio echoes are much like the radio forums on the major computer networks—loads of posted information and questions. The difference is that these forums are carried over many independent BBSs. One of the largest of these in the world is the FidoNet. The FidoNet has hundreds of different forums, including several different radio forums. These FidoNet forums can be subscribed to—as a result, some BBSs carry all of the FidoNet forums and others only carry three or four. This appendix covers hundreds of different FidoNet BBSs.

Radio-related BBSs specialize in radio files and information. In addition to the same types of Q&A information that fills the radio echoes and radio forums on the major computer networks, radio-related BBSs also feature demo copies of different radio software, scanned photos of QSLs and radio artwork, radio shareware, etc. This appendix covers several dozen different radio-related BBSs. "Radio related" in this case, means that the BBSs feature either shortwave or amateur radio files.

Using BBSs

BBSs change regularly; different FidoNet forums are added and dropped, BBSs move and phone numbers change, and new BBSs open while others close. This appendix lists some of the many FidoNet and radio-related BBSs in existence around the world. Remember that some of this information will have surely changed by the time you read this.

If you are planning a trip in advance, be sure to check into a BBS to make sure that you can receive access. Some BBSs require extensive user information before they allow access; it's best to get this out of the way before making the trip. Also, some BBSs have subscription fees. If a particular BBS would be handy during a trip, but there is a fee, try contacting the BBS system operator and letting him or her know that you will be in town. The sysop might be willing to give you a temporary free or reduced-rate BBS membership.

If you plan to call BBSs for radio info while traveling, make sure that, if you are calling from a motel, there is no extra fee for local calls. Most charge—either per call or per minute, and the results can be quite expensive.

Radio-related Internet forums

    alt.radio.digital
    alt.radio.pirate
    rec.radio.amateur.equipment
    rec.radio.amateur.misc
    rec.radio.broadcasting
    rec.radio.info
    rec.radio.noncom
    rec.radio.shortwave

Radio-oriented BBSs

Name	Location	Phone
ANARC/ACE BBS	Kansas City	913-345-1978
America Archive BBS	Cook County	708-426-8903
American Silver Dollar	Alexandria, LA	318-443-0271
Bad to the Bone	Casco, MI	810-749-3581
Brian's Brainy BBS	Danville, IL	217-442-9818
Chi.-Area Comp. Center	Bolingbrook, IL	708-230-9068
FBN BBS	Champaign, IL	217-359-2874
The File Bank	Denver, CO	303-534-4646
The /Flux/ Line	Bloomfield Hills	810-851-3509

Free Radio Network	Joplin, MO	417-624-1809
Grove BBS	Brasstown, NC	704-837-9200
HAM >>link<< RBBS	St. Paul, MN	612-HAM-0000
Ham It Up Data Systems	Moriches, NY	516-878-4906
KD3BJ Usenet BBS	Sellersville, PA	215-257-2744
Microcomputer Tech. BBS	Harrison, AR	501-365-7392
NAMU	Topeka, KS	913-273-1550
Radio Active	Atlanta, GA	404-437-1555
Radio Daze BBS	South Bend, IN	219-257-2138
The RF Deck	Baraboo, WI	608-356-4777
Saguaro Station	Phoenix, AZ	602-846-2318
Trance BBS	Torino, Italy	+39-11-482-751
WB3FFV Radio BBS	Baltimore, MD	410-661-2475

FidoNet BBSs

The following list contains some of the FidoNet-member BBSs from around the world. BBSs marked with an asterisk are verified as a carrier for the FidoNet Shortwave Echo (SW Echo) forum. Those not marked with an asterisk might carry the SW Echo or some of the other radio-related FidoNet forums (Ham Forsale, Ham-Tech, etc.). If no radio-related files are carried by the FidoNet member BBS, it is possible that they will be picked up later on or that you might be able to convince the sysop to join.

Alabama

* The ACCESS System	Huntsville	
The Anchor Inn	Mobile	334-675-8406
The Computrion BBS	Birmingham	205-595-0183
*The Dateline BBS	Cullman	205-747-4194
The Edge of Cyberspace	Albertville	205-891-3403
*The Jack of All BBS	Hartselle	205-773-2859
Shoal Creek eXpress	Florence	205-757-3355
*3 Nodes	Huntsville	205-551-9004

Alaska

TC's Byte Bank	North Pole	907-488-3751
Scriptorium	Anchorage	907-248-5985

Arizona

CNV	Phoenix	602-488-0656
The Desert Reef	Tucson	602-624-6386
Sunwise	Phoenix	602-584-7395
Union Jack	Phoenix	602-274-9921
The Wild Side BBS	Phoenix	602-258-8351

Arkansas

The Deserted Island	Little Rock	501-224-1605
Global Gateways	Conway	501-329-7508
Ozark Mt. BBS	Damascus	501-335-7871

California

*The Cat's Meow	Union City	510-471-0603
Crossroads	Irvine	714-724-1041
*Crossroads	Fairfield	707-426-3119
*Gandalf's	San Diego	619-466-9505
*The Electronic Grapevine	Napa	707-257-2338
Happy Trails BBS	Orange	714-547-0719
Manetheran BBS	Irvine	714-509-9276
*9-1-1 BBS	San Diego	619-669-0385
Online Resource	San Diego	619-793-8360
Silver Cactus BBS	Lancaster	805-949-7703
Sleuth BBS	Chatsworth	818-727-7639
The Solar System BBS	Mission Viejo	714-837-9677
*Ursa Major	Manhattan Beach	310-545-5611
*Zooming	San Diego	619-277-4140

Colorado

Access Denied	Arvada	303-431-6796
Colorado Connection	Arvada	303-423-9775
The Dart Board	Dolores	303-882-2360
Dimensions Online	Aurora	303-363-0910
*The File Bank, Inc.	Denver	303-534-4646
The Forum	Ft. Collins	303-226-4218
Grotto Lounge	Arvada	303-421-6965

Connecticut

Applause BBS	Waterbury	203-754-9598
Erasmus BBS	Newington	203-666-5113
*Hatters Park	Danbury	203-744-0179
*The Oracle	Bethel	203-794-8675
Shotgun BBS	Stamford	203-969-0825
USS Enterprise	Monroe	203-261-6863

Delaware

Baseline BBS	Newark	302-834-1089
The DVUG BBS	New Castle	302-324-8091
Imperialink BBS	Wilmington	302-892-9953

Florida

*Bullwinkle's Corner	Orlando	407-896-5772
Civil Air Patrol	Bradenton	813-750-9051
*Cornucopia TBBS	Winter Park	407-645-4929
CyberNexus	Merritt Island	407-459-9100
Esoteric Oracle	Gainesville	904-332-9547
Field of Dreams	Jacksonville	904-241-0735
*The Firehouse BBS	Lake Buena Vista	904-934-8678
Genesis BBS	Margate	305-753-5033

*Ground Zero	St. Petersburg	813-849-4034
*LASER!	Orlando	407-647-0031
Last Days	Mt. Dora	904-735-0531
Longbow BBS	Tampa	813-961-3653
*Mercury Opus	St. Petersburg	813-321-0734
Neon Knight's BBS	Crestview	904-689-3692
Nightmare Cafe	Panama City	904-874-2296
Parados	Jacksonville	904-260-3172
Prime Time BBS	Inverness	904-637-3713
Ups & Downs	Cooper City	305-434-8403
USS Enterprise EBBS	Englewood	813-473-1246
Whistler's Hollow	Jacksonville	904-727-9289

Georgia

American Connection	Savannah	912-927-7323
Crack of Dawn BBS	Ft. Stewart	912-369-7023
The Hotel California	LaGrange	706-845-7102
Planet X BBS	Gainesville	404-536-1652
* Total Recan	Douglasville	770-920-0621

Hawaii

Color Computer Library	Honolulu	808-545-8368

Idaho

*The Bodhi Tree	Boise	208-327-9916

Illinois

Boomtown BBS	McConnell	815-868-2422
*CfC	Chicago	708-362-7875
Creative Thoughts BBS	Barrington	708-382-3904
Dark Elf BBS	Cary	708-516-8953
East Village BBS	Chicago	312-777-2574
Gateway Elite BBS	Rockford	815-398-4678
Hook & Slice Club BBS	Danville	217-446-0105
*JPUSA BBS	Chicago	312-878-6030
National Islamic	Chicago	312-274-8136
RuneQuest	Chicago	312-728-7784
Shadowgate BBS	Sterling	815-622-9639
Speedy's Dilemma	Murphysboro	618-687-4806
Squirrel's Nest	Marseilles	815-795-6371
Steve's Maildrop & BBS	Columbia	618-281-8702
Uncle Bob's BBS	Lake Villa	708-265-0695
Under the Influence	Morris	815-942-2930
Village Oak BBS	Oak Lawn	708-430-7732

Indiana

ArcadiaVision BBS	Kouts	219-766-2378

Computer Support BBS	South Bend	219-272-8129
Electronic Warfare BBS	Vincennes	812-882-0644
4th Dimension BBS	Terre Haute	812-299-2983
The Flaming Star BBS	Terre Haute	812-898-2561
Graffiti o/t BBS Wall	Lafayette	219-448-2842
I.O. Board	Anderson	317-644-3039
Ivy Tech St. College	Terre Haute	812-299-9306
*Lakeland BBS	LaGrange	219-463-2484
The Software Zone	Martinsville	219-342-2094
The Sometimes BBS	Portage	219-763-2031

Iowa
Helium High-Grounds	Ames	515-232-0969

Kansas
*ANARC/ACE BBS	Kansas City	913-345-1978
The Quelle	Kansas City	913-287-2600

Kentucky
The Barbarian's Hut	Ft. Knox	502-942-8046
The ScrapYard	Ft. Knox	502-942-0864
U.S. Aviators Domain	Erlanger	502-342-6554
VACIS	Worthington	606-836-1267

Louisiana
Black Boar Inn	Metairie	504-885-3960
*The Digital Cottage	New Orleans	504-897-6614
Heavy Metal BBS	Houma	504-872-3043
N.Louisiana Connect	Ruston	318-255-4497
Southern Star BBS	New Orleans	504-885-5928

Maine
The Wizards Guild	Lewiston	207-783-0874

Maryland
*_ACE_ ONLINE	Silver Spring	301-942-2218
Bifrost	Mt. Ranier	301-779-9381
Econo-RBBS	Catonsville	410-747-3619
The Emerald Mines BBS	Largo	301-499-0326
G-Comm	Warren	410-436-7638
*Hafa Adai Exchange	Great Mills	301-994-9460
*The Last DX Connection	Bowie	301-805-8921
*MetroNet	Columbia	410-720-5506
MirageMecca RBBS	Baltimore	410-426-5097
Ron's Room	Severn	410-969-2241

Massachussetts
Computer Castle Link	Haverhill	508-521-6941
*Tom's BBS	Braintree	617-356-3538
*World of Burgers	Worcester	508-753-6969

Michigan
*The Air Studio BBS	Livonia	313-522-5349
Chess Players Forum	Lincoln Park	313-386-7054
Eau Claire Connection	Eau Claire	616-461-6801
The Programmer's Edge	Livonia	810-477-6695
The Rainbow Bridge BBS	Coldwater	517-278-7029
The Raven!	Redford	313-937-2016

Minnesota
The Black Hole BBS	Chaska	612-442-5682
*HAM>>link<< RBBS	Saint Paul	612-HAM-0000
Infiniti	Richfield	612-861-7460
SpareCom	Shakopee	612-445-5755

Mississippi
After Hours BBS	Jackson	601-371-0423
The Eagle's Eye	Vicksburg	601-636-3212
The Gateway	Biloxi	601-374-2697
Mid-South BBS	Amory	601-256-1494
StarNet BBS	Jackson	601-981-9228

Missouri
The Batboard TBBS	Columbia	314-446-0475
B.S. Box	St. Louis	314-861-3902
*Cork's Place BBS	Joplin	417-659-8666
*Hard Rock BBS	Nixa	417-725-8003
The Lake BBS	Osage Beach	314-348-9050
Mega-Link BBS	St. Joseph	816-364-1035
*Sound Advice	Gladstone	816-436-4516
St. Louis Online	St. Louis	314-561-3874
*Too Tall'S	Ladue	314-997-7060
*The Village Crier	St. Joseph	816-387-8881

Montana
*pC-Montana	Manhattan	

Nebraska
Castle Keep BBS	Bellevue	402-292-0789
MacNet Omaha	Elkhorn	402-289-2899

Nevada
ICON BBS	Las Vegas	702-256-4107

New Hampshire

Botnay Bay	Portsmouth	603-431-7090
The Cereal Port BBS	Rindge	603-899-3335
The Computer Castle	Newton	603-642-5949
EMail Connection	Rochester	603-335-0213
The Outpost	Nashua	603-888-3840

New Jersey

Alchemiga	Sparta	201-398-6360
Altered Illusions	Dayton	908-329-3216
The Armory BBS	Phillipsburg	908-859-0162
Beacon Studios BBS	Union City	201-863-5253
Beyond Reality	Fords	908-417-2175
The Bitter End BBS	Browns Mills	609-893-2874
Blackstar	Boonton	201-335-6132
The Black Tower	Rockaway	201-361-6651
The Boss BBS	Tenafly	201-568-7293
Dmaster's Den	Madison	201-301-9583
The Flight Deck	Ft. Monmouth	908-389-3202
The GearBox	New Milford	201-692-1110
Hard Drive Cafe BBS	Wayne	201-790-6300
Harry's Place	Mahwah	201-934-0861
HQ Informational Service	Orange	201-672-8969
In the Wind BBS	Hackensack	201-646-0227
It's All Rock'n'Roll	Trenton	609-695-9319
Just Programs BBS	Roselle Park	908-298-9098
The Licking Factory	Colonia	908-815-3146
MicroFone	Metuchen	908-494-8666
Mt. HED BBS	Parsippany	201-625-1519
Nightlink BBS	Manahawkin	609-597-4290
The Null Pointer BBS	Madison	201-301-2182
The Perverts BBS	Union Beach	908-888-0176
*Pics Online Multiuser System		609-753-2540
*Plain Brown Wrapper BBS	Cresskill	201-569-6685
Planet Shadowstar TBBS	Edison	908-494-3417
The Roach Motel	Nutley	201-667-3326
Shockwave Rider	Freehold	908-294-0659
The Sound Connection	Scotch Plains	908-322-0131
S.Coast Online Services	Ocean City	609-399-5708
The Tammy Board BBS	Mount Hope	201-361-5954
Union Lake BBS	Millville	609-327-5553
The Vortex BBS	Belleville	201-751-5608
The Waterside BBS	Little Ferry	201-641-5375

New Mexico

Construction Net #6	Los Alamos	505-662-0659
The Dragon's Lair	Portales	505-359-1770

Selective Online	Sante Fe	505-473-9765
Tel-Us Computer BBS	Las Vegas	505-425-6995

New York

ABC Online	Queens	718-446-2157
The Aquarium/AWE BBS	Buffalo	716-885-8363
The Belfry	New York	718-793-4796
The Boiler Room	Brooklyn	718-265-2532
The Brewster BBS	Brewster	914-279-2514
Coffee Clutch BBS	Clifton Park	518-383-3156
Control Central	Islandia	516-342-0382
The DarkLands BBS	Herkimer	315-866-8187
Dark Side of the Moon	Crompond	914-736-3186
Dragon's Cove	Rome	315-339-0208
East End Computer	Manorville	516-395-1033
The Enigma	Oswego	315-343-6706
*Ham-Net BBS	Manlius	315-682-1824
Harbringer BBS	Holbrook	516-472-1036
The Hidden Pyramid	East Syracuse	315-433-5398
Holman's World	New York	718-529-8890
HTBBS Network Services	Kew Gardens	718-380-8003
Kraut Haus	Rochester	716-359-0871
The Link BBS	Staatsburg	914-889-8379
The Machine BBS	Modena	914-883-6612
Medina Online	Medina	716-798-5549
Meng's Madhouse	Hilton	716-964-8323
The Messed BBS	Freeville	607-844-9216
MicroQuick	Plainview	516-935-5704
*The Night Owl BBS	Saratoga Springs	518-581-1797
The Pier BBS	Brooklyn	718-253-3583
The Promised Land	Bronx	718-295-3266
Quantum 2000 BBS	Queens	718-740-8411
Real Exposure	New York	212-691-2679
*Red Onion, ExPress	Wawayanda	914-342-4585
SMBBS Network	Wappingers Falls	914-297-2915
Tree Branch	Briarwood	718-739-5845
The United Front BBS	Flushing	718-886-6797
*Unity BBS!	Syracuse	315-488-0679

North Carolina

ANSI-Mation Alley	Jacksonville	910-346-6543
*Borderline! BBS	Concord	704-792-9241
Free Advice	Smithfield	919-934-1002
Hawaiian Hang Time	Hubert	910-326-5098
Hobbies Unlimited	Raleigh	919-821-4354
The Leap BBS	Cerro Gordo	910-654-5593

The Maniac's Den	Carthage	910-947-3077
Patriot BBS	Graham	910-222-8524
The Pig Pen BBS	Fayetteville	910-487-0363
Terminal Entry BBS	Rockingham	910-895-0368
The Vault BBS	Raleigh	919-821-4049
Wilmington Online	Wilmington	910-763-3849

North Dakota

The D-Generation	Bismarck	701-258-0872

Ohio

Bordertown/TexMex	Woodville	216-862-3423
Bulldog's Lounge	Wheelersburg	614-574-6590
The Consciousness BBS	Sciotoville	614-776-2618
The Crystal Palace	Port Clinton	216-797-4719
Freedom BBS	Streetsboro	216-655-9626
The Mother Board Club	Buckeye Lake	614-928-2259
The Mountain Lair	Belpre	614-423-0567
The New Frontier	Akron	216-733-0979
*PC-OHIO	Cleveland	216-381-3320
Spock's Adventure BBS	Cincinnati	513-779-9717
Twisted Reality	North Olmstead	216-779-4113
*WLIO TBBS	Toledo	419-228-8227

Oklahoma

*The Ham Radio Emporium	Owasso	918-272-4327
Light Speed][	Tulsa	918-299-4811
The Northern Exchange	Tulsa	918-747-2273

Oregon

A bit of Aloha	Aloha	503-591-7542
*Chemeketa Online	Chemeketa	503-393-5580
Hell Above Ground	Portland	503-284-3569
The Inferno	Eagle Creek	503-637-3178
*NWCS Online	Oregon City	503-655-3927
Ultimate BBS	Eugene	503-461-1148

Pennsylvania

Alpha Beta	Pittsburgh	412-683-4327
The Annex BBS	Pittsburgh	412-635-9165
Antarctica BBS	York	717-755-2440
Bart & Homer's BBS	Wernersville	610-678-2431
*BlinkLink	Pittsburgh	412-766-0732
Cheers BBS	Johnstown	814-539-6648
The Circuit Board	Greentown	717-676-9339
Cybernetics BBS	Ephrata	717-738-1976

The DataWorks	Dickson City	717-489-0862
The Eagle's Lair	Bellwood	814-742-9364
Edge of Chaos	Philadelphia	215-564-4208
Freddies' Playhouse BBS	Tarentum	412-226-3129
Freeland BBS	Freeland	717-636-0131
Gateway to End of Time	Collegeville	610-454-9862
The Hotline BBS	Lansdale	215-393-8594
The Keep BBS	Lansdale	215-855-0401
Keystone Amiga BBS	Allentown	610-770-0774
Memory Prime BBS	Waynesburg	412-627-3227
Milliways	Pittsburgh	412-766-1086
The Northeast Filebank	Jermyn	717-876-0152
Operation Mindcrime	Altoona	814-742-9885
Pennsylvania Online	Harrisburg	717-657-8699
Pirate's Den BBS	New Castle	412-652-1694
RBAS Unlimited	Trevose	215-357-8177
The Solution II	Richlandtown	215-529-9501
Tim's BBS	Erie	814-825-8660

Rhode Island

*Dimensions	Westerly	401-596-3502

South Carolina

Chaos	Sumpter	803-469-9267
Charleston Police Dpt.	Charleston	803-763-0846
Chris & Nancy's Place	Sumpter	803-494-2082
Custom Computers BBS	Anderson	803-375-0074

South Dakota

*Ducky's	Rapid City	605-393-9626

Tennessee

*The GOOD News BBS	Chattanooga	615-698-0407
Joe's Garage BBS	Bartlett	901-382-0268
The Hotel California	Johnson City	615-928-5704
*Music City Archives	Nashville	615-952-2254
*The Nashville Exchange BBS	Nashville	615-383-0727
The POST BBS	Knoxville	615-691-1887
The Shoreline BBS	Memphis	901-775-3190
Sam's Smalltown Diner	Greenbrier	615-643-7562
The Tea Room BBS	Maryville	615-681-2387
The Vision BBS	Murfreysboro	615-896-7949

Texas

The Asylum BBS	San Antonio	210-637-5670
The Black Gate	El Paso	915-585-3701

Botany Bay	Tyler	903-509-8518
C-Link Online Services	Desoto	214-223-8363
*Dallas Remote Imaging Group BBS	Carrollton	214-394-7438
The Digital Outlet	Borger	806-273-2834
Ft. Worth Online	Ft. Worth	817-735-8166
JAXBBS	Dallas	214-823-1579
Loengrube Mailbox	Odessa	915-550-5122
MacEndeavor	Houston	713-640-1298
Neon Jungle	San Antonio	210-656-0109
Pawn to King's Four	Huntsville	409-291-3322
Poseidon	El Paso	915-593-0639
The Raster Line	Houston	713-568-4128
Recompression Chamber	Brownfield	806-637-8113
*Resonant Frequency	Mesquite	214-686-0427
*The RF-BaKer	Borger	806-273-2407
Savage Jungle	Hurst	817-268-1914
*User-To-User	Dallas	214-393-9317
*WBBS	Conroe/Houston	409-447-4267
Ziggyuken	Cleburne	817-558-4600

Utah

The Cedar Chest BBS	Cedar City	801-586-8751
The Iron Grid	Salt Lake City	801-486-0929
Love Data	Syracuse	801-776-3459
South Valley BBS	West Jordan	801-567-0036

Virginia

The COMA BBS	Charlottesville	804-293-2400
The Elusive Diamond	Burke	703-323-6423
The Genesys BBS	Virginia Beach	703-499-9101
Kerry's Place	Moneta	703-297-8458
Mid-Atlantic OS/2 Group	Virginia Beach	804-422-8462
*No. VA Astronomy Club	Falls Church	703-256-4777
Shareware Solutions BBS	Virginia Beach	804-490-9630
Sparkies Machine BBS	Roanoke	703-366-4299
Virginia Data Exchange	Newport News	703-877-3539
Virginia Shareware Net	Dale City	703-730-8731
Woody's Warehouse	Woodbridge	703-878-3664

Washington

Cameron's Railroad	Arlington	206-659-2132
Columbia Basin BBS	Moses Lake	509-766-2867
The Dark Masters' BBS	Spanaway	206-846-8312
Dino's Doghouse	Oak Harbor	206-679-6971
Extensions BBS	E. Wenatchee	509-886-0306
The Final Frontier	Everett	206-303-9153

Four Aces	LaConner	206-466-ACES
More than Meets the Eye	Everett	206-787-5339
Night Voyager BBS	Veradale	509-926-1686
NW Disability Access	Tacoma	206-539-0704
*NorthWest Online	Spokane	509-244-DOOM
OS/2 Northwest Group	Bellevue	206-957-4513
Search BBS	Vancouver	206-253-5213
Starwest BBS	Clarkston	509-758-6248
Wally World BBS	Walla Walla	509-529-3726

West Virginia

The Bit Bank	Charlestown	304-728-0884
The Black Hole	St. Albans	304-727-5711
Coal County BBS	Hurricane	304-562-2263
The Empire BBS	Kincaid	304-465-5223
Killarney Narrows	Huntington	304-523-8643
The Mindless Ones	Weirton	304-723-2133
Seneca Station	Elkins	304-636-9592
Starbase 90 BBS	Crab Orchard	304-252-6390

Wisconsin

The Decker's BBS	Eau Claire	715-839-0942
*Exec-PC BBS	Milwaukee	414-789-4360
*Radio Free Milwaukee	Milwaukee	414-351-1823
The Shire	Sparta	608-269-7019

Australia

*Ancient Evenings	Perth	+61-9-250-7533
*Computer Connection	Adelaide	+61-08-326-2388
*Heaven's Door	Sydney	+61-2-415-6079
*Hot-Line BBS	Adelaide	+61-8-373-5136
*Melbourne PC User Group	Melbourne	+61-3-699-6788
*SATCOM_Australia	Sydney	61-2-905-0849
*Short Circuit BBS	Sydney	+61-02-569-6259
*Shortwave Possums BBS	Sydney	+61-2-651-3055
*Cross Facts BBS	Russell Lea	+61-2-712-3910
*S. Pacific Union Of Dxer's	Melbourne	

Belgium

*DXA-<Shortwave>-BBS	Antwerp	+32-3-8253613

Brazil

Sherwood BBS	Sao Paulo	+55-011-889-9677

Canada

*ABS International	Montreal	514-937-7451
*Astron BBS	St. Johns	506-652-8999
Calanost BBS/2	Edmonton	403-468-1741

*Channel-23	Orleans	613-830-8147
CompuBBS	Edmundston	506-735-3831
*Coven's Den	Vanier	613-746-3584
*CRS Online	Toronto	416-467-4975
Eco Communications BBS	Winnipeg	204-253-6711
*.\\iramax BBS	Vancouver	604-323-9698
*One Stop BBS	Vanier	613-746-3584
*The ODXA Listening Post	Aurora	905-841-6490
Pro Link BBS	Amos	819-732-7710
Quebec Online	Montreal	514-486-8959
SAUG	Saskatoon	306-242-6936
*SLIP	Alberta	403-299-9900
*Space Lodge BBS	Bowmanville	905-697-2935
The Staff Room BBS	Tecumseh	519-979-4208
TOTSE/2	Spencerville	613-658-5331
*Voice of Windsor BBS	Windsor	519-969-4292

Mexico
Pasito Tun Tun	Monterrey	+52-8-356-8446

The Netherlands
*Alexander BBS	Rotterdam	+31-10-4565600/2201454
*AINEX-BBS		+31-33-633916/653078
McBaud! BBS	Vlaardingen	+31-10-475-2961
*Omnix BBS		+31-0-1680-26819
*Scoop BBS		+31-3499-96366

New Zealand
*SunCity BBS	Nelson	+03-548-9171
*TV Mania	Christchurch	+64-3-352-8586

Puerto Rico
The Glass BBS	Mayaguez	809-832-4984
The Tropical Island BBS	Hormigueros	809-849-5921

Sweden
*GET	Lidingo	+46-8-7655670
*SweDX (Swedish DX Federation)		+46-853034727

United Kingdom
*The Control Center		+44-161-7079221
*The Crooked Spire Mult-BBS		+44-1246-551626/554626
*Pauls Point		+44-0191-4569135
*The Radio Shack		+44-1947-897551
*Golly! on Shortwave	TwyFord	+44-1734-320812
The Rock of Gibralter		+44-0181-6786087

APPENDIX C

Utility frequencies

The following list was provided by Sander Schimmelpenninck. Sander Schimmelpenninck lives in Canada. He became a shortwave listener in the Netherlands in 1937.

1,674.0	Scheveningen Coast D	2,192.0	Simplex calling
1,713.0	Scheveningen Coast F	2,203.0	Marine Simplex Gulf of Mexico
1,716.0	Scheveningen Coast B	2,214.0	Marine Simplex Public US
1,890.0	Scheveningen Coast I	2,260.0	AUS Flying MDs Sec
1,995.0	Scheveningen Ship C	2,261.0	USCG Air
2,003.0	Marine Simplex Gt Lakes	2,280.0	AUS Flying MDs Sec
2,045.0	Scheveningen Ship I	2,287.5	DOE
2,048.0	Scheveningen Ship I	2,298.0	Gander
2,051.0	Scheveningen Ship I	2,300.0	USACE
2,054.0	Scheveningen Ship I	2,326.0	USACE
2,057.0	Scheveningen Ship I	2,345.0	USACE
2,060.0	Scheveningen Ship B	2,348.5	USACE
2,065.0	Marine Simplex Public US	2,350.0	USACE
2,079.0	Marine Simplex Public US	2,350.0	USCG
2,086.0	Marine Simplex publ Mississippi	2,360.0	AUS Flying MDs Sec
2,093.0	Marine Simplex Public US	2,360.0	AUS Radphone Simplex
2,099.0	Scheveningen Ship D	2,366.0	Dutch Navy
2,103.0	USCG Internal	2,371.0	CAP all regions
2,138.0	Scheveningen Ship F	2,374.0	CAP all regions
2,142.0	Marine Simplex Pac Coast <42 N	2,428.0	RAF
		2,445.0	SHARES
2,182.0	Marine Calling/Distress	2,500.0	Time sigs VNG Penrith NSW

2,500.0	WWV	2,971.0	ICAO NAT-D
2,520.0	Scheveningen Ship A Call	2,980.0	WX Argentina
2,582.0	Bermuda Harbour Radio	2,980.0	WX Honolulu
2,582.0	CCG Sydney	2,980.0	WX Oakland
2,591.0	RAF Ascot	2,980.0	WX Tokyo
2,598.0	Canadian CG	2,992.0	ICAO MID-1
2,600.0	Scheveningen Coast C	2,994.0	ICAO MID-3
2,602.0	USACE	2,998.0	ICAO CWP
2,605.0	USACE	3,004.0	Flight Test
2,613.0	NMFS, KAC	3,004.0	ICAO NCA-3
2,622.5	DOE	3,007.0	LDOC Auckland
2,638.0	Intership Safety	3,007.0	LDOC Jakarta
2,641.0	RAF	3,007.0	LDOC Mexico City
2,648.0	Alaska	3,007.0	LDOC Rangoon
2,656.0	AUS Flying MDs Sec	3,007.0	LDOC Sydney
2,658.0	SHARES	3,010.0	LDOC Amsterdam
2,670.0	USCG w/ships, public	3,010.0	LDOC Lima
2,691.0	USCG 7th Dist	3,010.0	LDOC Trinidad
2,694.0	USCG Group Moriches	3,010.0	WW Berna
2,702.0	RN Coastal	3,013.0	LDOC Johannesburg
2,714.0	Harbour Control	3,013.0	WW HON SFO
2,716.0	US Navy Harbour	3,013.0	WW SFO
2,738.0	Intership exc Gt Lakes	3,016.0	ICAO EA-1
2,749.0	CCG E Coast wx	3,016.0	ICAO NAT-A
2,782.0	Intership Mississippi	3,019.0	ICAO NCA-1
2,799.8	Canadian Red Cross	3,019.0	Offshore Drilling
2,808.5	US Customs Channel XA	3,023.0	Search & Rescue
2,824.0	Scheveningen Coast A Call	3,026.0	RAF
2,830.0	Marine Simplex Gulf of Mexico	3,026.0	USAF ALE
		3,029.0	USAF
2,836.0	Harbour Control	3,032.0	RAAF
2,851.0	Flight Test	3,032.0	USAF
2,851.0	ICAO NCA-2	3,032.0	USAF Andrews
2,854.0	ICAO SAT-2	3,035.0	USN
2,869.0	Australian Flight Info	3,038.0	USN
2,869.0	ICAO CEP-1 CEP-2	3,041.0	USAF
2,875.0	Domestic AK	3,044.0	USAF-Canforce Primary
2,878.0	ICAO AFI-4	3,047.0	USN-Canforce primary
2,878.0	Offshore Drilling	3,050.0	USN
2,887.0	ICAO CAR-A	3,051.0	RAF
2,899.0	ICAO NAT-B	3,053.0	USCG
2,911.0	Domestic AK	3,056.0	USCG
2,932.0	ICAO NP	3,059.0	USAF
2,944.0	ICAO SAM-1	3,060.0	TAC
2,962.0	ICAO NAT-E	3,062.0	USAF
2,970.2	Canadian Red Cross	3,065.0	USAF

3,068.0	USAF		3,218.0	RAF
3,071.0	USAF		3,221.0	Air-Dale Flying Svc Ontario
3,074.0	USAF		3,225.0	Air-Dale Flying Svc Ontario
3,077.0	USAF		3,250.0	NATO Bosnia Dutch AF
3,080.0	USAF		3,281.0	Flight Test
3,083.0	USN		3,287.0	USACE
3,086.0	USN		3,290.0	USACE
3,089.0	USN		3,295.0	USAF S303
3,092.0	Can AF Edm/Tren/StJ		3,302.0	USACE
3,092.0	USN		3,303.0	DOT Emergency
3,095.0	USN		3,303.9	NATO Bosnia
3,098.0	USN		3,304.5	FHWA
3,101.0	USN		3,305.0	USACE
3,102.0	RAF		3,330.0	CHU Time
3,104.0	USN		3,336.6	DOE
3,107.0	USN		3,341.0	LDOC Stockholm
3,109.0	RAF		3,341.0	SHARES
3,110.0	USAF		3,345.0	RAF
3,113.0	USAF		3,357.0	Dutch CG
3,116.0	RAF		3,369.0	US Customs Channel YK
3,116.0	USAF		3,376.0	Ontario Natural Resources
3,119.0	USCG Air		3,378.0	Rainbow Radio
3,120.0	RAF		3,379.0	SHARES
3,122.0	USCG		3,380.0	RAF
3,125.0	USN		3,387.0	USCG Cutters
3,126.0	RAF		3,388.0	SHARES
3,128.0	NATO Bosnia naval		3,407.0	Hurricane Ctr Miami
3,128.0	USN		3,413.0	ICAO CEP-1 CEP-2
3,130.0	RAF		3,413.0	WW NY
3,131.0	USN		3,413.0	WX Shannon
3,133.0	USAF		3,419.0	ICAO AFI-2
3,134.0	USAF		3,428.0	US Customs Channel YA
3,137.0	USAF		3,432.0	WX Bangkok
3,140.0	USAF		3,434.0	Offshore Drilling
3,143.0	USAF		3,443.0	Flight Test
3,146.0	USAF		3,446.0	Rainbow Radio
3,149.0	USN		3,452.0	Australian Flight Info
3,152.0	Canadian Forces Vancouver		3,452.0	ICAO AFI SAT-1
3,152.0	USN		3,455.0	ICAO CAR-B
3,155.0	Bell Telephone		3,458.0	Rainbow Radio
3,158.0	NATO Bosnia naval		3,461.0	Australian Flight Info
3,166.0	Bell Can Quebec 1		3,467.0	ICAO AFI-3 MID-2
3,173.0	Wataway Natives Ontario		3,467.0	ICAO SP
3,182.0	NATO Bosnia naval		3,470.0	ICAO SEA-1 SEA-3
3,199.5	FHWA		3,476.0	ICAO INO-1
3,203.0	NATO Bosnia Dutch AF		3,476.0	ICAO NAT-E

Freq	Description
3,476.0	ICAO NAT-F
3,479.0	ICAO EUR-A
3,479.0	ICAO SAM-2
3,479.0	LDOC London
3,482.0	LDOC Portishead
3,485.0	ICAO SEA-1 EA-2
3,485.0	WX Gander
3,485.0	WX New York
3,493.0	WW NY
3,494.0	LDOC Stockholm
3,494.0	WW San Juan
3,495.0	WX Brazzaville
3,691.0	Hurricane TX H&W
3,695.0	Hurricane AL
3,707.0	WX Dakar
3,747.0	LDOC Portishead
3,776.0	RussAv A Kazakhstan, S Siberia
3,801.5	American Red Cross
3,808.0	Hurricane Carib
3,815.0	Hurricane Natl Net
3,818.0	Hurricane Antilles
3,845.0	Rainbow Radio
3,860.0	NACEC N Am Ctr Emerg Comm
3,864.0	RAF
3,867.0	RAF
3,880.0	RAF
3,910.0	Hurricane MS ARES
3,910.0	Hurricane TX H&W
3,915.0	Hurricane Carolinas
3,916.0	RAF
3,921.0	LDOC Amsterdam
3,923.0	Hurricane MS ARES
3,935.0	Hurricane Gulf Centr Net
3,939.0	RAF
3,943.0	Hurricane W Gulf
3,947.0	RAF
3,948.0	Hurricane CA
3,955.0	Hurricane S TX
3,967.0	Hurricane Gulf outgoing
3,975.0	Hurricane GA
3,987.5	Hurricane Mexican
3,993.0	Hurricane Gulf H&W
4,010.0	AUS Flying MDs Prim
4,010.0	AUS Flying MDs Sec
4,010.0	AUS Radphone Simplex
4,030.0	AUS Flying MDs Prim
4,030.0	AUS Radphone Simplex
4,030.0	AUS School of the Air
4,041.0	LDOC Stockholm
4,043.0	RAF
4,045.0	AUS Flying MDs Prim
4,045.0	AUS Radphone Simplex
4,045.0	AUS School of the Air
4,055.0	AUS Flying MDs Prim
4,055.0	AUS Radphone Simplex
4,055.0	DOT Emergency
4,077.0	Scheveningen Radio Ship
4,116.0	Ships CCG Halifax 4,408
4,125.0	Marine Calling/Distress
4,148.0	Ship/Ship
4,164.0	Ship/Ship
4,273.0	CAP Packet RTTY prim
4,275.0	Swedish AF
4,350.0	AUS Flying MDs Sec
4,365.0	AUS School of the Air
4,369.0	USCG WLC Rogers City MI
4,376.0	Can CG
4,376.0	DEA
4,376.0	USCG
4,387.0	Wellington NZ Radio
4,397.0	RAF
4,408.0	CCG Halifax Ship 4,116
4,426.0	USCG Portsmouth wx
4,429.0	Dutch Navy den Helder
4,438.0	Bell Telephone
4,438.5	NATO Bosnia naval
4,439.0	Bell Telephone
4,448.0	USAF Andrews
4,449.0	USAF
4,460.0	Ontario Natural Resources
4,464.0	RAF
4,466.0	CAP NE prim SE sec
4,469.0	CAP SE prim NE sec
4,478.0	ICAO NCA-2
4,480.5	DOE
4,484.0	RAF
4,485.0	CAP VA
4,495.0	USAF S304
4,500.0	US Customs Channel ZA
4,506.0	CAP N Cent prim

Freq	Station
4,509.0	CAP N Cent sec
4,510.0	USAF Coordination
4,520.0	Ontario Natural Resources
4,521.0	LDOC Portishead
4,535.0	Ontario Natural Resources
4,540.0	RAF Ascot
4,546.0	Lab(rador) Air
4,547.0	NATO Bosnia naval
4,550.0	Gulf of Mexico
4,555.0	NATO Bosnia naval
4,565.0	Ontario Natural Resources
4,580.0	Ontario Natural Resources
4,582.0	CAP Emergency ME PAC sec
4,585.0	CAP ME PAC prim
4,594.0	LDOC Stockholm
4,601.0	CAP MTN prim Gt Lakes sec
4,604.0	CAP Gt Lakes prim MTN sec
4,606.0	AUS Radphone Tx w/4,935
4,610.5	Bell Telephone
4,627.0	CAP SW prim
4,630.0	CAP SW sec
4,637.5	Gulf Oil Rigs
4,640.0	Canadian Army
4,645.0	Alaska
4,645.0	WX Tallinn
4,647.0	USAF
4,651.0	Bradley Air Svc Ontario
4,654.0	LDOC Stockholm
4,654.0	WW Berna
4,665.0	MAF Net Lat Am
4,666.0	ICAO CWP
4,666.0	LDOC Elite Canada 3000 ?
4,666.0	LDOC Sydney
4,668.0	WX Argentina
4,669.0	ICAO MID-1
4,669.0	ICAO MID-3
4,669.0	ICAO SAM-1
4,669.5	DOE
4,670.0	WW Berna
4,672.0	Offshore Drilling
4,675.0	ICAO NAT-D
4,678.0	Australian Flight Info
4,684.0	Australian Flight Info
4,687.0	LDOC Amsterdam
4,687.0	LDOC Rangoon
4,687.0	LDOC Sydney
4,690.0	LDOC Stockholm
4,693.0	Australian Flight Info
4,700.0	USN
4,702.0	WW Berna
4,703.0	Can AF Edm/Tren/StJ
4,703.0	USN Canforce sec
4,706.0	USN
4,707.0	RAF
4,709.0	USN
4,710.0	RAF
4,711.0	NATO Bosnia
4,712.0	USN
4,715.0	USN
4,716.0	RAF
4,718.0	USAF
4,719.0	RAF
4,721.0	CANFOR
4,721.0	USAF
4,724.0	USAF GHFS
4,727.0	USAF
4,730.0	Canada MACS
4,730.0	RAF
4,730.0	USCG
4,733.0	USCG
4,736.0	USN
4,739.0	Canada Search & Rescue
4,739.0	USAF
4,739.0	WX RAF UK
4,742.0	RAF Ascot
4,742.0	USAF
4,745.0	Hurricane hunter aircraft
4,749.0	RAF
4,752.0	Canada MACS
4,760.0	USAF Andrews
4,763.0	USAF tactical
4,763.4	Canadian hunters
4,775.0	Ontario Natural Resources
4,780.0	FEMA Ch 14
4,800.0	AUS School of the Air
4,807.0	LDOC Portishead
4,810.0	LDOC Portishead
4,822.0	RAF
4,826.0	Lab(rador) Air
4,840.0	National Guard
4,845.0	RAF
4,850.0	USACE

Frequency	Description
4,855.0	USAF Coordination
4,856.0	NASA Cape Radio
4,860.0	AUS School of the Air
4,872.0	NORAD
4,880.0	AUS Radphone Tx w/4,926
4,880.0	AUS School of the Air
4,880.0	Ontario Natural Resources
4,893.0	Medical Owings Mill MD
4,902.0	FHWA
4,910.0	Wataway Natives Ontario
4,926.0	AUS Radphone Rx w/4,880
4,926.0	AUS School of the Air
4,935.0	AUS Radphone Rx w/4,606
4,938.0	USAF
4,947.5	Alaska
4,977.0	Ontario Natural Resources
4,980.0	AUS Flying MDs Prim
4,989.0	Wataway Natives Ontario
4,991.0	US Customs Channel XB
4,992.0	NASA Cape Radio
5,000.0	MSF Time/
5,000.0	Time sigs VNG Penrith NSW
5,000.0	WWV
5,005.0	Bell Telephone
5,008.0	DOT Emergency
5,010.0	AUS School of the Air
5,011.0	USACE
5,015.0	USACE
5,019.0	Wataway Natives Ontario
5,027.0	RussAv C Volga
5,036.0	Gulf of Mexico
5,047.0	Norwegian AF
5,050.0	Norwegian AF
5,058.5	SHARES
5,061.0	NATO Bosnia Dutch AF
5,085.0	US Customs Channel XC
5,095.0	NATO Bosnia RAF
5,095.0	RAF
5,099.1	Bell Telephone
5,110.0	AUS Flying MDs Prim
5,122.5	Alaska
5,130.0	AUS School of the Air
5,145.0	AUS Flying MDs Prim
5,145.0	AUS School of the Air
5,164.0	French Canadian fixed
5,167.5	Alaska Emergency
5,170.0	Ontario Natural Resources
5,171.0	NATO Bosnia French AF
5,178.0	RAF
5,190.0	ETR PRI Night Channel
5,190.0	Universal Helicopters (Canada)
5,191.0	NATO Bosnia Dutch AF
5,196.4	Mississippi Emergency/RACES
5,211.0	FEMA/DOE/NRC Ch 15
5,227.0	AUS School of the Air
5,227.0	US Customs
5,230.0	AUS School of the Air
5,255.0	FHWA
5,260.0	AUS School of the Air
5,270.0	NATO Bosnia Dutch AF
5,277.0	DEA
5,295.0	DEW Line
5,297.0	NORAD
5,300.0	AUS Flying MDs Prim
5,300.0	AUS Radphone Simplex
5,300.0	AUS School of the Air
5,308.5	French Air Force
5,309.5	DOE
5,310.0	Alaska
5,310.0	NATO Bosnia naval coord
5,320.0	USCG Air
5,325.0	NATO Bosnia naval
5,327.0	USACE
5,335.0	USN FT
5,340.0	AUS School of the Air
5,346.0	USACE
5,350.0	USAF Coordination
5,360.0	AUS Flying MDs Prim
5,360.0	AUS Radphone Simplex
5,361.0	RAF
5,370.0	AUS School of the Air
5,383.0	NATO Bosnia Dutch AF
5,390.0	Bell Can Quebec 3 prim
5,400.0	USCG San Juan
5,403.0	RAF
5,410.0	AUS Flying MDs Prim
5,410.0	AUS Radphone Simplex
5,420.0	RAF
5,424.0	FHWA
5,430.0	Bell Can Quebec 4 sec

Frequency	Station
5,436.0	NASA
5,436.0	RAF
5,437.0	USACE
5,441.0	RAF
5,445.0	AUS School of the Air
5,447.0	RAF
5,450.0	RN Yeovilton Ops
5,451.0	Flight Test
5,462.0	NATO Bosnia air bridge
5,462.0	RAF
5,463.0	Domestic AK
5,463.0	Offshore Drilling
5,469.0	Flight Test
5,470.0	L2NA ITU propagation test
5,471.0	RAF
5,472.0	Domestic AK
5,475.0	LDOC Elite Canada 3000 ?
5,475.0	Ontario Natural Resources
5,477.0	WX Argentina
5,484.0	Domestic AK
5,490.0	Domestic AK
5,490.0	LIAT Antigua
5,490.0	Norontair S Ste Marie ON
5,493.0	ICAO AFI-4
5,496.0	Domestic AK
5,505.0	WX Shannon
5,508.0	Offshore Drilling
5,519.0	WX Hong Kong
5,519.0	WX Honolulu
5,519.0	WX Oakland
5,519.0	WX Tokyo
5,520.0	ICAO CAR-B
5,526.0	Australian Flight Info
5,526.0	ICAO SAM-2
5,526.0	Myanmar Pathein
5,529.0	WW HOU
5,532.0	LDOC Amsterdam
5,532.0	LDOC Johannesburg
5,532.0	LDOC Prague
5,535.0	LDOC Lima
5,535.0	LDOC London
5,535.0	LDOC London
5,535.0	LDOC Trinidad
5,538.0	Middle East Airlines
5,541.0	LDOC Stockholm
5,544.0	LDOC Mexico City
5,547.0	ICAO CEP-1 CEP-2
5,550.0	ICAO CAR-A
5,562.0	Hurricane Ctr Miami
5,565.0	ICAO SAT-2
5,567.0	RussAv B W Kazakh, Uzbekistan
5,571.0	Flight Test
5,571.0	US Customs Channel YB
5,574.0	ICAO CEP-1 CEP-2
5,577.0	Canadian Helicopters
5,589.0	LDOC Stockholm
5,598.0	ICAO NAT-A
5,601.0	WX Buenos Aires
5,604.0	Rainbow Radio
5,610.0	LDOC Portishead
5,616.0	ICAO NAT-B
5,628.0	ICAO NP
5,631.0	Domestic AK
5,634.0	ICAO INO-1
5,643.0	ICAO SP
5,646.0	ICAO NCA-1
5,649.0	ICAO NAT-C
5,649.0	ICAO SEA-1 EA-2
5,652.0	ICAO AFI-2
5,655.0	ICAO SEA-1 EA-2
5,658.0	ICAO AFI-3 MID-2
5,661.0	ICAO EUR-A
5,664.0	ICAO NCA-3
5,667.0	ICAO NP
5,673.0	WX Beijing
5,680.0	Air/Sea SAR worldwide
5,680.0	Canada North Civil Av
5,680.0	Ontario Natural Resources
5,684.0	Halifax Military
5,684.0	USAF ALE
5,685.0	RAF
5,687.0	USAF
5,690.0	Canada MACS VOLMET
5,690.0	Carib interisland police
5,690.0	Dutch Navy Valkenburg
5,690.0	LDOC Portishead
5,690.0	USAF
5,691.0	German AF
5,693.0	USN-pri/USCG-sec (Shared)
5,696.0	USCG Air prim
5,699.0	USCG

Frequency	Station
5,702.0	USAF
5,703.0	Canada MACS
5,705.0	Dutch Navy Valkenburg
5,705.0	USAF
5,708.0	USAF wx
5,711.0	USAF
5,712.0	RAF
5,714.0	RAF Ascot
5,714.0	USN
5,716.0	Swedish AF
5,717.0	RAF
5,717.0	USN/Canforce rescue
5,720.0	USN
5,721.0	RAF
5,723.0	USN
5,726.0	USN
5,729.0	RAF
5,731.0	AUS School of the Air
5,737.0	Chalk Airlines
5,747.0	RAF
5,751.0	CANFOR
5,752.5	DOE
5,760.0	USAF Andrews
5,800.0	USAF Andrews
5,810.0	USAF Coordination
5,820.0	USAF Andrews
5,841.0	DEA
5,845.0	AUS School of the Air
5,850.0	AUS School of the Air
5,887.5	Alaska
5,948.5	DOE
5,961.0	SHARES
6,049.0	NRC/FEMA Ch 19
6,107.0	NRC/FEMA Ch 20
6,108.0	NRC/FEMA Ch 21
6,151.0	NRC/FEMA Ch 22
6,176.0	NRC/FEMA Ch 23
6,209.0	Ships CCG Halifax 6,513
6,215.0	Marine Calling/Distress
6,224.0	Ship/Ship
6,224.0	Southbound 2 Atlantic wx
6,227.0	Ship/Ship
6,230.0	Ship/Ship
6,236.0	USCG Haitian Ops
6,266.0	USCG Cutters
6,405.0	USCG WLC Rogers City MI
6,477.0	UNHCR Bosnia
6,501.0	USCG Portsmouth wx
6,504.0	RAF
6,507.0	WX Sydney
6,513.0	CCG Ship 6,209
6,513.0	DEA
6,518.0	USCG Miami COMSTA
6,526.0	Gulf Air
6,526.0	WW Berna
6,526.0	WW NY
6,532.0	ICAO CWP
6,532.0	LDOC Sydney
6,535.0	ICAO AFI SAT-1
6,541.0	Australian Flight Info
6,549.0	ICAO SAM-1
6,550.0	Dutch CG a/g
6,550.0	Flight Test
6,555.0	Bradley Air Svc Ontario
6,556.0	ICAO SEA-1 SEA-3
6,556.0	LDOC Portishead
6,559.0	ICAO AFI-4
6,562.0	ICAO CWP
6,565.0	Australian Flight Info
6,571.0	ICAO EA-1
6,577.0	ICAO CAR-A
6,580.0	Australian Flight Info
6,580.0	Domestic AK
6,586.0	ICAO CAR-B
6,589.0	ICAO AFI-1
6,592.0	ICAO NCA-2
6,598.0	ICAO EUR-A
6,603.0	WX Brazil
6,604.0	Australian Flight Info
6,604.0	Domestic AK
6,604.0	WX Gander
6,604.0	WX New York
6,610.0	Australian Flight Info
6,616.0	Australian Flight Info
6,617.0	WX Brazzaville
6,622.0	ICAO NAT-F
6,629.0	ICAO NAT-E
6,631.0	ICAO MID-1
6,631.0	ICAO MID-3
6,634.0	LDOC Portishead
6,637.0	LDOC Auckland
6,637.0	LDOC Hong Kong

6,637.0	LDOC Jakarta	6,730.0	DOE
6,637.0	LDOC Singapore	6,730.0	RAF
6,637.0	WW HOU	6,730.0	USAF
6,640.0	WW HON SFO	6,731.0	German Navy Wilhelmshaven
6,640.0	WW NY	6,733.0	RAF
6,640.0	WW San Juan	6,733.0	USAF
6,640.0	WW SFO	6,736.0	USAF
6,643.0	WW Berna	6,739.0	RAF Ascot
6,646.5	DOE	6,739.0	USAF GHFS
6,655.0	ICAO NP	6,739.0	USCG
6,673.0	Hurricane Ctr Miami Delta	6,740.0	RAF
6,673.0	ICAO AFI-1	6,740.0	USAF Coordination
6,676.0	WX Bangkok	6,742.0	RAF
6,676.0	WX Singapore	6,742.0	USCG
6,676.0	WX Sydney	6,742.0	WW Berna
6,685.0	USAF	6,745.0	Canada MACS VOLMET
6,688.0	USN	6,745.0	USN
6,689.0	Dutch Navy Valkenburg	6,747.0	NATO Bosnia naval
6,690.0	RAF	6,748.0	RAF
6,691.0	USN	6,748.0	USN
6,693.0	Halifax Military	6,750.0	USAF
6,693.0	RAF	6,751.0	Hurricane hunter aircraft
6,694.0	Canforce wx	6,751.0	USAF
6,694.0	USN	6,754.0	USAF/Canforce wx
6,697.0	RAF	6,757.0	RAF
6,697.0	USN	6,757.0	USAF
6,700.0	USN	6,760.0	RAF
6,703.0	USN	6,760.0	USAF
6,705.0	WW Berna	6,763.0	Bell Telephone
6,706.0	USN/Canforce sec	6,765.0	RAF
6,708.0	Canada MACS	6,773.0	NATO Bosnia RAF
6,708.0	NORAD	6,785.0	AUS Radphone Rx w/6,866
6,709.0	USN	6,785.0	USACE
6,712.0	French Air Force	6,790.0	USAF Andrews
6,712.0	USAF GHFS	6,803.0	USAF
6,715.0	RAF	6,812.0	USAF Andrews
6,715.0	USAF/Canforce sec	6,817.0	USAF Andrews
6,718.0	USN	6,822.0	RussAv C Eur Russia, Urals, W Sib
6,720.0	RAF		
6,721.0	USN	6,825.0	AUS Flying MDs Sec
6,724.0	USN	6,830.0	USAF
6,726.0	Dutch Navy Valkenburg	6,837.0	NASA Cape Radio
6,727.0	RAAF	6,840.0	AUS Flying MDs Prim
6,727.0	USAF	6,840.0	AUS Radphone Simplex
6,729.0	Dutch Navy Valkenburg	6,845.0	AUS Flying MDs Sec
6,729.0	NATO air		

6,846.0	RussAv A Kazakhstan, S Siberia	7,262.0	NACEC N Am Ctr Emerg Comm
6,854.0	LDOC Portishead	7,268.0	Hurricane Waterway
6,866.0	AUS Radphone Tx w/6,785	7,283.0	Hurricane Gulf outgoing
6,879.0	Tactical, CHOKER	7,290.0	Hurricane TX H&W
6,880.0	AUS Flying MDs Sec	7,300.0	Bell Telephone
6,880.0	AUS Radphone Simplex	7,300.0	DEA
6,896.0	NASA Cape Radio	7,307.0	AUS Radphone Tx w/7,410
6,918.0	USAF Andrews	7,330.0	USAF S307
6,920.0	AUS Flying MDs Sec	7,335.0	CHU Time
6,920.0	AUS Radphone Simplex	7,340.0	AUS School of the Air
6,925.0	AUS School of the Air	7,348.0	DOE/FEMA
6,925.0	ICAO Iran	7,357.0	AUS School of the Air
6,927.0	USAF Andrews	7,373.5	DOT Emergency
6,945.0	AUS Flying MDs Sec	7,392.0	AUS Radphone Rx w/7,475
6,945.0	AUS Radphone Simplex	7,410.0	AUS Radphone Rx w/7,307
6,945.0	AUS School of the Air	7,419.5	FHWA Main Ch F3
6,950.0	AUS Flying MDs Sec	7,428.0	DOE
6,950.0	AUS Radphone Simplex	7,433.0	LDOC Stockholm
6,960.0	AUS Flying MDs Sec	7,439.0	USCG Cutters
6,960.0	AUS Radphone Simplex	7,453.0	LDOC Stockholm
6,960.0	AUS School of the Air	7,461.0	NASA Cape Radio
6,965.0	AUS Flying MDs Sec	7,465.0	AUS Flying MDs Sec
6,965.0	LRCN	7,465.0	Bell Can Quebec 2
6,970.0	Tactical, CHOKER	7,467.0	USN FACSFAC Oceana
6,982.5	DOE	7,475.0	AUS Radphone Tx w/7,392
6,993.0	USAF Andrews	7,475.0	DOT Emergency
6,997.0	NATO Bosnia ATC	7,475.0	FAA KIA21
6,998.0	Canadian Red Cross	7,475.0	FAA main
6,998.5	Intl Red Cross Switzerland	7,475.0	SHARES Emergency Net
7,055.0	Hurricane Puerto Rico Net	7,475.0	USAF W104
7,070.0	Hurricane Manana	7,500.0	WX Dakar
7,073.0	Hurricane Puerto Rico Net	7,507.0	Hurricane Warn
7,115.0	Hurricane Carib maritime	7,527.0	US Customs Channel ZB
7,165.0	Hurricane Antilles	7,533.5	FEMA
7,168.0	Caribbean Maritime	7,535.0	COMNAVFORCARIB
7,232.0	Hurricane South Carolina	7,540.0	SHARES main
7,235.0	Hurricane Baja California	7,546.0	RussAv B S Russia, Caucasus
7,235.0	Hurricane Gulf	7,549.1	Bell Telephone
7,240.0	East Coast Emergency net	7,550.0	AUS Radphone Tx w/8,144
7,243.0	Hurricane South Carolina	7,552.0	Bell Telephone Bellcore
7,245.0	Hurricane S AL	7,552.1	SHARES main
7,245.0	Hurricane W Gulf	7,553.0	Sunair KK2XCE
7,246.0	Hurricane Gulf H&W	7,565.0	AUS School of the Air
7,250.0	Hurricane S TX	7,582.0	DEA

Freq	Station
7,582.0	DOT Emergency
7,609.0	USCG
7,626.0	Special Patrol Ops
7,633.5	MARS
7,635.0	CAP SHARES main
7,635.0	CAP/USAF all regions 1615
7,657.0	DEA
7,657.0	USCG
7,690.0	USAF Andrews
7,697.1	Bell Telephone
7,701.5	DOE
7,726.5	FHWA
7,735.0	US Navy Norfolk
7,735.0	USAF Andrews
7,743.0	FHWA
7,765.0	USAF Andrews
7,778.5	US Customs Channel XD
7,790.0	LRCN
7,803.0	AUS School of the Air
7,813.0	USAF Andrews
7,820.0	LRCN
7,821.0	FHWA
7,840.5	DOE
7,858.0	USAF Andrews
7,870.0	L2NA ITU propagation test
7,880.0	NOAA
7,903.0	FAA Hampton GA
7,920.0	CAP/USAF all regions
7,952.0	Medical Owings Mill MD
7,961.0	USAF LSB
7,975.0	AUS Flying MDs Sec
7,975.0	AUS Radphone Simplex
7,997.0	USAF Andrews
8,014.0	AUS School of the Air
8,025.0	DOE
8,026.0	USAF Mystic
8,035.0	AUS School of the Air
8,040.0	USAF Andrews
8,060.0	USAF Andrews
8,091.0	ICAO Iran
8,095.0	LDOC Silvair Miami
8,125.0	FAA SHARES main
8,140.0	LDOC Sydney
8,144.0	AUS Radphone Rx w/7,550
8,150.0	AUS School of the Air
8,162.0	USAF Andrews
8,165.0	AUS Flying MDs Sec
8,165.0	AUS Radphone Simplex
8,170.0	LDOC Portishead
8,180.0	LDOC Portishead
8,185.0	LDOC Portishead
8,190.0	RAF Ascot
8,255.0	USCG Comsta Key West
8,261.0	Ships CCG Halifax 8,785
8,270.0	Scheveningen Ship Main
8,291.0	Marine Calling/Distress
8,294.0	Ship/Ship
8,297.0	Ship/Ship
8,364.0	Survival Craft
8,560.0	USN FX
8,565.0	AUS School of the Air
8,638.0	Time sigs VNG Penrith NSW
8,684.0	USCG Air
8,688.0	RAAF
8,760.0	USCG Comsta N Orleans
8,764.0	USCG New Orleans wx
8,764.0	USCG Portsmouth wx
8,794.0	Scheveningen Coast H05
8,806.0	St Lys Coast
8,819.0	Rainbow Radio
8,819.0	WX Krasnodar
8,822.0	Australian Flight Info
8,822.0	Flight Test
8,825.0	ICAO NAT-E
8,828.0	Honolulu WX
8,831.0	Australian Flight Info
8,831.0	ICAO NAT-F
8,837.0	Ben Gurion
8,839.0	WX Rostov
8,843.0	Australian Flight Info
8,843.0	ICAO CEP-1 CEP-2
8,843.0	WW NY
8,846.0	ICAO CAR-B
8,855.0	ICAO SAM-2
8,858.0	Australian Flight Info
8,861.0	ICAO AFI SAT-1
8,861.0	Wx Khabarovsk
8,864.0	ICAO NAT-B
8,867.0	ICAO SP
8,873.0	ICAO AFI-4
8,876.0	Australian Flight Info
8,876.0	Domestic AK

Freq	Station
8,876.0	Hurricane Ctr Miami Echo
8,879.0	ICAO INO-1
8,879.0	ICAO NAT-C
8,885.0	LDOC Lima
8,888.0	WX Syktyvar Novosibirsk
8,891.0	Australian Flight Info
8,891.0	ICAO NAT-D
8,894.0	ICAO AFI-2
8,897.0	ICAO EA-1
8,900.0	Australian Flight Info
8,900.0	LDOC Elite Canada 3000 ?
8,903.0	ICAO AFI-4
8,903.0	ICAO CWP
8,903.0	LDOC Sydney
8,903.0	WX Hong Kong
8,903.0	WX Honolulu
8,903.0	WX Oakland
8,903.0	WX Tokyo
8,906.0	ICAO NAT-A
8,906.0	ICAO NAT-E
8,912.0	US Customs Channel YC
8,913.0	Kinshasa Zaire
8,915.0	ICAO NP
8,918.0	ICAO CAR-A
8,921.0	Atlanta Flight Support
8,921.0	Dragon Radio
8,921.0	LDOC Jakarta
8,921.0	LDOC London
8,924.0	LDOC Amsterdam
8,924.0	LDOC Lima
8,924.0	LDOC Trinidad
8,927.0	LDOC Mexico City
8,930.0	LDOC Karachi
8,930.0	LDOC Singapore
8,930.0	LDOC Stockholm
8,930.0	LDOC Sydney
8,933.0	Egyptair
8,933.0	LDOC Collins Cedar Rapids
8,933.0	LDOC Johannesburg
8,933.0	WW NY
8,936.0	LDOC Sydney
8,936.0	WW Berna
8,938.0	Avianca
8,938.0	WX Argentina
8,939.0	WX Moscow
8,942.0	ICAO SEA-1 EA-2
8,946.0	ICAO India Pakistan
8,948.0	ICAO India Pakistan
8,951.0	ICAO MID-1
8,951.0	ICAO MID-3
8,951.0	ICAO NP
8,955.0	Bradley Air Svc Ontario
8,957.0	WX Shannon
8,960.0	LDOC Portishead
8,962.0	ICRC Bosnia transports
8,965.0	RAF
8,965.0	USAF
8,967.0	RAF
8,968.0	USAF GHFS
8,968.0	USAF main
8,970.0	Dutch Navy Valkenburg
8,970.0	Kuwaiti AF
8,971.0	USN
8,972.0	LDOC Stockholm
8,972.0	LDOC Sydney
8,974.0	USN
8,976.0	RAAF
8,977.0	RN Yeovilton Ops
8,977.0	USN
8,980.0	USCG
8,983.0	USCG Air prim
8,984.0	RAF
8,985.0	Swedish AF
8,986.0	USAF ALE
8,987.0	RAF
8,989.0	Belgian Air Force
8,989.0	Edm/Tren Mil
8,989.0	RAAF
8,989.0	USAF
8,990.0	RAF
8,992.0	Canada Search & Rescue
8,992.0	Hurricane hunter aircraft
8,992.0	USAF
8,993.0	Portland Ops UK
8,995.0	USN
8,996.0	RAF
8,997.0	Antarctica US bases
8,998.0	USN
9,001.0	USN
9,004.0	USN
9,006.0	NASA Cape Radio
9,006.0	RAAF

Freq	Station
9,006.0	UNHCR Bosnia
9,007.0	USN/Canforce sec
9,010.0	Canada MACS
9,010.0	USN
9,011.0	RAF
9,013.0	RN Prestwick Ops
9,013.0	USAF
9,016.0	USAF
9,017.0	SHARES
9,019.0	USAF
9,020.0	Hurricane Ctr Miami
9,022.0	RAF
9,022.0	USAF
9,023.0	NORAD
9,023.0	RAAF
9,024.0	RAF
9,025.0	RAF
9,025.0	USAF
9,028.0	USN
9,031.0	NATO Bosnia other
9,031.0	RAF Ascot
9,031.0	USN
9,032.0	RAF
9,034.0	USN/USCG sec
9,036.0	RAF
9,037.0	USN/USCG sec
9,040.0	LDOC Sydney
9,043.0	NATO Bosnia nav air recon
9,043.0	USAF Andrews
9,057.0	USAF Andrews
9,074.5	DOT Emergency
9,106.0	Antarctica
9,115.5	DOE
9,120.0	USAF Andrews
9,122.5	USACE Fridays 15 UTC
9,158.0	USAF Andrews
9,169.0	FHWA
9,172.0	Ontario Natural Resources
9,180.0	USAF Andrews
9,197.0	FHWA Main Ch F4
9,211.0	WW Berna
9,238.0	US Customs Channel XE
9,270.0	USAF
9,305.0	Sunair, KK2XCE
9,315.0	USAF Coordination
9,316.0	USAF Coordination
9,320.0	USAF Andrews
9,380.0	Hurricane Warn
9,497.0	DEA
9,793.0	NORAD
9,802.0	US Customs Channel ZC
9,806.0	USAF W107
9,919.5	DOE
9,958.0	USAF Andrews
9,963.5	DOE
9,974.0	USAF Coordination
9,991.0	USAF Andrews
10,000.0	WWV
10,002.0	WX Jeddah
10,015.0	Hurricane Ctr Miami Fox
10,018.0	ICAO AFI-3 MID-2
10,024.0	ICAO SAM-1
10,033.0	LDOC Mexico City
10,033.0	WW HON SFO
10,039.0	ICAO NCA-3
10,042.0	ICAO EA-1
10,045.0	Flight Test
10,048.0	ICAO NP
10,051.0	WX Gander
10,051.0	WX New York
10,057.0	WX Brazil
10,066.0	ICAO SEA-1 SEA-3
10,069.0	WW Berna
10,072.0	LDOC Auckland
10,072.0	LDOC Jakarta
10,072.0	LDOC London
10,072.0	LDOC Sydney
10,073.0	WX Brazzaville
10,075.0	LDOC Collins Cedar Rapids
10,075.0	Middle East Airlines
10,075.0	WW HOU
10,078.0	LDOC Singapore
10,084.0	ICAO EUR-A
10,093.0	Gulf Air
10,096.0	ICAO NCA-2
10,096.0	ICAO SAM-2
10,112.0	USAF Andrews
10,159.0	NATO Bosnia Belgian AF
10,165.0	LDOC Stockholm
10,194.0	Canada MACS
10,194.0	FEMA
10,195.0	SHARES main

10,215.0	WX Jeddah	11,193.0	USN
10,232.0	NATO Bosnia Belgian AF	11,196.0	USCG
10,242.0	SHARES	11,199.0	USCG
10,242.0	US Customs Channel TA	11,202.0	USCG prim
10,264.0	Rainbow Radio	11,205.0	RAF Ascot
10,291.0	LDOC Portishead	11,205.0	USN
10,365.0	Bosnia Assoc Hum Equilibr	11,208.0	RAF
10,407.0	L2NA ITU propagation test	11,208.0	USN
10,427.0	USAF Andrews	11,211.0	USN
10,493.0	FEMA Ch 28	11,212.0	RAF
10,530.0	USAF Andrews	11,214.0	CFB Trenton patches
10,575.0	LDOC Stockholm	11,214.0	USAF
10,583.0	USAF Andrews	11,217.0	German AF
10,588.0	FEMA main Ch 29	11,217.0	USAF
10,588.0	USCG Miami	11,220.0	USAF
10,780.0	NASA Cape Radio	11,222.0	LDOC Stockholm
10,788.0	COMSTA Charleston	11,223.0	French AF
10,805.0	LDOC Stockholm	11,223.0	USAF/Canforce Alaska
10,881.0	USAF Andrews	11,225.0	LDOC Stockholm
10,884.0	LDOC Portishead	11,225.0	Swedish AF
10,884.0	LDOC Stockholm	11,226.0	SHARES
10,891.0	FHWA Main Ch F5	11,226.0	USAF
10,918.0	FHWA	11,229.0	USAF
10,960.0	LDOC Portishead	11,232.0	Canforce/USAF sec
11,028.0	DOT/FBI main	11,234.0	RAF
11,035.0	USAF Andrews	11,235.0	USAF
11,053.0	USAF Andrews	11,237.0	RAAF
11,073.0	US Customs Channel XF	11,238.0	USAF/Canforce Alaska
11,076.0	DEA	11,241.0	USAF
11,124.0	TAC	11,244.0	USAF GHFS
11,126.5	DOE	11,246.0	DEA
11,132.0	LDOC Sydney	11,247.0	Canforce/USAF sec
11,141.0	NORAD	11,247.0	RAF Ascot
11,175.0	USAF GHFS	11,250.0	Canforce/USAF ALE sec
11,178.0	Dutch Navy Suffisant	11,250.0	Hurricane hunter aircraft
11,178.0	USAF	11,250.0	RAF
11,178.0	WX RAF UK	11,253.0	USN
11,181.0	USAF	11,256.0	USN
11,182.0	Dutch Navy Valkenburg	11,257.0	RAF
11,184.0	USN	11,259.0	USN
11,185.0	RAF	11,262.0	USN
11,186.0	NAS Key West	11,265.0	USN
11,186.0	RAAF	11,267.0	RAAF
11,187.0	German Air Force	11,268.0	Belgian Air Force
11,187.0	USN	11,268.0	USN
11,190.0	USN	11,270.0	RAF

Frequency	Description
11,271.0	Canforce/USAF sec
11,279.0	ICAO NAT-B
11,279.0	ICAO NAT-D
11,282.0	ICAO CEP-1 CEP-2
11,284.0	NORAD
11,288.0	Flight Test
11,288.0	LDOC Cairo/Jeddah
11,291.0	ICAO SAT-2
11,297.0	WX Kiev St Petersburg
11,300.0	ICAO AFI-3 MID-2
11,306.0	Flight Test
11,306.0	LDOC Portishead
11,306.0	Lima Radio
11,306.0	Portishead/Panama
11,309.0	ICAO NAT-E
11,318.0	WX Syktyvar Novosibirsk
11,319.0	WX Argentina
11,330.0	ICAO CAR-B
11,334.0	ICAO CWP
11,342.0	WW HON SFO
11,342.0	WW NY
11,342.0	WW San Juan
11,345.0	Atlanta Flight Support
11,345.0	LDOC Lima
11,345.0	LDOC Stockholm
11,345.0	LDOC Trinidad
11,348.0	WW SFO
11,351.0	LDOC Karachi
11,351.0	LDOC Rangoon
11,354.0	Gulf Air
11,354.0	LDOC Johannesburg
11,356.0	Italian AF
11,357.0	Domestic AK
11,360.0	ICAO SAM-1
11,363.0	Domestic AK
11,363.0	LDOC Portishead
11,375.0	ICAO MID-1
11,375.0	ICAO MID-3
11,381.0	USAF
11,387.0	WX Bangkok
11,387.0	WX Singapore
11,387.0	WX Sydney
11,390.0	LDOC Elite Canada 3000 ?
11,393.0	LDOC Lima
11,396.0	ICAO CAR-A
11,396.0	ICAO SEA-1 EA-2
11,396.0	ICAO SEA-1 SEA-3
11,397.5	DOE
11,398.0	Hurricane hunter aircraft
11,407.0	USAF Andrews
11,413.0	USAF Andrews
11,414.0	NASA Cape Radio
11,425.0	Hurricane hunter aircraft
11,440.0	USAF Rescue
11,441.0	USAF Andrews
11,451.0	SHARES
11,460.0	USAF Andrews
11,466.0	USAF Andrews
11,470.0	LDOC Silvair Miami
11,480.0	USAF P382
11,484.0	SHARES
11,484.0	USAF Andrews
11,494.0	US Customs Channel VF
11,494.0	USAF
11,498.0	USAF Andrews
11,501.2	Mission Av Guelph Ontario
11,518.7	FHWA
11,545.0	USAF Andrews
11,548.0	NASA Cape Radio
11,556.5	DOE
11,576.5	SHARES
11,596.0	USAF Andrews
11,615.0	NASA Cape Radio
11,615.0	USAF Andrews
11,627.0	USAF Andrews
11,634.0	USAF Andrews
11,643.0	USAF
11,643.0	USAF Discrete Ascension
11,721.0	NRC/FEMA Ch 30
11,800.0	NATO Bosnia Dutch AF
11,801.0	DOT/NRC/FEMA Ch 31
11,898.0	Hurricane Ctr Miami
11,910.0	LDOC Elite Canada 3000 ?
11,957.0	NRC/FEMA Ch 32
11,975.0	CAP/USAF all regions
11,994.0	SHARES
12,070.0	FEMA Mt Weather VA
12,070.0	USAF W108
12,072.0	WX Brazzaville
12,107.0	USAF
12,133.0	LDOC Portishead
12,158.0	FHWA

Frequency	Station
12,168.0	LDOC Portishead
12,178.7	FHWA
12,216.0	FEMA Ch 34
12,220.0	DEA
12,222.0	US Customs Channel ZD
12,246.0	Hurricane hunter aircraft
12,266.0	Scheveningen Ship Main
12,266.0	Ships CCG Halifax 13,113
12,290.0	Marine Calling/Distress
12,317.0	USAF Andrews
12,324.0	USAF Andrews
12,327.0	Medical Owings Mill MD
12,353.0	Ship/Ship
12,356.0	Ship/Ship
12,359.0	Ship/Ship
12,761.0	USN HICOM WESTPAC
12,984.0	Time sigs VNG Penrith NSW
13,089.0	USCG Portsmouth/SFO wx
13,113.0	CCG Halifax Ship 12,266
13,113.0	Scheveningen Coast H05
13,150.0	DEA
13,172.0	ATL Fleet Ships, A/C
13,181.0	US Navy HICOM
13,200.0	USAF GHFS
13,203.0	USAF
13,205.0	WW Berna
13,206.0	USAF
13,209.0	USAF ALE
13,211.0	Portishead
13,212.0	USAF
13,215.0	USAF
13,218.0	USCG
13,221.0	USCG
13,224.0	Hurricane hunter aircraft
13,224.0	USN
13,227.0	USN
13,230.0	USN
13,233.0	USN
13,236.0	USN
13,237.0	RAF
13,239.0	USN
13,242.0	USAF
13,245.0	Hurricane hunters
13,245.0	RAF
13,245.0	USAF
13,247.0	FEMA
13,248.0	USAF
13,248.6	Rainbow Radio
13,250.0	Swedish AF
13,251.0	USN
13,254.0	USN
13,257.0	RAF Ascot
13,257.0	USN
13,257.0	USN/Canforce sec
13,261.0	ICAO SP
13,261.0	WW CEP
13,264.0	WX Shannon
13,266.0	WW Berna
13,267.0	Hurricane Ctr Miami
13,267.0	WX Khabarovsk
13,270.0	WX Gander
13,270.0	WX New York
13,273.0	ICAO AFI-2
13,273.0	ICAO NP
13,282.0	WX Auckland
13,282.0	WX Hong Kong
13,282.0	WX Honolulu
13,285.0	Bradley Air Svc Ontario
13,285.0	Rainbow Radio
13,288.0	ICAO AFI-3 MID-2
13,288.0	ICAO CEP-1 CEP-2
13,288.0	ICAO EUR-A
13,291.0	ICAO NAT-B
13,291.0	ICAO NAT-D
13,291.0	ICAO NAT-F
13,294.0	ICAO AFI-4
13,297.0	ICAO CAR-A
13,297.0	ICAO SAM-2
13,300.0	ICAO CWP
13,300.0	LDOC Stockholm
13,300.0	WW NY
13,303.0	ICAO NCA-3
13,304.0	LDOC Tel Aviv
13,306.0	ICAO INO-1
13,306.0	ICAO NAT-A
13,306.0	ICAO NAT-C
13,309.0	ICAO SEA-1 EA-2
13,312.0	Flight test
13,312.0	Flight Test
13,312.0	US Customs Channel YE
13,315.0	ICAO NCA-1
13,315.0	ICAO SAT-2

13,318.0	ICAO SEA-1 SEA-3	13,348.0	LDOC N Am
13,324.0	WW Berna	13,348.0	WW HON SFO
13,330.0	LDOC Africa	13,348.0	WW SFO
13,330.0	LDOC Beirut	13,351.0	LDOC Asia
13,330.0	LDOC Johannesburg	13,351.0	LDOC Bangkok
13,330.0	LDOC N Am	13,351.0	LDOC Beijing
13,330.0	LDOC Stockholm	13,351.0	LDOC Bombay
13,330.0	LDOC Sydney	13,351.0	LDOC Dublin
13,330.0	Middle East Airlines	13,351.0	LDOC Europe
13,330.0	WW HON SFO	13,351.0	LDOC Jakarta
13,330.0	WW HOU	13,351.0	LDOC Manila
13,330.0	WW San Juan	13,351.0	LDOC Ostende
13,333.0	LDOC Asia	13,351.0	LDOC Paris
13,333.0	LDOC Auckland	13,352.0	WX Brazil
13,333.0	LDOC Europe	13,352.0	WX S Am
13,333.0	LDOC Hong Kong	13,354.0	Hurricane Ctr Miami
13,333.0	LDOC London	13,354.0	ICAO CEP-1 CEP-2
13,333.0	LDOC Seoul	13,354.0	ICAO NAT-E
13,333.0	LDOC Singapore	13,354.0	WW HON SFO
13,336.0	LDOC Amsterdam	13,354.0	WW NY
13,336.0	LDOC Bogota	13,356.0	LDOC Kingston
13,336.0	LDOC Europe	13,356.5	Rainbow Radio
13,336.0	LDOC Lisbon	13,357.0	ICAO AFI SAT-1
13,336.0	LDOC Rome	13,373.5	Rainbow Radio
13,336.0	LDOC S Am	13,412.0	USAF Andrews
13,336.0	LDOC Trinidad	13,420.0	Rainbow Radio
13,339.0	ICAO NP	13,432.5	DOT Emergency
13,339.0	LDOC Africa	13,440.0	USAF Andrews
13,339.0	LDOC Bahrain	13,455.0	USAF Andrews
13,339.0	LDOC Havana	13,457.0	FAA
13,339.0	LDOC Jeddah	13,457.0	USAF Andrews
13,339.0	LDOC Mexico City	13,485.0	USAF Andrews
13,339.0	LDOC N Am	13,565.0	USAF Andrews
13,339.0	LDOC Nairobi	13,576.0	LDOC Stockholm
13,342.0	LDOC Asia	13,585.0	USAF Andrews
13,342.0	LDOC Europe	13,593.0	WW Berna
13,342.0	LDOC Karachi	13,630.0	FAA main
13,342.0	LDOC Rangoon	13,630.0	SHARES Emergency Net
13,342.0	LDOC Stockholm	13,710.0	USAF Andrews
13,344.0	WX Hong Kong	13,777.0	USAF
13,344.0	WX Honolulu	13,803.5	DOE
13,344.0	WX Oakland	13,823.0	USAF Andrews
13,344.0	WX Tokyo	13,865.0	LDOC Portishead
13,345.0	LDOC Europe	13,878.0	USAF Andrews
13,348.0	LDOC Africa	13,881.0	Sunair, KK2XCE
13,348.0	LDOC Collins Cedar Rapids	13,907.0	US Customs Channel TB

Frequency	Station
13,915.0	Canadian Red Cross
13,915.0	Intl Red Cross Switzerland
13,942.0	LDOC Stockholm
13,960.0	USAF Andrews
13,962.0	Swiss embassy Cairo
13,965.0	Canadian Red Cross
13,965.0	Intl Red Cross Switzerland
13,970.0	Canada MACS
13,973.0	Canadian Red Cross
13,973.0	Intl Red Cross Switzerland
13,993.0	SHARES main
13,997.0	Intl Red Cross Switzerland
13,998.0	Canadian Red Cross
13,998.5	American Red Cross
14,185.0	Hurricane Carib
14,235.0	Pan Am Health Org
14,265.0	IARN II, Red Cross Net
14,268.0	UN Disaster Net 4UNUN1 NY
14,270.0	American Red Cross
14,275.0	Hurricane IARN/Red Cross
14,280.0	Hurricane Puerto Rico Net
14,283.0	Hurricane Caribus
14,283.0	UN Relief Net
14,300.0	Maritime Mobile Net
14,316.0	Maritime Mobile Net
14,325.0	Hurricane Ctr/Hams Watch
14,340.0	Hurricane LA
14,347.0	NACEC N Am Ctr Emerg Comm
14,350.0	DEA
14,375.0	Canadian Red Cross
14,383.5	SHARES main
14,384.5	CFARS Can Forces Affil Radio Svc
14,396.5	FEMA
14,405.0	L2NA ITU propagation test
14,450.0	FEMA main Ch 35
14,461.0	FHWA
14,467.0	USN Ships
14,470.0	USN Ships
14,493.0	FBI/Utah Natl Guard
14,493.5	SHARES
14,495.0	FBI
14,500.0	NATO Bosnia Italian AF
14,501.2	Mission Av Guelph Ontario
14,545.0	Rainbow Radio
14,606.0	MARS
14,615.0	USAF Coordination
14,645.0	LDOC Stockholm
14,657.0	DOE
14,670.0	CHU Time
14,686.0	DEA P Alt day
14,690.0	DEA
14,715.0	USAF Andrews
14,776.0	FEMA Ch 36
14,818.5	USN Ships
14,832.0	SHARES
14,837.0	NRC/FEMA
14,886.0	NRC/FEMA
14,890.0	LDOC Portishead
14,894.0	NORAD
14,899.0	NRC/FEMA
14,901.2	Mission Av Guelph Ontario
14,902.0	CAP/USAF all regions 1615
14,902.0	USAF Andrews
14,905.0	CAP SHARES main
14,908.0	NRC/FEMA
14,913.0	USAF Andrews
14,991.2	Mission Av Guelph Ontario
15,000.0	WWV
15,010.0	Belgian Air Force
15,010.0	Canforce/USAF sec
15,013.0	USAF
15,016.0	USAF ghfs
15,019.0	USN
15,021.0	LDOC Stockholm
15,022.0	USN
15,025.0	USN
15,028.0	USN
15,031.0	Canforce/USAF sec
15,031.0	RAF Ascot
15,034.0	Canforce wx/USAF sec
15,037.0	USAF
15,040.0	USAF
15,043.0	USAF
15,044.0	SHARES
15,046.0	LDOC Stockholm
15,046.0	USAF
15,046.0	WW Berna
15,049.0	USN
15,050.0	WW Berna
15,052.0	USN

15,055.0	USN	16,465.0	Scheveningen Ship Main
15,058.0	USN	16,528.0	Ship/Ship
15,061.0	USN	16,531.0	Ship/Ship
15,064.0	USN	16,534.0	Ship/Ship
15,067.0	USN	16,590.2	Medical Owings Mill MD
15,070.0	USN	17,314.0	USCG Portsmouth wx
15,073.0	USN	17,347.0	Scheveningen Coast H05
15,076.0	USN	17,350.0	Hongkong Coast
15,079.0	USN	17,385.0	USAF Andrews
15,082.0	USCG	17,398.5	DOE
15,085.0	USCG	17,405.0	LDOC Portishead
15,088.0	USCG	17,415.0	WW Berna
15,091.0	USAF	17,421.0	DOT Emergency
15,094.0	USAF	17,440.0	LDOC Sydney
15,097.0	USAF	17,480.0	USAF Andrews
15,522.0	COMNAVFORCARIB	17,485.0	LDOC Barbados
15,685.0	LDOC Stockholm	17,525.0	FHWA
15,687.0	USAF Andrews	17,601.0	US Customs Channel XH
15,835.0	WW Berna	17,901.0	Hurricane Ctr Miami
15,867.0	SHARES	17,902.5	DOE
15,867.0	US Customs Channel ZE	17,904.0	ICAO CEP-1 CEP-2
15,910.0	FHWA	17,904.0	ICAO CWP
15,912.0	Irish Army UNIFIL Lebanon	17,904.0	ICAO NP
15,953.5	US Customs Channel XG	17,904.0	ICAO SP
15,962.0	USAF S315	17,907.0	ICAO CAR-A
15,964.0	LDOC Portishead	17,907.0	ICAO CAR-B
15,964.0	US Customs Channel VD	17,907.0	ICAO SAM-1
16,000.0	Time sigs VNG Penrith NSW	17,907.0	ICAO SAM-2
16,032.0	USAF Andrews	17,907.0	ICAO SEA-1 EA-2
16,065.5	DOE	17,907.0	ICAO SEA-1 SEA-3
16,080.0	USAF Andrews	17,910.0	Bradley Air Svc Ontario
16,117.0	USAF Andrews	17,910.0	LDOC Barbados
16,141.0	DEA	17,910.0	Rainbow Radio
16,201.0	FEMA	17,916.0	LDOC Asia
16,207.0	LDOC Stockholm	17,916.0	LDOC Athens
16,211.5	FHWA	17,916.0	LDOC Bombay
16,216.0	Ostende Radio	17,916.0	LDOC Dublin
16,320.0	USAF Andrews	17,916.0	LDOC Europe
16,348.0	FAA main	17,916.0	LDOC Jakarta
16,348.0	SHARES Emergency	17,916.0	LDOC Manila
16,370.0	LDOC Portishead	17,916.0	LDOC Paris
16,382.0	USACE	17,916.0	LDOC Seoul
16,407.0	USAF Andrews	17,916.0	LDOC Stockholm
16,420.0	Marine Calling/Distress	17,919.0	LDOC Buenos Aires
16,443.0	LDOC Portishead	17,919.0	LDOC Mexico City
16,450.0	Medical Owings Mill MD	17,919.0	LDOC N Am

17,919.0	LDOC S Am	17,940.0	LDOC Mexico City
17,920.0	LDOC Bangkok	17,940.0	LDOC N Am
17,922.0	LDOC Asia	17,940.0	LDOC Ostende
17,922.0	LDOC Europe	17,940.0	LDOC Rome
17,922.0	LDOC London	17,940.0	LDOC Sydney
17,925.0	LDOC Africa	17,940.0	WW HOU
17,925.0	LDOC Jeddah	17,946.0	ICAO NAT-A
17,925.0	LDOC Johannesburg	17,946.0	ICAO NAT-B
17,925.0	LDOC N Am	17,946.0	ICAO NAT-C
17,925.0	WW HON SFO	17,946.0	ICAO NAT-D
17,925.0	WW NY	17,946.0	ICAO NAT-F
17,925.0	WW San Juan	17,946.0	ICAO NP
17,925.0	WW SFO	17,950.0	Av Lat America?
17,928.0	LDOC Asia	17,955.0	ICAO AFI SAT-1
17,928.0	LDOC Bogota	17,955.0	ICAO SAT-2
17,928.0	LDOC Buenos Aires	17,958.0	ICAO EA-1
17,928.0	LDOC S Am	17,958.0	ICAO NCA-1
17,928.0	LDOC Tokyo	17,958.0	ICAO NCA-2
17,928.0	LDOC Trinidad	17,958.0	ICAO NCA-3
17,931.0	LDOC Africa	17,961.0	ICAO AFI-2
17,931.0	LDOC Bahrain	17,961.0	ICAO AFI-3 MID-2
17,931.0	LDOC Beirut	17,961.0	ICAO AFI-4
17,931.0	LDOC Cairo	17,961.0	ICAO EUR-A
17,931.0	LDOC Europe	17,961.0	ICAO INO-1
17,931.0	LDOC Frankfurt	17,961.0	ICAO MID-1
17,931.0	LDOC Nairobi	17,961.0	ICAO MID-3
17,931.0	WW Berna	17,964.0	Flight Test
17,934.0	LDOC Asia	17,970.0	USN
17,934.0	LDOC Havana	17,973.0	USAF
17,934.0	LDOC N Am	17,976.0	USAF GHFS
17,934.0	LDOC Rangoon	17,979.0	USN
17,934.0	LDOC Singapore	17,982.0	LDOC Stockholm
17,934.0	LDOC Sydney	17,982.0	USN
17,937.0	LDOC Africa	17,985.0	USN
17,937.0	LDOC Lima	17,988.0	USCG
17,937.0	LDOC S Am	17,991.0	USCG
17,940.0	Dragon Radio	17,992.0	German Air Force
17,940.0	LDOC Amsterdam	17,994.0	USAF
17,940.0	LDOC Asia	17,995.0	Edm/Tren/StJ Mil
17,940.0	LDOC Auckland	17,997.0	USAF
17,940.0	LDOC Europe	18,000.0	USAF
17,940.0	LDOC Hong Kong	18,003.0	USAF
17,940.0	LDOC Karachi	18,004.0	Belgian Air Force
17,940.0	LDOC Las Palmas	18,006.0	German Air Force
17,940.0	LDOC Lisbon	18,006.0	USAF
17,940.0	LDOC Madrid	18,009.0	USAF

Freq	Station
18,012.0	USN
18,015.0	USN
18,018.0	Hurricane hunter aircraft
18,018.0	RAF Ascot
18,018.0	USAF ALE
18,021.0	USAF ALE
18,023.0	FEMA
18,023.0	WW Berna
18,024.0	USAF
18,027.0	Canada MACS
18,027.0	Swedish AF
18,027.0	USAF
18,042.0	LDOC Stockholm
18,046.0	USAF
18,063.0	SHARES
18,091.0	Hurricane hunter aircraft
18,091.0	Hurricane USAF
18,205.0	CAP AK
18,210.0	LDOC Portishead
18,283.0	US Customs
18,390.0	USAF
18,397.0	USAF W112
18,480.0	WW Berna
18,594.0	SHARES
18,594.0	US Customs Channel VF
18,650.0	SHARES
18,666.0	DEA
18,990.0	USAF/USN Coord
19,131.0	US Customs Channel XI
19,223.0	FHWA
19,303.0	USAF Coordination
19,410.0	SHARES
19,510.0	LDOC Portishead
19,554.0	WW Berna
19,640.0	NASA Cape Radio
19,845.0	LDOC Ostende
19,984.0	USAF Coordination
20,000.0	WWV
20,016.0	USAF Andrews
20,027.0	NRC FEMA
20,035.0	WW Berna
20,053.0	USAF Andrews
20,065.0	LDOC Portishead
20,124.0	USAF W115
20,154.0	USAF Andrews
20,160.0	LDOC Sydney
20,167.0	USAF W116
20,191.0	USAF Coordination
20,192.0	NASA Jupiter Control
20,313.0	USAF Andrews
20,320.0	Belgian Air Force
20,390.0	NASA Cape Radio
20,407.0	USAF W117
20,475.0	USAF Coordination
20,600.0	USAF Lat Am Co-op
20,620.0	Belgian Air Force
20,631.0	US Customs Channel VB
20,665.0	LDOC Portishead
20,753.0	Canadian Red Cross
20,753.0	Intl Red Cross Switzerland
20,770.0	LDOC Stockholm
20,800.0	Canadian Red Cross
20,812.0	Intl Red Cross Switzerland
20,815.0	Intl Red Cross Switzerland
20,852.0	FAA
20,855.0	NORAD
20,870.0	WW Berna
20,873.0	CAP/USAF all regions
20,876.3	MIT Observatory
20,885.0	MAAG Bolivia, Panama
20,890.0	US Customs Channel TC
20,939.0	Intl Red Cross Switzerland
20,942.0	Canadian Red Cross
20,942.0	Intl Red Cross Switzerland
20,945.0	L2NA ITU propagation test
20,987.0	Intl Red Cross ?
20,993.5	American Red Cross
20,998.0	Canadian Red Cross
20,998.0	Intl Red Cross Switzerland
21,272.0	FEMA
21,390.0	Hurricane Interamericas
21,400.0	Hurricane Transatlantic
21,410.0	Intl Police Assn
21,765.0	LDOC Portishead
21,925.0	ICAO NP
21,925.0	LDOC Amsterdam ?
21,931.0	Flight Test
21,933.0	WW Berna
21,937.0	Hurricane Ctr Miami
21,940.0	LDOC Europe
21,940.0	LDOC Paris
21,943.0	Gulf Air

21,943.0	LDOC Africa	21,979.0	LDOC Frankfurt
21,943.0	LDOC Bahrain	21,982.0	LDOC Africa
21,943.0	LDOC Beirut	21,982.0	LDOC Cairo
21,943.0	LDOC Johannesburg	21,982.0	LDOC Nairobi
21,946.0	LDOC Europe	21,985.0	ICAO CWP
21,946.0	LDOC London	21,985.0	LDOC Havana
21,949.0	LDOC Asia	21,985.0	LDOC N Am
21,949.0	LDOC Bombay	21,988.0	LDOC Europe
21,949.0	LDOC Manila	21,988.0	WW Berna
21,949.0	LDOC Rangoon	21,994.0	LDOC Africa
21,949.0	LDOC Singapore	21,994.0	LDOC Jeddah
21,949.0	LDOC Tokyo	21,997.0	LDOC Europe
21,952.0	LDOC Dublin	21,997.0	LDOC Stockholm
21,952.0	LDOC Europe	22,012.0	Scheveningen Ship Main
21,952.0	LDOC Lisbon	22,015.0	Portishead Ships
21,952.0	LDOC Rome	22,159.0	Ship/Ship
21,955.0	LDOC Bogota	22,162.0	Ship/Ship
21,955.0	LDOC Buenos Aires	22,165.0	Ship/Ship
21,955.0	LDOC S Am	22,168.0	Ship/Ship
21,955.0	LDOC Trinidad	22,171.0	Ship/Ship
21,958.0	LDOC Europe	22,687.0	ATL Fleet Ships
21,961.0	LDOC Africa	22,708.0	Scheveningen Coast H05
21,964.0	LDOC N Am	22,711.0	Portishead Coast
21,964.0	WW HON SFO	22,722.0	Medical Owings Mill MD
21,964.0	WW HOU	22,723.0	USAF Andrews
21,964.0	WW NY	23,001.2	Mission Av Guelph Ontario
21,964.0	WW San Juan	23,040.0	LDOC Stockholm
21,964.0	WW SFO	23,070.0	LDOC Sydney
21,967.0	LDOC Europe	23,142.0	LDOC Portishead
21,967.0	LDOC Las Palmas	23,206.0	TAC
21,967.0	LDOC Madrid	23,210.0	LDOC Portishead
21,970.0	LDOC Asia	23,210.0	LDOC Stockholm
21,970.0	LDOC Auckland	23,214.0	US Customs Channel TD
21,970.0	LDOC Bangkok	23,220.0	USAF
21,970.0	LDOC Hong Kong	23,224.0	ATL Fleet Aircraft
21,970.0	LDOC Jakarta	23,226.0	RAF
21,970.0	LDOC Karachi	23,227.0	USAF
21,970.0	LDOC Seoul	23,236.0	RAF
21,970.0	LDOC Sydney	23,250.0	Canada MACS
21,973.0	LDOC Amsterdam	23,265.0	USAF Andrews
21,973.0	LDOC Europe	23,275.0	Belgian Air Force
21,976.0	LDOC Lima	23,285.0	WW Berna
21,976.0	LDOC S Am	23,287.0	USN HICOM LANT/CARIB
21,977.0	LDOC Stockholm	23,332.0	Belgian Air Force
21,979.0	LDOC Athens	23,403.0	DEA I Alt day
21,979.0	LDOC Europe	23,412.0	LDOC Portishead

23,413.0	NASA Cape Radio	25,578.0	USAF Andrews
23,675.0	DEA	26,471.0	USAF Andrews
24,655.0	LDOC Portishead	26,617.0	CAP/USAF all regions
24,860.0	USAF Lat Am Co-op	26,620.0	CAP/USAF all regions
24,865.0	USAF Lat Am Co-op	27,218.0	LDOC Portishead
25,035.0	LDOC Stockholm	27,625.0	FAA
25,109.0	LDOC Portishead	27,870.0	US Customs Channel VA
25,350.0	US Customs Channel TE	27,998.0	Intl Red Cross Switzerland
25,385.0	LDOC Stockholm	28,091.0	FEMA
25,433.0	USAF Andrews	28,450.0	Hurricane Puerto Rico Net
25,500.0	WW Berna	29,701.0	Intl Red Cross Switzerland

Index

A

A Voz da Resistencia do Galo Negro (Angola), 116
abbreviations, 329-331
Abu Dhabi, 341
accessories for radio receivers, 34-36
ACE, The, 199
Aden, 341
Admiralty Islands, 347
Adventist World Radio (Costa Rica), 128
Adventist World Radio (Guam), 138
Aero Marine Beacon Guide, 60
AF gain, 28
Afars, 338
Afghanistan, 201, 202, 341
Air Force One transmissions, 224
Albania, 116, 344
Algeria, 116, 337
All India Radio, 140-141, **141**
AM band radio, 23
amateur (ham) radio (see also CB radio), 4, 245-263
   Amateur Radio Relay League (ARRL), 245
   "amatirs" amateur (ham) radio and CB, 198-199
   antennas, 253
   beat frequency oscillator (BFO) mode, 250-252
   callsigns, 254-255, 263
   carrier wave (CW) mode, 251
   codes, 255
   country prefix codes, 254-255
   double sideband (DSB) transmissions, 252
   foxhunts, 260-261
   frequencies, 249-250, **250**
   functions of amateur radio, 246
   hamfests, 258-260, **259**
   history and development, 246-248
   license classes, 248-249
   longwave radio, 59
   military amateur radio service (MARS), 222
   Morse code transmission, 245, 251
   outbanders (freebanders), 241
   packet radio, 256-258
   QSL card, **251**, **254**, **262**
   QSL reports to hams, 261-262
   receivers, 250-252, **253**
   regulations governing amateur radio, 245-246, 248-249
   shortwave listening hits, 249-253
   single sideband (SSB) mode, 250-252
   telephone transmissions, 245
   television (SSTV) transmissions, 245, 258
   tuning in, 253
   VHF repeater operations, 256
   Web sites, 254, 258, 263
amplitude modulation (AM), 19, 63
AMTOR and SITOR data transmissions, 230-231
Andaman and Nicobar Islands, 141, 341
Andorra, 344
Anglo-Egyptian Sudan, 337
Angola, 116, 201, 202, 204, 337
Antarctica, 117, 341

Illustrations are in **boldface**.

antennas, 4, 29, 37-53
  amateur (ham) radio, 253
  Beverage antennas (see longwire)
  clip or test leads as connectors, 42
  coaxial cable, 47-48, 49
  connections to receivers, 40-42, **41**, 47-48, 49-50, **50**
  converters, 34
  corrosion/tarnish on antenna wire, 39
  curtain antenna, Swiss Radio International, **38**
  dipoles, 48-50, **49**
  directional antennas, 78
  doublets, 49
  fallacies about antennas, 39
  gauge of antenna wire, 49, 52
  grounding, 48
  height of antenna, 39, 52
  indoor antennas, 39
  inductive coupling and connections, 42
  insulation on antenna wire, 39
  insulators, 49, **51**
  longwave radio, 62
  longwire antennas, 42-44, 50-53, **51**, **52**, **53**, 77-79
  loop antennas, 39, 78-79, **78**
  mediumwave radio, 77-79
  noise, 52
  nulling effect, 52
  PL-259 connectors, 47-48
  portable receiver antennas, 37, 39
  reel antennas, 40-41, 44-46, **45**, 109
  remote or portable antennas, 107-109
  safety, 40-41, 109
  self-sticking screen antenna, 46-47,
  specific-frequency antennas, 39
  tabletop receiver antennas, 47-53
  temporary antenna stringing tips, 40-41
  terminal lugs, 48
  tree-mounted antenna, **51**
  tuners, 34
  TV antennas, 39
  VHF signal monitoring, 39, 270-272
  wave antennas (see longwire)
  whip antennas, 37, 39
  wire antennas, 42-44
Antigua, 117, 346
Antilles, 346
antique radios (see also receivers), **5**, 285-294
Arbenz, Jacabo, 206-215
Argentina, 117, 348
Argentine Antartica, 117
Armenia, 118, 341
Armstrong, Edwin, 2
Army Corps of Engineers QSL, **227**

artificially stimulated emissions, 61
Ascension Island, 118, 337
Asiatic R.S.F.S.R., 341
astronomy, radio-astronomy, 60
audion to detect radio signals, 2
Aum Shinrinkyo, 10-11
auroral skip, 268
Australia, **74**, 118, **119**, 347
Australian DXpedition results, 110-114
Austria, 120, 344
automatic gain control (AGF), 28
aviation transmissions, 222, 224
Azad Kashmir, 342
Azerbaijan, 120, 341
Azores, 337

B

Bahamas, 346
Bahrain, 121
Balearic Islands, 344
Bali, 341
Bangladesh, 121, 342
Barbados, 346
Basutoland, 337
battery-powered receivers, 26
Bayrak Radio (Cyprus), 131
beacons, 221
beat frequency oscillator (BFO) mode, 21, 250-252
Bechuanaland, 337
Belarus, 121
Belgian Congo, 337
Belgium, 121, 344
Belize, 346
Bell, Alexander G., 1
Bellville, Rob, 240
Benin, 122, 339
Bennett, Hank, 173
Beverage antennas (see longwire)
Beverage, H.H., 50-51
Bhutan, 122, 341
Biafra, 339
Black Liberation Radio pirate station, 191-192
Bolivia, 100, 122, 348
Bonaire, 346
Borneo, 341
Bosnia-Hercegovina, 122
Botswana, 122, 337
Bougainville, 202, 347
bounce, 55
Brannigan, Alice, 72
Brazil, 123, 348
Britain (see also United Kingdom)
Britain Radio pirate station, 196

Britain's Better Music Station pirate station, 196
British Broadcasting Corp. (BBC), 8, 68, 88, 117, 118, 178-179
broadcast radio, 2-3, 4, 56
Broadcasting Service of Kingdom of Saudi Arabia, 163
Brunei, 341
Brussels Calling, 121
Bryant, John H., 294
Bulgaria, 123, 344
Bunis, Marty and Sue, 285
Burkina Faso, 123, 339
Burma, 201, 202, 341
Burundi, 201
Byelorussia, 344

C

cable radio/TV, 12, 189
callsigns, 71-72, 254-255, 349-356
Cambodia, 123-124, 204, 342
Cameroon, 124, 338
Canada, 124-126, 346, 347
Canal Zone, 346
Canary Islands, 338
Capeverde, 338
CARACOL Columbia, 99, 100, 128
Carr, Joe, 4, 61, 79
carrier signals, 19
carrier wave (CW) mode, 251
CB Magazine, 237
CB radio (see also amateur (ham) radio), 235-244
   channels and frequencies, 238, **239**
   Community Radio Network (CRN), 240
   current status of CB, 237-238
   history and development of CB, 236-237
   international CB radio groups, 241-243
   outbanders (freebanders), 241
   REACT emergency monitoring group, 240
   restrictions to CB transmissions, 235-237
   Web sites, 242-244
CB Times, 237
Celebes, 341
Central African Republic, 127, 338
Ceylon, 341
CFRX (Canada), 124
CFVP (Canada), 124
Chad, 127, 338
Channel Africa (South Africa), 165
Channel Islands, 344
Chile, 127, 348
China, 9, 127, 149, 203, 341
CHNX (Canada), 124

Christian Science Monitor, 156
CKFX (Canada), 124
CKZN (Canada), 124
CKZU (Canada), 124
Clandestine Broadcasting Directory, 215
clandestine radio (see also pirate radio), 10, 200-216
   history of clandestine radio, 200
   jamming, 205
   pirate radio vs., 200
   schedules for broadcasts, 201-204
   source of information, 215-216
   Web sites, 216
clear-channel mediumwave radio stations, 64-65
clip or test leads as connectors, 42
coaxial cable, 47-49
Coburn, Robert A., 234
collectibles, 279-298
Collins, 292, 294, 292
Colombia, 99, 100, 128, 348
Community Radio Network (CRN), 240
Comoros, 338
Computer Aided Technologies, 303
computers, 34, 90, 299-306
   computerized receivers, 304
   packet radio, 256-258
   receiver-control programs, 301-304
   sources of information/software, 303
   SWECHO FidoNet BBS list, 357-371
   Web sites, 306
   WiNRADiO software, 302, **302**, 304-306, **305**
Cones, Harold N., 294
Congo, 128
connectors, 40-42, **41**, 47-48, 49-50, **50**
continuous wave (CW) transmission, 230
converters, 34
Cook Islands, 128, 348
Costa Rica, 99, 128-129, 346
Cote d'Ivoire, 130
coverage, 23-24, **267**
Cox, James, 3
CQ magazine, 26, 260
Croatia, 130
crystal receivers, 3
Cuacao, 346
Cuba, 101, 130-131, 200, 201, 202, 203, 204, 205, 346
Cumbre DX, 95, 103
Curry Communications, 58-59
curtain antenna, Swiss Radio International, **38**
Cyprus, 131, 342
Czech Republic, 131, 344

D

D layer propagation, 63-64
Dachis, Chuck, 294
Dahomey, 339
Dansk Shortwave Listener's Club International (DSWCI), 95
data decoders, 227-228
dawn chorus phenomenon, 61
daytime phenomena, 101
DBS satellite TV, 12
DD-1 Skyrider Diversity receiver, 90-92, **91**
de Leath, Vaughn, 4-5
DeForest, Lee, 2
Denmark, 132, 344
Deutch Welle, 88, 117, 137, 149, 150, 162
Dexter, Gerry, 215
digital signal processing (DSP), 34-36, **35**
dipoles, 48-50, **49**
Direct Broadcast Satellite (DBS), 189
directional antennas, 78
diversity reception (see stereo diversity reception)
Djibouti, 338
Dodecanese Islands, 342
Dominican Republic, 100, 132, 347
double sideband (DSB) transmissions, 252
doublets, 49
Drake receivers, **17**, 19, 21-24, **21**, 30
drift in frequency, 100
Dubai, 342
ducting, 267
DW Radio Tune In, **89**
DX Monitor, 80
DX test, 80-81
DXing basics, 6, 83-114
   amateur (ham) radio, 249-253
   computerized tracking of frequencies, 90
   DXpeditions, 105-114
   finding DX station locations, 97-103
   frequencies, 88, 90, 99, 101
   frequency drift, 100
   friends through DXing, 96-97
   guides to DXing, 94-95
   hour-of-day reception differences, 101
   identifying what you hear, 86, 94-95, 97-103
   improving audio quality, 90
   interference, 104-105
   language of broadcast, 102
   location for receiver, 83-86, **85**
   logsheets, 103-104
   music, 102
   program listening, 86-93
   radio frequency interference (RFI), 104-105
   regional indicator stations, 102, **102**
   reporting stations, 103-104
   schedules for programming, 88, 90
   stereo diversity reception, 90-93
   Web sites, 95
   WiNRADIO computerized tracker, 90
DXpeditions, 105-114
Dybka, Jill, 60

E

earthquake-related radio signals, 60-61
Ecos del Atrato (Columbia), 100
Ecos del Torbes (Venezuela), 99
Ecuador, 99, 132-133, 348
educational uses of shortwave, 8-9
Egypt, 133, 338
El Salvador, 205, 347
ELBC (Liberia), 148
Electric Radio, 92
electromagnetic spectrum, 3-4, **3**
emergency communications, 221
Emissora Cuidad de Montevideo (Uruguay), 179
England (see also United Kingdom), 178, 344
Equatorial Guinea, 133, 338, 340
Eritrea, 134, 203
Estonia, 134, 345
Ethiopia, 134, 203, 338
European DX Council, 97
European Music Radio pirate station, 197
Evans, Robert utility stations., 234
expanded band allotments, mediumwave radio, 66, **66-67**, 68
experimental stations/hours, DX tests, 81, **226**

F

F2 skip, 268
555 codes, 315
fading, 91
Falkland Islands, 349
Far East Broadcasting Co. (Philippines), 160
Faro del Caribe (Costa Rica), 99, 100, 128
fax transmissions, 224-225
FEBA Radio (Seychelles), 163
Federal Communications Commission (FCC), 4
   amateur radio, 245-246, 248-249
   CB radio, 235-237
   pirate radio, 195-196
Federal Radio Commission (FRC), 4
feeder broadcasts, 221
Feminist International Radio Endeavor (FIRE), 129
Fernando Poo, 338
Fessenden, Reginald, 2
Fidonet, 31, 357-371
filters, 19-20, 28, 34

Fine Tuning, 95
FineWare, 303-304, **304**
Finland, 134-135, 345
foreign language broadcasts (see also identifying), 319-326
foxhunts, amateur (ham) radio, 260-261
France, 135, 345
Free Hope Experience pirate station, 193
Free Radio Berkeley (pirate station), 191-192
Free Radio Network (FRN) web page, 199
Free Radio Service Holland pirate station, 197, **198**
Free Radio Weekly, 95
freedom of speech, 189
French Equatorial Africa, 338
French phrases, **324-325**
FRENDX, 96
frequency, 2, 3-4, **3**, 88, 90, 101
   amateur (ham) radio, 249-250, **250**
   drift in frequency, 100
   longwave radio, 62
   same-frequency stations, 99
   VHF, 265
frequency modulation (FM) (see also VHF), 2, 8, 12, 63
Fuji Islands, 348
future of shortwave, 12-13

G

Gabon, 136, 338
Gad, Robert, 234
gain, 28
Galapagos Islands, 349
Gambia, 339
gauge of wire for antennas, 49, 52
Georgia, 136, 201, 202, 203, 342
Germany, 136-137, 345
Ghana, 97, 137, 339
Gilbert Islands, 348
Globe Wireless, 221-222, **222**
Goa, 141, 342
Godwin, Larry, 69-70, 73
Gold Coast, 339
government activity on longwave radio, 59-60
graveyard mediumwave radio channels, 65
Greece, 137-138, 342, 345
Greenland, 347
Grenada, 347
Grey, Earl T., 199
grounding antennas, 48
groundwave signals, 55
Grove Communications Expo, 97
Grove SP-200 A signal enhancer, **90**
Grove, Bob, 234
Grundig receivers, 24, 25, **25**, 30

Guadeloupe, 347
Guam, 138-139, 348
Guatemala, 99, 139, 205, 206-215, 347
Guinea, 139, 339
Guinea Bissau, 340
Guisti, Gary, 61
Guyana, 139, 348

H

Haiti, 347
Hall, Robert, 222
Hallicrafters (see also antique radios), 23-24, 90-92, **91**, 285-287, **286**, **288**
Halligey, Geoff, 234
ham radio (see amateur radio)
Hammarlund SP-600 receiver (see also antique radios), 24
Harding, Warren G., 3
Harrington, Thomas, 234
HCJB (Ecuador), 132-133
height of antenna, 52
Hertz, Heinrich, 1
heterodynes, 35
High Adventure Broadcasting (Palau), 158
High Adventure Broadcasting Ministries (Lebanon), 148
hiss, 61
history of shortwave radio, 1-8, 88
Hobbs, Charles P., 88
Hollerman, Lynn, 80-81
Honduras, 99, 139-140, 347
Hong Kong, 342
hooks, 61
HRVC (Honduras), 139-140
Hungary, 140, 345

I

IBS, 26
Iceland, 140, 345
Icom R-71A receiver, 24
identifying what you hear, 86, 94-95, 97-103
image rejection, 20-21
India, 140-141, 201, 203, 341, 342
Indonesia, 142, 341, 342, 343, 344, 348
inductive coupling and connections, 42
insulators, 49, **51**
interference, 62, 104-105
International Radio Club of America, 80, 81
Internet and World Wide Web, 13, 31, 299, 300
   amateur (ham) radio sites, 254, 258, 263
   antique radios, 289-292
   callsigns and slogans, 72
   CB radio sites, 242-244
   clandestine radio sites, 216

Internet and World Wide Web *continued*
   collectibles sites, 282, 284
   computer/radio sites, 306
   DXing sites, 95
   FM sites, 276-277
   Globe Wireless site, 222
   Great Circle map site, 109
   longwave radio sites, 59, 60, 62
   mediumwave radio sites, 80
   pirate radio sites, 95, 199-200
   ship-to-shore site, 222
   SWECHO FidoNet BBS list, 357-371
   TV sites, 276-277
   utility stations sites, 234
   VHF signal monitoring sites, 276-277
ionospheric propagation, 63-64
Iran, 142, 203, 204, 342
Iraq, 143, 201, 202, 204, 342
Ireland, 143, 345
Israel, 143, 204, 343
Issas, 338
Italian Radio and Television Service, 144
Italian Radio Radio Relay Service, 144
Italy, 144-145, 345
Ivory Coast, 130, 339

J
Jamaica, 347
jamming, 205
Japan, 10-11, 145, 342
Java, 342
Johnson, Lance, 252
Jolly Roger Radio (Ireland), 143
Jordan, 146, 344
Journal, The, 103
JPS NTR-1 DSP, **35**

K
KA2XAU experimental broadcast station QSL, **226**
KAIJ (United States), 180
Kantanko, Mbanna, 192
KAPW pirate station, 191
Karelo-Finnish SSR, 345
Kashmir, 342
Katanga, 337
Kazakhstan, 146, 342
KC4ZGL Ham Software, 304
KCNA (North Korea), 224
KDED pirate station, 193
KDKA Pittsburgh, 3
Kenwood receivers, 24, **253**
Kenya, 146-147, 339
KFBS (Northern Mariana Islands), 156

KHAS-TV, **267, 271**
KHBI (Northern Mariana Islands), 156, **157**
KHBN (Palau), 158
KHLO, Hilo, Hawaii, 73
King of Hope (Lebanon), 147-148
Kingenfuss, Joerg, 234
Kirbati, 147
Kirghiz, 343
KIWI Radio (New Zealand), 153-154, 198
KJES (United States), 180
KNLS (United States), 180, **180**
Kol Israel, 143, **144**
KPD-581 Connecticut DOT QSL, **232**
KPRC (pirate station), 191
Kropf, Mathias, 215
KSDA (Guam), 138-139, **138**
KTBN (United States), 181
KTWR (Guam), 139
Kurdistan, 202, 203
Kuwait, 147, 343

L
La Voix du Zaire, 188
La Voz de Alpha 66, 205
La Voz de la Liberacion, 206-215
La Voz del CID (Cuba), 200
La Voz del Evangelica (Honduras), 139-140
La Voz del Napo (Ecuador), 133
La Voz Radio Centinela del Sur (Ecuador), 133
Laos, 147, 342
Laster, Clay, 245, 246-248
Latvia, 147, 345
layers of ionosphere, 63-64
Lebanon, 147-148, 343
Lesotho, 148, 337
Lesser Sunda Islands, 343
Levine, Joel, 92
Lewis, Tom, 2
LF Engineering Co., 59
Liberia, 97, 148, 339
Libya, 148
licensing of radio stations, 189
lightning-related radio signals, 61
Lilya, 339
line-of-sight reception, 55, 266
Lithuania, 149, 345
logsheets, 103-104
Long Wave Club of America (LWCA), 58
long-channel mediumwave radio stations, 65
longwave radio, 55-62
   amateur radio, 59
   antennas, 62
   astronomy, radio-astronomy, 60
   beacon tracking, 59-60

bounce, 55
broadcasting, 56
earthquake-related radio signals, 60-61
frequencies, 62
government activity on longwave radio, 59-60
hobbyist beacons, 57-59
interference, 62
Long Wave Club of America (LWCA), 58
LORAN navigation signals, 59
"lowfers," 58
Morse code transmissions, 58-60
natural radio signals, 60-62
noise, 60
propagation, 55-56
receivers, 56-57
select broadcasting stations, list, **57**
skip/skip distance, 55
transmitters, 57-59
tuning in to longwave, 62-63
Web sites, 59, 60, 62
longwire antennas, 42-44, 77-79
loop antennas, 78-79, **78**
LORAN navigation signals, 59, 221
Lowdown, The, 57
Lowe HF-150 receiver, 19
"lowfers," 58
luggables (semi-portable receivers), 15-18
Luxembourg, 345

M

M Street Radio Directory, 79
Madagascar, 149, 339
magazines and bulletins, 294-298, **296**, **297**
Magnavox, 30
Magne, Larry, 26
Malawi, 149, 340
Malaysia, 149, 341, 343, 344
Maldives, 343
Mali, 149, 339
Malta, 150, 345
man-made object skip, 269
Manchuria, 343
manufacturers of receivers, 29-33
Marianas, 348
Marshall Islands, 348
Martinque, 347
Mason, Simon, 217
Mauritania, 150, 339
Mauritis, 339
maximum usable frequency (MUF), 268
Maxwell, James C., 1
Mayotte, 338
McGreevy, Stephen P., 61-62

Medium Wave Circle club, 80
mediumwave radio, 63-81
  antennas, 77-79
  bounce, 64
  callsigns and slogans, 71-72
  clear-channel stations, 64-65
  D layer propagation, 63-64
  DX test, 80-81
  DXing mediumwave radio, 69-71
  E layer propagation, 64
  expanded band allotments, 66, **66-67**, 68
  F layer propagation, 64
  fading, 64
  graveyard channels, 65
  information on mediumwave radio, 79-80
  ionospheric propagation, 63-64
  listening patterns, 74-75
  long-channel stations, 65
  North American mediumwave, 64-68
  number of stations available, 73
  pirate radio, 69-71, 72
  propagation, 63-64
  receivers, 75-77, **76**
  regional-channel stations, 65
  requirements for good DXing, 73-80
  skip, 64
  Web sites, 72, 80
  worldwide mediumwave radio broadcasting, 68
memory (in receivers), 29
meteor scatter, 269
Mexico, 150, 347
Middle East, 204
military amateur radio service (MARS), 222
military communications, 222
mode selection, 28
Moldova, 150-151, 345
Molucca Islands, 343
Monaco, 151, 345
Mongolia, 151, 343
Monitor Radio International, 156
Monitoring Times, 26, 80, 88, 95, 205, 228, 234, 284
Montserrat, 347
Moon, Havana, 217
Moore, Don, 159, 205, 206
Moore, Raymond S., 289
Morocco, 151, 340
Morse code transmissions, 1-2, 58-60, 221, 230, 251, 333-334
Mozambique, 151, 339
multiplexing, 2
music via shortwave, 10
Myanmar, 152, 202, 203, 341

N

Namibia, 98, **98**, 152, 340
NASWA radio country list, 95, 334-356
National Broadcasting . . . Papua New Guinea, 98, **99**, 158
National Radio Club, 79
National Radio Company, 292, **293**
National Voice of Cambodia, 123
natural radio signals, 60-62
Nepal, 152, 343
Netherlands, 152-153, 345
New Caledonia, 10, 348
New Guinea, 98
New Hebrides, 348
New Ireland, 348
New Zealand, 153-154, 348
Newark News Radio Club, 6
news and information via shortwave, 8-9, 12
Nicaragua, 154, 347
Niger, 339
Nigeria, 97-98, 155, 339
night-related phenomena, 63-64, 101
noise, 28, 52, 60
noise blanker, 28
North American Pirate Radio Relay Service (NAPRS), 193
North Korea, 155-156, 205, 343
Northern Ireland, 345
Northern Mariana Islands, 156
Norway, 156, 346
notch filters, 28
Noumea Radio, 10, **10**
NRC AM Radio Log, 79
nulling effect, 52
numbers stations, 216-217, 220
Nusa Tenggara, 343
Nuts'n'Volts, 260
Nyasaland, 340

O

ODXA, 95
Office de Radiodiffusion et Television du Benin (Benin), 122
Oman, 157, 343
oscillators, 2
Oubangi-Chari, 338
outbanders (freebanders), 241
output (in receivers), 29
overloading, 20

P

P.J. Sparx, 193
packet radio, 256-258
Pakistan, 157-158, 342
Palau, 158
Palestine, 343
Panama, 347
Panasonic receivers, 24, 30
Papua New Guinea, 98, 158, 348
Paraguay, 159, 349
passband tuning, 28
Passport to World Band Radio, 26, 94, 103
pennants, 281-282, **283**
People's Republic of China, 341
Peoples' Republic of the Congo, 338
Personal Communications, 237
Peru, 10, 99, 100, 101, 159-160, 349
Philippines, 160, 343
Philips receivers, 24, **76**
photos, 284
Pirate Pages, 95
pirate radio (see also clandestine radio), 10, 69-72, 92,190-200
   "amatirs" amateur (ham) radio and CB, 198-199
   clandestine radio vs., 200
   enforcement of FCC regulations, 195-196
   European pirates, 196-198
   format, 193
   information sources on pirates, 199-200
   local broadcasters, 191
   music, 193
   outbanders (freebanders), 241
   time and place for listening, 192-194
   Voice of the Rock pirate station, **192**
   Web sites, 95, 199-200
   worldwide pirates, 198-199
PL-259 connectors, 47-48
pocket sized receivers, 24
Poland, 160, 346
polar flutter, 91
political and propaganda stations, 9-11, **11**, 200-216
Popular Communications, 26, 80, 88, 95, 200, 225, 228, 234, 237
Portugal, 161, 346
Portugese phrases, **322-323**
power supplies, 36, 107, 110
preamplifiers/preselectors, 36
Presidential radio transmissions, 224
press service transmissions, 12, 224, **224**
Print Disabled Radio (New Zealand), 300-301
Proceedings of Fine Tuning, The, 94, 95
propaganda broadcast (c. 1940), **7**
propagation of radio waves, 1, 3-4, **3**
   artificially stimulated emissions, 61
   astronomy, radio-astronomy, 60
   auroral skip, 268
   bounce, 55, 64

D layer propagatio, 63-64
dawn chorus phenomenon, 61
daytime phenomena, 64, 101
drift in frequency, 100
ducting, 267
DX test, 80-81
E layer propagation, 64
earthquake-related radio signals, 60-61
experimental hours, DX tests, 81
F layer propagation, 64
F2 skip, 268
fading, 64, 91
frequencies, 265-266
groundwave signals, 55
hiss, 61
hooks, 61
hour-of-day reception differences, 63-64, 101
interference, 104-105
ionospheric propagation, 63-64, 265-266
jamming, 205
layers of ionosphere, 63-64
lightning-related radio signals, 61
line-of-sight signals, 55, 266
longwave radio, 55-56
man-made object skip, 269
maximum usable frequency (MUF), 268
mediumwave radio, 63-64
meteor scatter, 269
natural radio signals, 60-62
night-related phenomena, 63-64, 101
noise, 60
polar flutter, 91
radio frequency interference (RFI), 104-105
selective fading, 91
skip, 55, 64, 268
spherics, 61
sporadic E skip, 268
static, 61
stereo diversity reception, 90-93
sunspot cycles, 268
triggered emissions, 61
tropospheric propagation, 266-268
tweeks, 61
VHF frequencies, 265-269
weather-related reception phenomena, 266-268
whistlers, 61
Pyongyang Radio (see North Korea)

Q

Q codes, 2, 6, 314-315, 332
Qatar, 161, 343

QSL cards, 6, **10**, 231-232, 279-281, 308, **309**, **310**
QST, 2, 26, 245, 246, 260

R

Radio Vilnius (Lithuania), 149
Radio 10th (Russia), 199
Radio 270 pirate station, 196
Radio Abidjan (Cote d'Ivoire), 130, **130**
Radio Acunta (Peru), 159
Radio Africa (Equatorial Guinea), 133, **133**
Radio Algiers International, 116
Radio Alianza (Ecuador), 199
Radio Alma Ata (Kazakhstan), 146
Radio Altura (Peru), 159
Radio Ancash, Peru, 10
Radio Anhanguera (Brazil), 123
Radio Apintie (Suriname), 169
Radio Aquarius pirate station, 197
Radio Arista (Indonesia), 199
Radio Atlanta pirate station, 196
Radio Atlantida (Peru), 159
Radio Aum Shinrinkyo, 10-11
Radio Australia, 118, **119**
Radio Austria International, 120, **120**
Radio Bahai (Ecuador), 133
Radio Bahrain, 121
Radio Bandeirantes (Brazil), 123
Radio Bangladesh, 121
Radio Bare (Brazil), 123
Radio Belarus, 121
Radio Botswana, 122
Radio Bras (Brazil), 123
Radio Budapest (Hungary), 140
Radio Buenas Nuevas (Guatemala), 139
Radio Bulgaria, 123
Radio Burkina (Burkina Faso), 123
Radio Cairo (Egypt), 133
Radio Canada International, 125-126, **125**, **126**
Radio Capital, Venezuela, 10
Radio Caroline pirate station, 196, **197**
Radio Catolica International (Columbia), 199
Radio Central (Papua New Guinea), 158
Radio China International, 127
Radio Chota (Peru), 159
Radio Christian Voice (Zambia), 188
Radio City pirate station, 196
Radio Clube do Para (Brazil), 123
Radio Continental (Venezuela), 100
Radio Cook Islands, 128
Radio Cora (Peru), 159
Radio Cristal International (Dominican Republic), 100, 132
Radio Cultural (Guatemala), 99, 139

Radio Dada Gorgud (Azerbaijan), 120
Radio Damascus (Syria), 173
Radio Database International, 26
Radio Denmark, 132
Radio Diffusion Television Congolaise (Congo), 128
Radio Dniester International (Moldova), 150
Radio Dubai (United Arab Emirates), 177
Radio Dublin (Ireland), 143, 197
Radio Educacion (Mexico), 150
Radio Enga (Papua New Guinea), 158
Radio England pirate station, 196
Radio Ethiopia, 134
Radio Europe (Italy), 144
Radio Exterior de Espana (Spain), 167
Radio Farabundo Marti, 206
Radio Fides (Peru/Bolivia), 99, 122
Radio For Peace International (Costa Rica), 128-129, **129**
Radio France International, 135, 136
Radio Free Noumea, 10, **10**
Radio Free Speech News, 193, **194**
radio frequency interference (RFI), 104-105
Radio from Abu Dhabi (United Arab Emirates), 177
Radio Georgia, 136
Radio Havana (Cuba), 127, 130-131
Radio Illimani (Bolivia), 122
Radio Internacional (Honduras), 99, 140
Radio Iraq International, 143
Radio Jamahiriya (Libya), 148
Radio Japan, 136, 145, **146**
Radio Jordan, 146
Radio K'ekchi (Guatemala), 99, 139
Radio Kiev (Ukraine), 177
Radio Kirbati, 147
Radio Korea (South Korea), 166-167
Radio Kuwait, 142, 147
Radio La Oroya (Peru), 100
Radio Latvia International, 147
Radio Lesotho, 148
Radio Liberacion, 206
Radio Liberia, 97, 205
Radio Lider (Peru), 99
Radio London pirate station, 196
Radio Madre de Dios (Peru), 159
Radio Maya (Guatemala), 139
Radio Mercur pirate station, 196
Radio Meteorologia (Brazil), 99
Radio Mexico International, 150
Radio Milne Bay (Papua New Guinea), 158
Radio Mogadishus (Somalia), 165
Radio Moldova International, 150-151, 150
Radio Monte Carlo (Uruguay), 179
Radio Moscow, 10, 162, 300

Radio Moyobamba (Peru), 160
Radio Mozambique, 151
Radio Myanmar, 152
Radio Nacional Amazonas (Brazil), 123
Radio Nacional de Angola, 116
Radio Nacional de Arcangel (Antarctica), 117
Radio Nacional de Colombia, 128
Radio Nacional de Venezuela, 187
Radio Nacional del Colombia, 99, 100
Radio Nacional del Paraguay, 159
Radio Nacional, Venezuela, 10
Radio National de Chile, 127
Radio Nederland (Netherlands), 149, 152-153
Radio Nepal, 152
Radio New Zealand International, 154, **154**
Radio Newyork International (pirate station), 191
Radio Nigeria, 97-98
Radio Nord pirate station, 196
Radio Northern (Papua New Guinea), 158
Radio Norway International, 132, 156
Radio Nuevo Cajamarca (Peru), 159
Radio Oman, 157
Radio Omdurman (Sudan), 169
Radio Onda Popular (Peru), 159
Radio Orbita (Russia), 199
Radio Pakistan, 157-158
Radio Pilipinas (Philippines), 160
Radio Polonia, 160
Radio Portugal, 161
Radio Prague (Czech Republic), 131
Radio Pyongyang (North Korea), 9, 127, 155-156, 167
Radio Quillabamba (Peru), 101
Radio Quince de Septiembre, 206
Radio Quito (Peru/Ecuador), 99, 133
Radio Rebelde (Cuba), 101, 177
Radio Republik Indonesia, 142
Radio Romania International, 161, **161**
Radio Rwanda, 162
Radio San Ignacio (Peru), 159
Radio Sana'a (Guyana), 139
Radio Santa Ana (Bolivia), 122
Radio Santa Cruz (Bolivia), 122
Radio Santa Maria (Chile), 127
Radio Satelite (Peru), 10, 99, **101**, 159
Radio Shack, 30
Radio Singapore International, 164
Radio Slovakia International, 9, 164-165, **166**
Radio Sofia (Bulgaria), 123
Radio South West Africa, 10
Radio St. Helena, 167-169, **168**
Radio Suara Kasih Agung (Indonesia), 199
Radio Sweden, 170-171, **171**
Radio Tanzania, 174

Radio Tashkent (Uzbekistan), 186
Radio Thailand, 174-175, **174**
Radio Tirana (Albania), 116
Radio Titanic pirate station, 197
Radio Tropical (Peru), 159
Radio Tunisia, 175
Radio Uganda, 177
Radio Ukraine International, 10, 177
Radio Ulaan Baator (Mongolia), 151
Radio Union (Peru), 159
Radio USA pirate station, 193
Radio Valentine pirate station, 197
Radio Vanuatu, 186
Radio Venceremos (El Salvador), 205
Radio Veneremos, 206
Radio Veritas Asia (Philippines), 160
Radio Veronica pirate station, 196
Radio Viking pirate station, 197
Radio Virus pirate station, 193
Radio Vlaanderen International (Belgium), 121
Radio Western (Papua New Guinea), 158
Radio Yerevan (Armenia), 118
Radio Yugoslavia (Serbia), 163
Radio-Television Malagasy (Madagascar), 149
Radiodiffusion Argentina al Exterior, 117
Radiodiffusion National Tchadienne (Chad), 127
Radiodiffusion Television Centrafricaine (Central African Rep), 127
Radiodiffusion Television Guineenne (Guinea), 139
Radiodiffusion-Television de Mauritanie (Mauritania), 150
Radiodiffusion-Television Gabonaise (Gabon), 136
Radiodiffusion-Television Malienne (Mali), 149
Radiodiffusion-Television Marocaine (Morocco), 151
radiotelephone transmissions, 231
radioteletype (RTTY) transmissions, 231
REACT emergency monitoring group, 240
Realistic DX-390 receiver, **30**, 30
receiver incremental tuning (RIT), 29
receivers (see also antique radios), 2, 3, 4, **5**, 15-36
   accessories, 34-36
   AF gain, 28
   AM band radio, 23
   amateur (ham) radio, 250-252, **253**
   antennas, 29, 34
   antique radios, 285-294
   automatic gain control (AGF), 28
   batteries, 26
   beat-frequency oscillators (BFO), 21
   checklist before purchasing, 18-19
   computerized, 34, 301-304, 304-306
   converters, 34
   coverage, 23-24
   data decoders, 227-228
   data-receiving station, 228-229
   digital signal processing (DSP), 20, 34-36, **35**
   digital vs. analog readouts, 22-23
   display, 27
   dual diversity receiver, U.S. Navy, **93**
   features found on receivers, 18-26
   filters, 19-20, 28, 34
   gain, 28
   heterodynes, 35
   image rejection, 20-21
   improving audio quality, 90
   location for receiver, 83-86, **85**
   longwave radio, 56-57
   luggables (semi-portable receivers), 15-18
   mail-order receivers, 32-33
   manufacturers of receivers, 29-33
   mediumwave radio, 75-77, **76**
   memory (in receivers), 29
   mode selection, 28
   modern and 1950s-era receiver mix, **17**
   new vs. used receivers, 15-18
   noise blanker, 28
   notch filters, 28
   output, 29
   overloading, 20
   passband tuning, 28
   pocket sized receivers, 24
   portable vs. table-top receivers, 15-16, 37, 39
   preamplifiers/preselectors, 36
   price vs. quality, 16, 30-31
   receiver incremental tuning (RIT), 29
   reviewing equipment before purchase, 26
   RF gain, 28
   RF notch filters, 28
   S meters, 28
   scanners, 29
   selectivity, 19-20
   sensitivity, 19
   signal enhancers, 90
   single-sideband (SSB) transmission, 21
   size and weight issues, 15-18, 24
   "smart" functions, 83
   squelch, 29
   station name tuning, 22-23
   stereo diversity reception, 90-93, **91**, **93**
   surge suppressors, 36
   tabletop receivers and antennas, 47-53
   tone, 28
   tuning in high-speed data transmissions, 229-231
   tuning knob, 27

receivers *continued*
  tuning rate, 23
  tuning speed, 27-28
  utility stations, 225
  VHF signal monitoring, 23, 269-270
  WiNRADiO software, 302, **302**, 304-306, **305**
reception reports, 6
reel antennas, 40-41, 44-46, **45**, 109
regenerative receivers, 2
regional indicator stations, 102, **102**
regional-channel mediumwave radio stations, 65
remote or portable antennas, 107-109
repeaters, amateur (ham) radio, 256
reporting and verification, 307-326
  555 codes, 315
  codes, 314-316
  data in report, 311-313, **313**, **314**
  enclosures, 326
  foreign-language reports, 319-326
  postage, 316-317
  Q codes, 314-315
  QSLs, 308, **309**, **310**
  RST codes, 315
  SINPO codes, 315-316
  tape-recorded reports, 318-319
  time, UTC, 317-318, 327-328
Reunion, 340
RF gain, 28
RF notch filters, 28
RFO Guyane (French Guiana), 136
RFO Tahiti, 173
Rhodes, 342
Rhodesia, 339, 340
Rikisutvarpid Reykjavik (Iceland), 140
Rio Muni, 340
Romania, 161, 346
RRI Sorong (Indonesia), 142
RRI Ujung Padang (Indonesia), 142
RST codes, 315, 331-332
RTTY Listener, The, **223**, 234
Ruanda-Urundi, 162, 340
Russia, 10, 162
RXKR pirate station, 196
Ryukyu Islands, 343

S

S meters, 28
73 Amateur radio, 26, 260
S9 Hobby Radio, 237
Sabah, 341
Samoa, 348
Sangean receivers, 24, 30
Sao Tome, 163, 340
Sarawak, 343

Sarnoff, David, 2
satellite transmissions, 12, 189
Saudi Arabia, 163, 203, 343
scanners, 29
schedules for programming, 88, 90
Schimmel, Don, 217
Scotland, 346
screen antenna, 46-47
selective fading, 91
self-sticking screen antenna, 46-47
Senegal, 339
sensitivity, 19-20
Sentech Shortwave Services (South Africa), 165-166
Serbia, 163
Seychelles, 163, 340
SGC Corp., 35
Sharjah, 343
ship-to-shore communications, 221
shore communications, 221-222, **222**
Shortwave Clandestine Confidential, 215
Shortwave Magazine, 95
Sicily, 346
sideband signals, 19
Sierra Leone, 163-164, 340
signal enhancers, 90
Singapore, 164, 344
single sideband (SSB) mode, 21, 250-252
SINPO codes, 315-316
skip, 55, 268
Slovakia, 9, 164-165
Soloman Islands, 10, 165, 348
Somalia, 165, 201, 202, 203, 338, 339
Sony receivers, 22, 24, 30
South Africa, 165-166, 340
South Korea, 166-167, 201, 202, 204, 205, 343
South Pacific Union of DXers Inc. (SPUD), 106
South West Africa Radio, 10
Southern Rhodesia, 340
Spain, 167, 346
Spanish phrases, 320-321
spark-gap transmitters, 3
spay radio stations, 10
Spectrum Systems, 304
spherics, 61
sporadic E skip, 268
spy stations (see numbers stations)
squelch, 29
Sri Lanka, 167, 341
SRS-News, 95
St. Helena, 167-169, 340
Stalin radio broadcast (c. 1940), **7**
static, 61
station name tuning, 22-23
Stellwag, August, **17**

stereo diversity reception, 90-93, **91**, **93**
stickers, 284
Stryker, Ken, 60
Sudan, 169, 201, 204, 337
Sulawesi, 341
Sumatra, 344
sunspot cycles, 268
superheterodyne receivers, 2
surge suppressors, 36
Surinam, 169, 349
Swan Island, 347
Swaziland, 169-170, 340
SWBC stations (NASWA), 335
SWECHO FidoNet BBS list, 357-371
Sweden, 170, 346
Swiss Radio International, 136, 172, **172**
Switzerland, 172, 346
SWL Winterfest, 97
Syria, 173, 344

T

T-shirts, 284
Tadzhik, 344
Tahiti, 173, 348
Taiwan, 173, 341
Tajikistan, 173, 201, 202
Tanganyika, 340
Tangier, 340
Tanzania, 174, 340
tapes or cassettes, 282, 284, 318-319
Taylor, Charles, 97
telegraph and Morse Code, 1-2
telephone, 1-2
television (see also VHF signal monitoring), 258
terminal lugs, 48
TGNA (Guatemala), 139
Thailand, 174, 344
TIAWR (Costa Rica), 128
Tibet, 344
TIFC (Costa Rica), 128
time, UTC, 317-318, 327-328
Timor, 343
Titanic disaster, radio as life-saver, 2
Togo, 175, **175**, 338
tone, 28
Tonga, 348
tools, 109-110
Trans World Radio (Guam), 139
Trans World Radio (Monaco), 151
Trans World Radio (Swaziland), 169-170
Trans World Radio, **69**
Trans-Jordan, 344
Transmitiendo Gratas Nuevas Alegres (TGNA) (Guatemala), 100

transmitters, 3
traveler's information stations (TIS), 220, **220**
tree-mounted antenna, **51**
triggered emissiions, 61
Trinidad, 347
Tristan Da Cunha, 340
tropical bands, 8
tropospheric propagation, 266-268
TRS Consultants, 303
tuners, antenna, 34
tuning rate, 23
tuning speed, 27-28
Tunisia, 175, 340
Turkey, 176, 344
Turkmen, 344
Turks and Caicos Islands, 347
tweeks, 61

U

U.S. Navy dual diversity receiver, **93**
Ubico, Jorge, 206
Uganda, 177, 340
Ukraine, 10, 177, 346
United Arab Emirates, 177-178, **178**, 341-343
United Kingdom, 178-179
United States, 180-186, 346, 347, 348
Universal M-7000 communications terminal, 228-229, **228**
Universal Time Coordinated (UTC), 317-318, 327-328
Up Against the Wall Radio pirate station, 193
Upper Volta, 339
Uruguay, 179, 349
utility stations, 12, 219-234, 373-395
    AMTOR and SITOR data transmissions, 230-231
    aviation transmissions, 222, 224
    beacons, 221
    commercial transmissions, 224
    continuous wave (CW) transmission, 230
    emergency communications, 221
    equipment necessary, 225
    fax transmissions, 224-225
    feeder broadcasts, 221
    Globe Wireless, 221-222, **222**
    high-speed data decoders, 227-228
    legality of listening, 225
    military amateur radio service (MARS), 222
    military communications, 222
    Morse code transmissions, 221, 230
    numbers stations, 220
    press service transmissions, 224, **224**
    QSLs from utility stations, 231-232
    radiotelephone transmissions, 231

utility stations *continued*
   radioteletype (RTTY) transmissions, 231
   setting up data-receiving station, 228-229
   ship-to-shore communications, 221
   shore communications, 221-222, **222**
   sources of information, 234
   traveler's information stations (TIS), 220, **220**
   tuning in high-speed data transmissions, 229-231
   U.S. Presidential transmissions, 224
   weather map faxes, 225
   Web sites, 234
Uzbekistan, 186, 344

V

Van Horn, Gayle, 234
Vanuatu, 186, 348
Vatican Radio, **70**, 187, 346
Venezuela, 10, 99, 100, 187, 349
VHF repeater operations, amateur (ham) radio, 256
VHF signal monitoring, 23, 63, 265-277
   antennas, 270-272
   auroral skip, 268
   ducting, 267
   DXing tips, 272-275
   equipment needed, 269-270
   F2 skip, 268
   frequencies, 265-266
   high- vs. low-band VHF, 273
   line-of-sight reception, 266
   man-made object skip, 269
   maximum usable frequency (MUF), 268
   meteor scatter, 269
   propagation, 265
   QSLs, 275, **275**
   skip propagation, 268
   sources of information, 275-277
   sporadic E skip, 268
   sunspot cycles, 268
   tropospheric propagation, 266-268
   weather-related reception phenomena, 266-268
   Web sites, 276-277
Vietnam, 127, 187, 342
Voice of America, 9, 86, 118, 151, 160, 163, 181-182, **182**, 185-186
Voice of Armenia, 118
Voice of Free China (Taiwan), 173
Voice of Greece, 137-138
Voice of Guyana, 139
Voice of Hope (Lebanon), 158
Voice of Indonesia, 142
Voice of Kenya, 146-147
Voice of Lebanon, 148
Voice of Malaysia, 149
Voice of Nigeria, 98, 155, **155**
Voice of Russia, 162
Voice of Sudan, 169
Voice of the Broad Masses of Eritrea, 134
Voice of the Grateful Dead pirate station, 193
Voice of the Islamic Republic of Iran, 142
Voice of the Mediterranean (Malta), 150
Voice of the Night pirate station, 195-196
Voice of the Rock pirate station, **192**
Voice of Turkey, 176, **176**
Voice of Venus (pirate station), 191
Voice of Vietnam, 127, 187
Voice of Voyager (pirate station), 191
VOLMET transmissions (aviation weather), 222

W

Wales, 346
WARR pirate station, 194
Watkins-Johnson, 35
wave antennas (see longwire)
WDTV QSL, **276**
weather map faxes, 225
weather-related reception phenomena, 266-268
West Coast Radio (Ireland), 143
Western Sahara, 204
Westside Radio pirate station, 197
WEWN (United States), 182
WFAT (pirate station), 191
WGTG (United States), 182-183
whip antennas, 37, 39, 37
whistlers, 61
WHRI (United States), 183, 205
Williams, Dallas, 234
WINB (United States), 183
Wings of Hope (Lebanon), 148
WiNRADiO, 90, 302, **302**, 304-306, **305**
wire antennas, 42-44
WJCR (United States), 183
WJLR pirate station, 92
WJR clear-channel mediumwave radio, Detroit, 65
WLIS (We Love Interval Signals) pirate station, 193
WMLK (United States), 183
World Parody Network pirate station, 194
World Radio TV Handbook, 26, 79, 94, 103
World War II and shortwave listening, 6-8, **7**
World Wide Christian Radio (WWCR), 184-185, 195, 300

World Wide Web (see Internet and World Wide Web)
Worldradio, 260
Worldwide Pirate Radio Station, 194
Worldwide UTE News Club (WUN), 234
WPN pirate station, 194
WPRS pirate station, 194
WREC pirate station, 193
WRMI (United States), 183-184, 205
WRNO (United States), 184, 300
WRTH, 26
WRV pirate station, 193
WSHB (United States), 184
WUB4 Army Corps of Engineers QSL, **227**
WUMS pirate station, 196
WVHA (United States), 184
WWCR (United States), 184-185, 195, 300
WYFR (United States), 185

X
XERK pirate station, 195

Y
Yaesu FRG-100 receiver, 24
Yaesu FRG-7 receiver, 16-17
Yemen, 139, 187, 341, 344
YLE Radio Finland, 134-135, **135**
Yugoslavia, 346

Z
Zaire, 188, 337
Zambia, 188, 339
Zanzibar, 340
Zeller, George, 199
Zimbabwe, 188, **188**, 340
ZXLA: Radio for the Print Disabled (New Zealand), 154, 300-301

About the Author

One of the "brightest stars" in hobby writing, according to *Monitoring Times*, Andrew Yoder is the author or coauthor of nine books about audio and radio, including the extremely well-received, *Build Your Own Shortwave Antennas*, also published by McGraw-Hill. He is an expert on pirate radio and TV stations and has contributed articles to *Radio!*, *Popular Electronics*, *Radio World*, *New Jersey Monthly*, and *Popular Communications*.

Grove: Your Global Communications Source

Shortwave radio has made incredible strides in recent years. Keeping up with broadcast schedule changes is hard enough—but how do you keep up with the changes in technology?

That's where Grove Enterprises comes in, providing both the information and the equipment for the successful SWL.

As the publisher of *Monitoring Times* and *Satellite Times* magazines, Grove provides you with all the resources you need to keep abreast of developments in broadcasting, as well as in the emerging technologies of satellite and Internet communications. *MT* and *ST* remain the only full-spectrum publications in the world, reporting on any communications media that travel by radio wave or wired electrical impulse!

Of course, many SWLs already depend on us for our renowned "Shortwave Guide," published each month in *MT*. We pride ourselves in providing you with the most timely and accurate programming data in the world. In addition to the "Guide," we also publish longwave, mediumwave and VHF/UHF frequencies—literally everything from "DC to daylight."

Want to know more about AMSAT's Phase IIID amateur satellite technology? In *ST* you will find up-to-date reports on everything from LEO to GEO and beyond—including satellite launches, new products, domestic and international TVRO, personal communications, weather satellites, and much more!

Subscribe to *MT* for as little as $12.95 for six months or $23.95 for one year (US). *ST* rates start at $19.95 for one year (US).

Want even more current news? Then come to our comprehensive site on the World Wide Web (see URLs below) for all the late-breaking developments and frequency updates.

SHORTWAVE PRODUCTS

Renowned for its great prices and quality service, Grove Enterprises is your source for top-of-line shortwave receivers (such as the Drake R-8 and Sony ICF-2010 and quality accessories such as the Grove SP-200A Sound Enhancer (shown above with the ICF-2010)—the best performance boost you can get for under $200!

We also carry a full line of antennas, books, filters and pre-amps. Request a free catalog today, or visit our on-line catalog at the address below!

Grove Enterprises, Inc.
7540 Highway 64 West
Brasstown, NC 28902

1-800-438-8155 (US & Can.); 704-837-9200
FAX 704-837-2216; E-mail: order@grove.net

On the Web:
Grove Home Page: www.grove.net
MT: www.grove.net/hmpgmt.html
ST: www.grove.net/hmpgst.html
Catalog: www.grove.net/hmpgcat.html

WiNRADiO™

The World's Most Surprising Communications Receiver

**WiNRADiO card.
A new look in radios.**

"The sensitivity seems to be pretty good across the whole range... ...unique and useful monitoring product...worth a serious look."
Monitoring Times Magazine

"...I don't know of any scanner, where I succeeded instantly in successful reception without studying the handbook..."
RadioScanner Magazine

"Of all the cool PC cards you could stick in your computer, WiNRADiO takes the cake."
internet.au Magazine

"...high quality workmanship, good reception and easy usage." Chip Magazine

"...a must-have for hackers. A scanner user's dream." Radio & Communications Magazine

"The most innovative new product we saw at Dayton HamVention..." W5YI Report

**WiNRADiO software.
Virtual front panel on your PC.**

"WiNRADiO has enticing possibilities...The manual is an exciting book not only because of its beautiful cover, high quality paper, and easy instructions, but also because it contains a mix of operating and technical informations about various aspects of radio you might have forgotten or never knew."
World Scanner Report, Volume 6, No. 7

What are the advantages of having a PC-based receiver compared to a stand-alone one?

1. Communications receivers are similar to test instruments - the trend is towards PC-based instrumentation which allows many of the traditional front-panel functions to be more flexible and informative compared to a traditional, dedicated control panel.

2. The PC-based software controls all the ancillary functions such as scanning parameters, memories, logging and various operation modes. Compared to hardware or ROM-based firmware control, this gives the receiver a greater flexibility, greater number and sophistication of ancillary functions, practically unlimited memory capacity, and the ability to customize the receiver for special applications.

3. Without the constraints of a fixed control panel, a receiver can have different "personalities" depending on the user's applications and preferences. New functions, for example frequency databases, can be easily added and integrated with the receiver.

4. A number of independent WiNRADiO receivers can be controlled by a single PC. This is very useful if you need to monitor a large range of frequencies on a continuous basis, or where various methods of multi-channel transmissions are employed.

5. A PC-based receiver allows the user to take advantage of the digital signal processing capabilities of the PC. Modern PCs are fast enough to do such signal processing, decoding and display in real time, as well as provide mass storage for received signals.

For your nearest distributor please contact:

WiNRADiO Communications
222 St. Kilda Road St. Kilda, 3182　　Phone:　+61 3 9525 5300　　Email:　info@winradio.net.au
Australia　　　　　　　　　　　　　　Fax:　　+61 3 9525 3560　　Web:　　www.winradio.net.au

You've tried listening
Try Broadcasting!

Radio is a lot of fun to listen to, but have you ever considered broadcasting: owning and operating your own radio station or producing a syndicated program?

It's easier than you might think! With the more-sophisticated, yet less-expensive technology of the 1990s, you'll find that your broadcasting dreams are within your reach!

Hobby Broadcasting is a quarterly magazine, dedicated to hobby radio and television operations. Whether college, low-power, pirate, syndicated, or community-access radio or TV, it's all here.

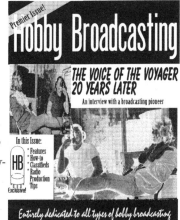

Hobby Broadcasting is a must-have for anyone involved in radio or TV operations. Some of the regular topics include:

* Broadcasting news
* How to get on the air
* How to broadcast successfully
* How-to articles about Internet radio, Part-15, Pirate, Syndicated, Carrier-current, Community-access, and International shortwave broadcasting.
* Features about various successful broadcasters
* Programming and production tips
* Technical features
* Music features and reviews
* Equipment reviews

Yes! Please sign me up!
HB Hobby Broadcasting

☐ Send an SASE (United States) or send $1 (elsewhere) for subscription information and the availability of back issues

John Doe
123 Any Street
Anytown, USA 01234

Hobby Broadcasting
PO Box 642
Mont Alto, PA 17237

...ORE with advanced equipment from Universal!

...S RECEIVERS

Universal Radio carries an excellent selection of new and used communications receivers. Japan Radio NRD-535D shown.

PORTABLE RECEIVERS

ICOM SONY GRUNDIG SANGEAN

Universal offers over 40 portable receivers from $50 to over $500. Our **free** catalog fully describes and prices all models.

COMMUNICATIONS BOOKS

● **Passport To Worldband Radio** *By L. Magne*
Graphic presentation of all shortwave broadcast stations. Plus exhaustive receiver reviews. A *must have* book. $19.95 (+$2)

● **World Radio TV Handbook**
All shortwave broadcast stations organized by country with schedules, addresses, power, etc. $24.95 (+$2)

● **Worldwide Aeronautical Frequency Directory**
By R. Evans. The definitive guide to commercial and military, shortwave and VHF/UHF aeronautical communications including ACARS. Second Edition. 260 pages. $19.95 (+$2)

● **Shortwave Receivers Past & Present** *By F. Osterman*
Covers over 500 receivers made in the last 50 years with new and used values, photos, specs. and features. $19.95 (+$2)

● **Discover DXing!** *By J. Zondlo*
An introduction to DXing AM, FM and TV. $4.95 (+$2)

● **The World Below 500 KiloHertz** *By P. Carron*
An introduction to the world of longwave DX. $4.95 (+$2)

● **Guide to Utility Station** *By J. Klingenfuss*
Unquestionably the best directory of *non*-broadcast stations on shortwave. Covers SSB, CW and RTTY modes. $39.95 (+$2)

● **Shortwave Radio Listening Guidebook** *By H. Helms*
This noted author provides a readable book on all aspects of shortwave listening. Second Edition. $19.95 (+$2)

● **Comprehensive Guide to Military Monitoring** *By Douglass*
Covers monitoring equipment, shortwave and VHF/UHF frequencies, identifiers, bases, black projects, etc. $19.95 (+$2)

● **Easy-Up Antennas for Listeners & Hams** *By E. Noll*
Learn about low cost, easy to erect antennas for LW, MW, SW, FM and ham radio that really work. $16.95 (+$2)

● **National Radio Club AM Radio Log**
All US & Canadian AM stations with addresses! $22.95 (+$2)

• Please include $2 per title for bookrate shipping.

COMMERCIAL RECEIVERS

The improved Watkins-Johnson HF-1000 is *the* ultimate receiver! D.S.P. technology, 58 bandwidths, 1Hz display. Under $4000.

RTTY AND FAX DECODERS

■ **Universal M-450 Reader**

The self-contained Universal M-450 is a sophisticated decoder *and* tone reader. The shortwave listener can decode: **Baudot, SITOR, FEC-A, ASCII, SWED-ARQ** and weather **FAX**. The VHF-UHF listener can copy the **ACARS** VHF aviation teletype mode plus **GOLAY** and **POCSAG** digital pager modes plus **DTMF, CTCSS** and **DCS**. Big two-line, 20 character LCD. Operates from 12 VDC or with the supplied 110 VAC adapter. No PC is required, but a serial port is provided. $399.95 (+$6)

■ **Universal M-1200 Decoder Card**

Turn your *PC* into a powerful intercept device! Modes include: **Morse, Baudot RTTY, SITOR, FEC-A, ARQ 6-90, ARQ-M2, ARQ-E/E3, ASCII, Packet, FAX, ACARS, POCSAG** and **GOLAY**. Requires a *PC* with VGA monitor. $399.95 (+$5)

■ **Universal M-8000 Decoder**

The professional choice. The Universal M-8000 decodes: **CW, Baudot, SITOR, ARQ-M2/4, ARQ-E/E3, ARQ6-90, ARQ-S, SWED-ARQ, FEC-A, FEC-S, ASCII, Packet, Pactor, Piccolo, POL-ARQ, GMDSS, VFT, ACARS, POCSAG** and **GOLAY**. Plus display modes for: Cyrillic, Literal and Databit analysis. Breathtaking **FAX** images to VGA monitor or printer. Monitor and printer optional. 115/230 AC 50/60 Hz. $1399.00 (+$12)

VISIT UNIVERSAL RADIO ON THE INTERNET
http://www.universal-radio.com

• Visa
• MasterCard
• Discover

• Prices and specifications are subject to change.
• Used equipment list available on request.
• Returns are subject to a 15% restocking fee.

Universal Radio, Inc.
6830 Americana Pkwy.
Reynoldsburg, OH 43068
☎ 800 431-3939   Orders & Prices
☎ 614 866-4267   Information
→ 614 866-2339   FAX Line
dx@universal-radio.com

FREE 100 PAGE CATALOG
Our informative catalog covers **everything** for the shortwave, amateur and scanner enthusiasts. With photos and informative descriptions.   **Free** in North America. (Five I.R.C.s elsewhere).